Electrified Aircraft Propulsion

What are the benefits of electrified propulsion for large aircraft? What technology advancements are required to realize these benefits? How can the aerospace industry transition from today's technologies to state-of-the-art electrified systems? Learn the answers with this multidisciplinary text, combining expertise from leading researchers in electrified aircraft propulsion. The book includes broad coverage of electrification technologies – spanning power systems and power electronics, materials science, superconductivity and cryogenics, thermal management, battery chemistry, system design, and system optimization – and a clear-cut road map identifying remaining gaps between the current state-of-the-art and future performance technologies. Providing expert guidance on areas for future research and investment and an ideal introduction to cutting-edge advances and outstanding challenges in large electric aircraft design, this is a perfect resource for graduate students, researchers, electrical and aeronautical engineers, policymakers, and management professionals interested in next-generation commercial flight technologies.

Kiruba Haran is Professor of Electrical and Computer Engineering at the University of Illinois at Urbana–Champaign, where he is also Director of the Grainger Center for Electric Machinery and Electromechanics. He is a fellow of the IEEE.

Nateri Madavan is Deputy Director for the Transformative Aeronautics Concepts Program, NASA Aeronautics Mission Directorate, NASA.

Tim C. O'Connell is Senior Lead Engineer for P. C. Krause & Associates and Adjunct Research Assistant Professor of Electrical and Computer Engineering at the University of Illinois at Urbana–Champaign.

Electrified Aircraft Propulsion

Powering the Future of Air Transportation

Edited by

KIRUBA HARAN
University of Illinois at Urbana–Champaign

NATERI MADAVAN
NASA Aeronautics Mission Directorate, NASA

TIM C. O'CONNELL
P. C. Krause & Associates

CAMBRIDGE
UNIVERSITY PRESS

CAMBRIDGE
UNIVERSITY PRESS

University Printing House, Cambridge CB2 8BS, United Kingdom

One Liberty Plaza, 20th Floor, New York, NY 10006, USA

477 Williamstown Road, Port Melbourne, VIC 3207, Australia

314–321, 3rd Floor, Plot 3, Splendor Forum, Jasola District Centre, New Delhi – 110025, India

103 Penang Road, #05–06/07, Visioncrest Commercial, Singapore 238467

Cambridge University Press is part of the University of Cambridge.

It furthers the University's mission by disseminating knowledge in the pursuit of education, learning, and research at the highest international levels of excellence.

www.cambridge.org
Information on this title: www.cambridge.org/9781108419345
DOI: 10.1017/9781108297684

© Cambridge University Press 2022

This publication is in copyright. Subject to statutory exception and to the provisions of relevant collective licensing agreements, no reproduction of any part may take place without the written permission of Cambridge University Press.

First published 2022

A catalogue record for this publication is available from the British Library.

Library of Congress Cataloging-in-Publication Data
Names: Haran, Kiruba, editor.
Title: Electrified aircraft propulsion : powering the future of air transportation / edited by Kiruba Haran, University of Illinois at Urbana-Champaign, Nateri Madavan, NASA Aeronautics Mission Directorate, NASA, Tim C. O'Connell, P.C. Krause & Associates.
Description: Cambridge, United Kingdom ; New York, NY, USA : Cambridge University Press, 2021. | Includes bibliographical references and index.
Identifiers: LCCN 2021012177 (print) | LCCN 2021012178 (ebook) | ISBN 9781108419345 (hardback) | ISBN 9781108297684 (epub)
Subjects: LCSH: Electric airplanes. | BISAC: TECHNOLOGY & ENGINEERING / Engineering (General) | TECHNOLOGY & ENGINEERING / Engineering (General)
Classification: LCC TL683.3 .E44 2021 (print) | LCC TL683.3 (ebook) | DDC 629.134/35–dc23
LC record available at https://lccn.loc.gov/2021012177
LC ebook record available at https://lccn.loc.gov/2021012178

ISBN 978-1-108-41934-5 Hardback

Cambridge University Press has no responsibility for the persistence or accuracy of URLs for external or third-party internet websites referred to in this publication and does not guarantee that any content on such websites is, or will remain, accurate or appropriate.

To all scientists and engineers working to make electrified aircraft propulsion a reality

Contents

Contributors

Thanatheepan Balachandran
Department of Electrical and Computer Engineering,
University of Illinois at Urbana-Champaign, Urbana, IL

Robert Bayles
Collins Aerospace, Rockford, IL

Ruirui Chen
University of Tennessee–Knoxville, Knoxville, TN

Rodger Dyson
NASA Glenn Research Center, Cleveland, OH

Jonathan C. Gladin
Georgia Tech University, Marietta, GA

Handong Gui
University of Tennessee–Knoxville, Knoxville, TN

Kiruba Haran
Department of Electrical and Computer Engineering,
University of Illinois at Urbana-Champaign, Urbana, IL

Timothy Haugan
Air Force Research Laboratory, Wright-Patterson AFB, OH

Ralph Jansen
NASA Glenn Research Center, Cleveland, OH

Charles E. Lents
Raytheon Technologies Research Center, Amston, CT

Nateri Madavan
NASA Ames Research Center, Moffett Field, CA

Ajay Misra
NASA Glenn Research Center, Cleveland, OH

Tim C. O'Connell
PC Krause & Associates, Inc., Savoy, IL

Fei (Fred) Wang
University of Tennessee–Knoxville, Knoxville, TN

Patrick Wheeler
Department of Electrical and Electronic Engineering, University of Nottingham, University Park, Nottingham

Xiaolong Zhang
University of Illinois at Urbana-Champaign, Urbana, IL 61801

Zheyu Zhang
Clemson University, Charleston, SC

Preface

The genesis of this book was two special sessions on electrified aircraft propulsion (EAP) at the Institute of Electrical and Electronics Engineers (IEEE) Energy Conversion Conference and Exhibition (ECCE), held in Montreal, Canada, in September 2015. At that time, small two- to four-passenger electric aircraft were being developed, and the circumnavigation by the solar-powered, single-seat, all-electric *Solar Impulse 2* was underway, catching the imagination of the public. Although small electric planes could operate using 2015 technology, the consensus was that disruptive, breakthrough technologies in several areas would be required to enable electrified propulsion for the vast fleet of large commercial transport aircraft that accounted for most of the aviation industry's carbon emissions. To spark innovation, the US National Aeronautics and Space Administration (NASA) Aeronautics division, through its Advanced Air Transport Technology Project (AATT), had recently launched several significant NASA Research Agreements (NRAs) to develop technologies enabling megawatt-class electric propulsion systems. The ECCE sessions were organized to bring together technology experts and raise awareness of the significant challenges addressed by these NRAs. They were well attended and well received, revealing great enthusiasm in the technical community to overcome the challenges ahead.

The success at ECCE 2015 spawned numerous similar sessions at related conferences, and in April 2016, a technology road map for large electrical machines was drafted at a workshop hosted at the University of Illinois at Urbana–Champaign by the Grainger Center for Electric Machinery and Electromechanics and NASA. Joint workshops between the American Institute of Aeronautics and Astronautics (AIAA) and IEEE were then held, fostering stronger collaboration between the electrical engineering and aerospace communities and culminating in the creation of the annual Electric Aircraft Technologies Symposium (EATS), collocated with the popular AIAA Propulsion and Energy Forum. Seeing the momentum building in this burgeoning field, Cambridge University Press approached Nateri Madavan and Kiruba Haran to distill the important concepts discussed at these events into an edited book. Contributions were solicited from a large team of diverse technology experts to create a compendium of the emerging field. Tim O'Connell then joined the team of editors to perform detailed copyediting of the contributed works and help push the book across the finish line.

Electrified Aircraft Propulsion: Powering the Future of Air Transportation is intended for engineers, scientists, advanced undergraduate or graduate students, management professionals in aerospace-related fields, and policymakers who want to understand the numerous obstacles to electrifying the propulsion systems of large, passenger-class (or similar-sized) aircraft. As these obstacles are many, the book necessarily spans a wide range of topics, including power systems and electronics, materials science, superconductivity/cryogenics, thermal management, battery chemistry, system design, and optimization. Rather than "deep-diving" into any one area, it contains general concepts, tools, and information, providing the reader with a solid top-level understanding of the material. Thus, while it may be used as a textbook for a university-level survey course, this book's primary function is as a desk reference and launching point to deeper study.

In nine comprehensive chapters, this book addresses three main questions:

(1) What are the benefits of electrified propulsion for large aircraft?
(2) What technology advancements are required to realize these benefits?
(3) How can the aerospace industry transition from today's state of the art to these advanced technologies?

To address these in a logical way, the book has been arranged to create a natural progression through its wide range of topics. While every attempt has been made to ensure the consistent use of notation, variables, acronyms, etc., because each chapter has been written by a separate technology area expert (or group of experts), there may be some discrepancies. For example, for practical or historical reasons, it is common to find the same engineering concept taught and used slightly differently across disciplines, each with its own unique notations, equations, variables, etc. In such cases, no attempt has been made to unify the notations. Rather, it is the editors' belief that each subject area should be presented in its "native" format so that the reader will more easily be able to apply the knowledge gained herein to other related sources.

Further, because this field is rapidly evolving, some of its key identifying terms and definitions have undergone several iterations over the years. For example, when this book was first proposed, its working title was *Hybrid Electric Aircraft Propulsion*. The term "hybrid" was originally used in the community to describe systems that used any combination of traditional and nontraditional propulsion. When these concepts were new, hybrids were a natural stepping stone between traditional turbofan propulsion and fully electric propulsion. Over the years, as propulsion systems have become more fully electrified, terms such as "electric/electrified aircraft," "electric propulsion," and "propulsion electrification" have become more prevalent. At the time of this writing, a consensus seems to have been reached for the all-encompassing term "electrified propulsion," which can include architectures that derive some or all of their propulsion energy from electricity. This includes those architectures previously called "hybrid." The book's final published title reflects this consensus. While we have made every attempt at a consistency of notation throughout, the legacy of diverse terminology lives on, and there are likely a few cases where alternative terms are used. In the context of this text, the reader can safely assume that all of these terms are essentially synonymous with each other.

Back to the three main questions: in Chapter 1, "Benefits of Electrified Propulsion for Large Aircraft," Rodger Dyson, Ralph Jansen, and Nateri Madavan address the first one, making the case for EAP through numerous trade studies and the analysis of several concept vehicles. Because it introduces the architectures that are the focus of the text and lays the groundwork for the chapters that follow, Chapter 1 is recommended as a prerequisite to the other chapters. Beyond this, the chapters largely stand alone and can be studied in any order that suits the reader.

The subsequent eight chapters, while remaining focused on question 1, collectively address questions 2 and 3 in technical detail. In Chapter 2, "Aircraft Electric Power System Design, Control, and Protection," Robert Bayles presents a modern aircraft electric power system (EPS), summarizing its design, control, and protection functions. Several key EPS components and features are described, which provide context to the rest of the book.

Electric machines (EMs) are the topic of Chapters 3 and 4. In Chapter 3, "Megawatt-Scale Electric Machines for Electrified Aircraft Propulsion," Tim C. O'Connell and Xiaolong Zhang focus on the large, conventional (i.e., non-cryogenic), megawatt-scale EMs that are required to facilitate electric propulsion in large passenger-class aircraft. The material provides a comprehensive overview, with a focus on methods for mass reduction and specific power improvement. In Chapter 4, "Superconducting Machines and Cables," Thanatheepan Balachandran, Timothy Haugan, and Kiruba Haran propose options for drastically increasing machine specific power and efficiency based on superconducting (SC) technology. The authors present several SC machine topologies being pursued by different research groups, followed by the physics and advantages of SC cables, and a look forward based on current SC technology trends.

Chapters 5 and 6 focus on power electronic circuits. In Chapter 5, "Conventional Power Electronics for Electrified Aircraft Propulsion," Patrick Wheeler introduces the basic concepts of power electronics, focusing on those circuits and devices that are crucial for EAP. As with EMs, the push for ever-higher specific power requires power circuits capable of power levels and efficiencies well beyond the current state of the art. This motivates Chapter 6, "Cryogenic Power Electronics," by Zheyu Zhang, Fei (Fred) Wang, Ruirui Chen, and Handong Gui. In it, the authors describe the development of cryogenic power electronics, from the component up to the converter level, highlighting their massive potential for high specific power.

In Chapter 7, "Electrochemical Energy Storage and Conversion for Electrified Aircraft," Ajay Misra provides an overview of the electrochemical energy storage and conversion systems for electrified aircraft, including batteries, fuel cells, supercapacitors, and multifunctional structures with energy storage capability. He highlights the extremely high energy storage requirements of fully electric passenger-class aircraft and points to some promising technologies on the horizon that may help us get there.

Electric drivetrains for EAP have unique thermal management requirements. Charles E. Lents addresses these in Chapter 8, "Thermal Management of Electrified Propulsion Systems," in which he walks through the fundamental equations and calculations for designing a notional TMS that includes both a liquid- and oil-cooling loop and multiple heat exchangers.

The book concludes with Chapter 9, "Performance Assessment of Electrified Aircraft," in which Jonathan C. Gladin ties the material together by presenting a systematic performance assessment process for EAP concept architectures. Gladin calls upon concepts and conclusions from earlier chapters to present this comprehensive method for electric aircraft design.

This book represents years of work by a diverse community of engineers and scientists who are dedicated to making electrified propulsion a reality. As with any compendium of state-of-the-art technology, it provides a snapshot of a quickly evolving field and makes projections using the best knowledge available today. It is our hope that it will provide inspiration and stimulate continued research and rapid innovation. Paradoxically, this book will have successfully done its job if the material herein becomes dated quickly, meriting a second edition. We hope that you will enjoy reading and using it as much as we have enjoyed producing it.

Finally, we would like to thank all of our authors for taking time out of their busy schedules to contribute their knowledge to this collection, and Thanatheepan Balachandran for preparing many of the high-quality figures for production.

1 Benefits of Electrified Propulsion for Large Aircraft

Rodger Dyson, Ralph Jansen, and Nateri Madavan

Introduction

Three main questions are addressed in this book:

(1) What are the benefits of electrified propulsion for large aircraft?
(2) What technology advancements are required to realize these benefits?
(3) How can the aerospace industry transition from today's state of the art to these advanced technologies?

This chapter addresses the first question, making the case for electrified aircraft propulsion (EAP) through numerous trade studies and the analysis of several concept vehicles. This will lay the groundwork for the chapters that follow, which – while remaining focused on question 1 – collectively address questions 2 and 3 in technical detail.

There is substantial interest in the investigation of improvements to aircraft efficiency through the introduction of electrical components into the propulsion system. In the case of turboelectric and hybrid electric aircraft, the electrical systems provide unmatched flexibility in coupling the power generation turbines to the fan propulsors. This flexibility facilitates tight propulsion–airframe integration and can result in reduced noise, emissions, and fuel burn. However, the greatly expanded electrical system incurs substantial weight and efficiency penalties at odds with its benefits. A promising intermediate step between a conventional turbofan aircraft and a fully turboelectric or electric aircraft is an aircraft with a partially turboelectric or hybrid electric propulsion system. Initial studies show that significant aerodynamic benefits can be achieved by sourcing just a small fraction of the propulsive power electrically. However, it is difficult to arrive at authoritative conclusions since the concept aircraft configurations thus far considered and many of the major electrical system components have yet to be built or verified.

In this chapter a breakeven analysis is presented to elucidate the electrical power system performance requirements necessary to achieve electrified aircraft propulsion – specifically fully turboelectric, partially turboelectric, and parallel hybrid electric. This first-order analysis provides a framework for comparing electric drive system performance factors, such as the electrical efficiency, in the context of aircraft propulsion systems. The value of this analysis is both to guide electrical system component research and to provide aircraft configuration researchers with reasonable component expectations.

Similar parametric analyses were presented previously for a fully turboelectric propulsion system [1] and a partially turboelectric system [2]. The study summarized here investigates a broader array of aircraft types, including the fully and partially turboelectric aircraft already addressed, as well as parallel hybrid electric aircraft. In the cases of partially turboelectric and hybrid electric systems, the fraction of thrust power will be varied between the turbofan engines and electric distribution to additional propulsors. A key difference between this study and the prior studies is in the breakeven analysis assumptions. Here the input power and ratio of operating empty weight to aircraft initial weight are held constant among the aircraft types, in addition to equating the range and payload weight. The other studies held either the initial aircraft weight or the fuel weight, as well as the operating empty weight, to be the same.

1.1 Benefits and Costs of Electrified Aircraft Propulsion

1.1.1 Benefits of Electrified Aircraft Propulsion

The turboelectric aircraft propulsion-derived system benefits have been described in previous papers by Jansen et al. [1, 2], and the main points are now summarized. Higher propulsive efficiency due to increased bypass ratio (BPR), higher propulsive efficiency due to boundary layer ingestion (BLI), and lift-to-drag ratio (L/D) improvements are facilitated by EAP.

The introduction of an electric drive system between the turbine and fan enables the decoupling of their speeds and inlet/outlet areas. With this approach, high BPR can be achieved since any number and size of fans can be driven from a single turbine. Increasing BPR results in improved propulsive efficiency. Also, the speed ratio between the turbine and the fan can be arbitrarily set and varied during operation, thereby removing the physical constraint levied by either direct shaft or geared coupling. As a result, the fan pressure ratio and the turbine/compressor ratios can be optimized independently. The propulsive efficiency benefits due to higher BPR could be as high as 4–8 percent [3, 4].

BLI increases propulsive efficiency by ingesting lower velocity flow near the airframe into the propulsors, reenergizing the wake, and thereby reducing drag. BLI can be implemented on both conventional tube-and-wing and hybrid wing body (HWB) aircraft. The propulsor is mounted such that the slow-moving flow near the aircraft is ingested, reenergized, and exhausted where the aircraft wake would have been. The BLI benefits to propulsive efficiency are expected to be 3–8 percent [4, 5]. Combining BPR and BLI propulsive efficiencies listed here yields improvements of 7–17 percent.

Distributed propulsion is expected to improve both lift and L/D ratio through wing flow circulation control. The propulsors can be distributed above or below, or embedded in the traditional tube-and-wing configuration. Likewise, HWB configurations can employ fans distributed across or embedded within the upper surface. Improvements in L/D ratio may result in smaller wing area and reduced drag and weight. The benefits of lift augmentation can be taken in reduced wing area for a given load capacity or shorter takeoff distances. Reduction in wing area lowers wing weight and drag, thereby imparting fuel savings. Alternatively, the improved lift could be focused on increased climb rate and reduced takeoff distance in order to decrease the noise footprint around the airfield. The L/D ratio could be improved by 8 percent [6] to 16 percent [5].

1.1.2 Costs of Electrified Aircraft Propulsion

Introducing an electric drive system, with or without batteries, into the aircraft propulsion system will add weight and reduce efficiency. Here, the electric drive system includes the electric machines, the power management and distribution system, and the thermal system related to heat removal in the two prior systems. Specifically, the electric drive system could include generators, rectifiers, distribution wiring, fault protection, inverters, motors, and the thermal control for those components.

The United States National Aeronautics and Space Administration (NASA) is investigating high-performance motors and batteries that could make electrified aircraft propulsion viable. With regard to the electric drive components, NASA is looking to improve both the efficiency and the specific power of generators, motors, inverters, and rectifiers. A NASA research announcement has a goal of developing technologies and demonstrating an MW-class motor with efficiency greater than 96 percent and power density of greater than 13 kW/kg. This is just one component of the electric drive system. The partially turboelectric STARC-ABL (single-aisle turboelectric aircraft with aft boundary layer propulsor) aircraft concept assumes those values for the motors and generators, as well as rectifiers and inverters with 19 kW/kg and 99 percent efficiency. Stacking up all the components for this aircraft – including cables, circuit protection, and thermal management – yields an electric drive efficiency of 89.1 percent [7].

With regard to batteries, current state-of-the-art lithium-ion batteries have a specific energy on the cell level of up to 200 Wh/kg. Projected values in 15 and 30 years are 650 and 750 Wh/kg, respectively, for lithium–sulfur (LiS), and 950 and 1,400 Wh/kg, respectively, for lithium–air (Li-Air) [8]. These values have to be de-rated based on depth of discharge, battery structure, and battery management. For comparison, the specific energy of aviation fuel is approximately 12,000 Wh/kg.

Clearly, the benefits of improved propulsive efficiency from high BPR and BLI, as well as increased L/D, must be greater than the costs of electrified aircraft propulsion, and the balance of these benefits and constraints are presented here.

1.1.3 Aircraft Concepts with Electrified Propulsion

NASA has been investigating several different EAP systems for aircraft, including fully turboelectric, partially turboelectric, and parallel hybrid electric systems.

The N3-X concept shown in Figure 1.1 is a 300-passenger, HWB aircraft with a fully turboelectric propulsion system and a design range of 7,500 nautical miles (NM).

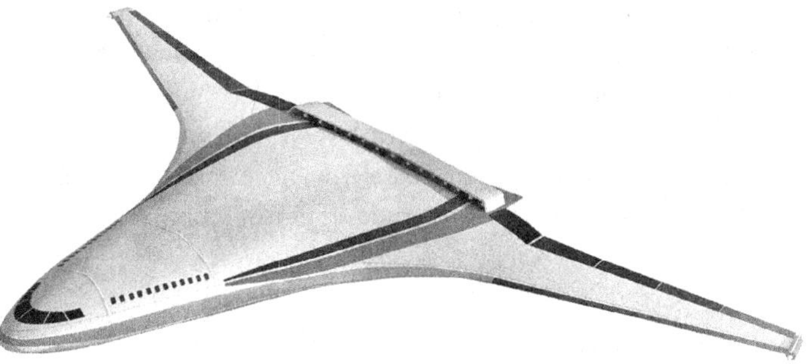

Figure 1.1 N3-X concept vehicle.

Figure 1.2 STARC-ABL concept vehicle.

Turbine engines are located at the wing tips, powering generators. Electric power is then transmitted through cables to a series of motor-driven fans located near the trailing edge of the aircraft. This configuration allows for a higher lift-to-drag ratio due to the hybrid wing body, as well as higher propulsive efficiency due to the increase in fan bypass ratio and boundary layer ingestion. This concept, described by Felder et al. [5], was conceived as a future-generation aircraft to meet NASA's goal of 70 percent fuel burn reduction. Out of the 70 percent overall improvements, 18–20 percent of fuel burn reduction was attributed to the turboelectric propulsion system architecture.

Figure 1.2 shows the partially turboelectric concept STARC-ABL, which is a 154-passenger aircraft with a design range of 3,500 NM. This commercial transport concept was developed for notional entry into service in 2035 and was compared to a conventional configuration using similar technology by Welstead and Felder [9]. The propulsion system consists of two underwing turbofans with generators extracting power from the fan shaft and transmitting it to a rear fuselage, axisymmetric, boundary layer ingesting fan. The power to the tailcone fan is constant and contributes approximately 20 percent of the thrust at takeoff and about 45 percent of the thrust at cruise. Analysis in [9] indicates that the partially turboelectric concept has an economic mission fuel burn reduction of 7 percent, and a design mission fuel burn reduction of 12 percent compared to the conventional configuration. It should be noted that subsequent studies have

Figure 1.3 PEGASUS concept vehicle.

predicted fuel burn reductions that are in the range of 3–4 percent, but they were not available for referencing at the time of this publication.

The Parallel Electric-Gas Architecture with Synergistic Utilization Scheme (PEGASUS) concept is shown in Figure 1.3, which is a 48-passenger parallel hybrid electric aircraft. This concept is described by Antcliff and Capristan [10]. A detailed analysis of an intermediate parallel hybrid electric concept was performed by Antcliff et al. [11], which was based on the ATR-42-500 conventional fuel-based aircraft with a range of 600 NM. The analysis included various levels of battery specific energy, which is a critical parameter as battery weight has been shown to be a significant penalty for these types of aircraft. They found that a specific energy of 750 Wh/kg was required to break even on total energy, even as the aircraft weight increased over the baseline value.

The N3-X, STARC-ABL, and PEGASUS concepts will be used as case studies for the breakeven analysis in this study.

1.2 Breakeven Analysis

1.2.1 Key Performance Parameters and Key Assumptions

In order to conduct the breakeven analysis, we first define the key performance parameters (KPPs), the key assumptions, and the electrical power system boundary. Then we will formulate range equations for each aircraft type. Finally, we find the breakeven relationship by implicitly solving for the electric drive specific power and efficiency while holding constant the ratio of operating empty weight to initial weight, payload weight, input energy (from fuel and/or batteries), and aircraft flight range. The resulting parametric curves can be used as the top-level requirements for the electrical power system and bounding guidelines for further aircraft exploration.

Specifically, the key performance parameters (KPPs) are as follows:

- Electric drive system efficiency η_{elec}, expressed as a percentage.
- Electric drive system specific power Sp_{elec}, in kW/kg.
- Electric propulsion fraction ξ for partially turboelectric and parallel hybrid electric aircraft.

The breakeven assumptions in this analysis used to determine the values of the KPPs include the following:

- The ranges of the conventional and electrified aircraft are equal.
- The input energy (fuel and/or battery energy) of the conventional and electrified aircraft are equal.
- The payload weights of all the aircraft are equal.
- The ratios of operating empty weights (OEW) to initial aircraft weights are equal, where OEW does not include the weights of the electric drive and batteries.

1.2.2 Electrified Aircraft Propulsion System Definitions

Each EAP system will now be described, along with the boundaries of the electric drive system for each case. In Figures 1.4–1.7 are shown simplified diagrams of the conventional (fuel-based) turbofan, fully turboelectric, partially turboelectric, and parallel hybrid electric aircraft propulsion systems, respectively. The conventional turbofan system is considered the baseline aircraft system for comparison. The building blocks of the systems are the energy source (fuel and/or battery), the turbine engine, the propulsor, and the electric drive for the EAP cases. We denote the *conventional turbofan aircraft*, *fully turboelectric*, *partially turboelectric*, and *parallel hybrid electric* parameters with the subscripts *AC*, *TE*, and *PE*, and *HE*, respectively. *Power* is denoted by the letter *P*, *efficiency* by η, *specific energy* by *Se*, and *specific power* by *Sp*.

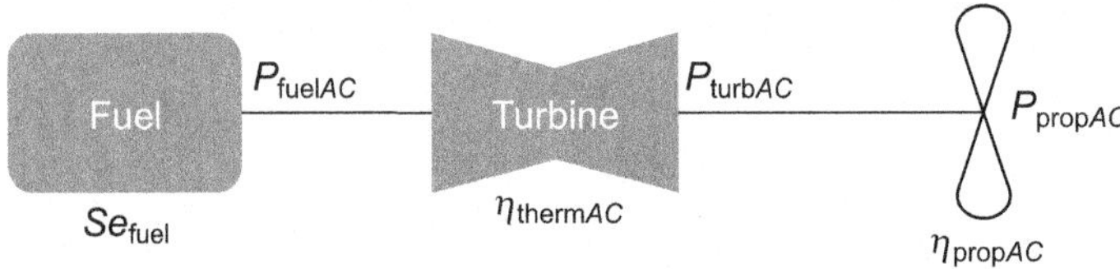

Figure 1.4 Conventional, fuel-based aircraft propulsion system (AC).

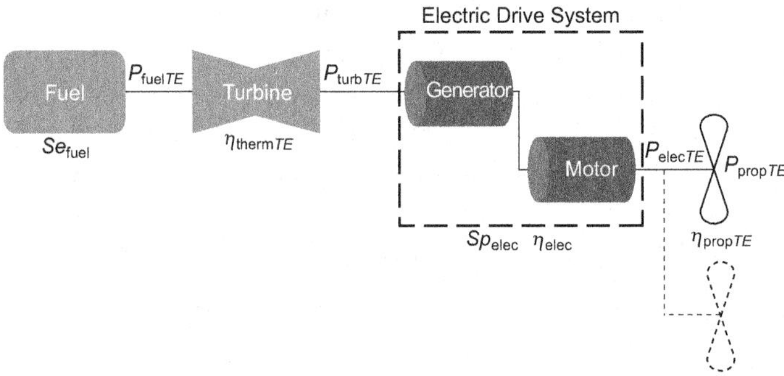

Figure 1.5 Fully turboelectric aircraft propulsion system (TE).

The turbine, propulsor, and electric drive have associated thermal (η_{therm}), propulsive (η_{prop}), and electrical (η_{elec}) efficiencies, expressed throughout this chapter as percentages. The fuel power P_{fuel}, battery power P_{batt}, turbine engine power P_{turb}, electrical power P_{elec}, and propulsive power P_{prop}, all in kW, are defined as output power of the fuel, battery, turbine engine, electric drive, and propulsors, respectively. The variables in Figures 1.4–1.7 illustrate the association between the propulsive subsystems, powers, and efficiencies for each propulsion system. In the partially turboelectric and parallel hybrid electric cases, we must introduce the electrical propulsion fraction ξ, defined as the fraction of total aircraft thrust at cruise produced by electrically driven propulsors. When the electrical propulsion fraction is equal to one, all the thrust during cruise is provided by electrically driven propulsors. The fully turboelectric system is one in which all the thrust throughout the mission, including takeoff and cruise, is provided by electrically driven propulsors. Therefore, the electric drive system will need to be sized accordingly.

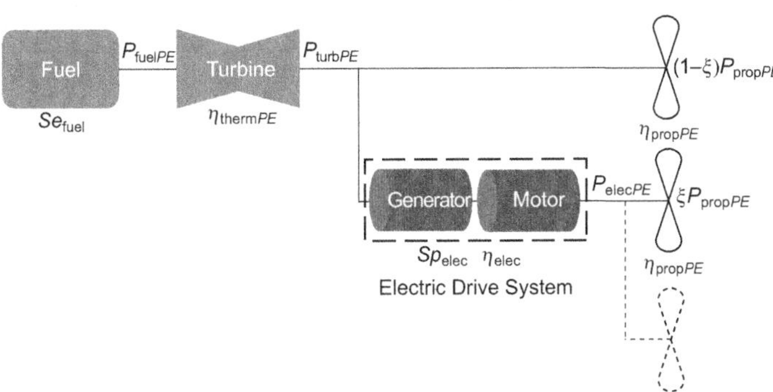

Figure 1.6 Partially turboelectric aircraft propulsion system (PE).

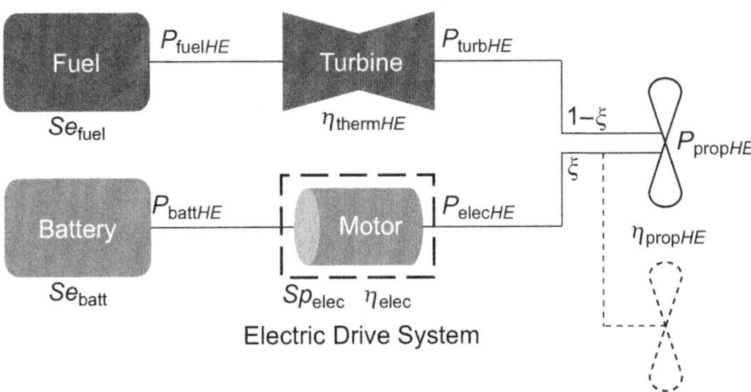

Figure 1.7 Parallel hybrid electric propulsion system (HE).

The electric drive specific power, efficiency, and electrical propulsion fraction are the electrified aircraft electric drive system KPPs. Specific power is the ratio of the rated electric drive output power to its mass. Efficiency is the ratio of the output power to the input power of the electric drive system, multiplied by 100 percent. Electrical propulsion fraction is the fraction of total aircraft thrust at cruise produced by electrically driven propulsors. These three KPPs will be used to describe electrical power system performance and establish necessary levels of performance.

The boundary of the electric drive system is defined to lend meaning to the KPPs. In the analysis presented here, the boundary will include generators, rectifiers, distribution wiring, fault protection, inverters, motors, and the thermal control for those components. The parallel hybrid electric system does not require generators. Some variants of the electrical drive system may use a subset of these components or alternative layouts. The specific power and electrical efficiency analyzed in this study include all of the components inside the boundary. Notably, the turbine engine and the propulsors are outside the electric drive boundary.

A simplified assessment of the relationship of the electric drive system KPPs, aircraft range, and input energy is proposed for top-level aircraft performance comparisons. The range equations are discussed first, followed by the input energy, and, finally, the component weights. The breakeven equations are derived for fully turboelectric, partially turboelectric, and parallel hybrid electric aircraft.

1.2.3 Breakeven on Range

The basis of the analysis is an expansion of the traditional terms in the Breguet range equation for fuel-based aircraft to include the efficiency and weight of the electric drive system. The range equation for battery-powered aircraft from Hepperle [12] is expanded in a similar way. These equations apply to situations where overall aerodynamic efficiency, L/D, and flight velocity are constant over the duration of cruise. Although this is not true for the entire flight envelope, this description is a reasonable approximation for cruise conditions.

We develop range equations of the typical form, representing the conventional and EAP aircraft configurations concurrently for comparison. The range equations for fuel-based and battery-based aircraft are, respectively,

$$R_{\text{fuel}} = \frac{Se_{\text{fuel}}}{g}\frac{L}{D}\eta_{\text{o}}\ln\left(\frac{W_{\text{i}}}{W_{\text{f}}}\right) \tag{1.1}$$

and

$$R_{\text{batt}} = \frac{Se_{\text{batt}}}{g}\frac{L}{D}\eta_{\text{o}}\left(\frac{W_{\text{batt}}}{W_{\text{i}}}\right), \tag{1.2}$$

where R is the range, in m; Se_{fuel} and Se_{batt} are the specific energies of the fuel and battery, respectively, in J/Kg; and η_{o} is the overall efficiency percentage of the propulsion system.

For fuel-based aircraft, the final aircraft weight, W_f, is equal to the initial aircraft weight, W_i, minus the fuel weight, W_{fuel}, where all are expressed in N. Thus, the fuel-based range equation is

$$R_{fuel} = \frac{Se_{fuel}}{g}\frac{L}{D}\eta_o \ln\left(\frac{1}{1 - W_{fuel}/W_i}\right). \tag{1.3}$$

Note that for small values of W_{fuel}/W_i,

$$\ln\left(\frac{1}{1 - W_{fuel}/W_i}\right) \approx \frac{W_{fuel}}{W_i}, \tag{1.4}$$

which shows that Equations (1.1) and (1.2) have a similar form. Thus, the range is approximately proportional to the ratio of the energy source weight to the aircraft initial weight. Since $Se_{batt} \ll Se_{fuel}$, battery weight for the same range will be much larger than fuel weight.

The overall efficiency of each aircraft type is defined in Equations (1.5)–(1.8) in Table 1.1 as functions of propulsive, thermal, and electric drive efficiency. Note that the propulsive efficiency defined here is actually the product of transfer efficiency and propulsive efficiency.

To see how adding the electric drive system affects overall efficiency, the ratio of electrified aircraft to baseline conventional overall efficiency is plotted in Figure 1.8 as a function of electric propulsion fraction. Here it is assumed that the thermal efficiency is 55 percent and the electric drive efficiency is 90 percent. Increasing ξ decreases overall efficiency for the turboelectric cases, since the electric drive system is in series with the turbine engine. Since η_{elec} is larger than η_{therm}, the hybrid electric system has increasing overall efficiency compared to the baseline. However, the battery weight required for hybrid electric will be a significant penalty in the breakeven analysis.

1.2.4 Breakeven on Input Energy

The input energy of fuel is simply the product of the specific energy of the fuel and the fuel mass. Similarly, the input energy of the battery is simply the product

Table 1.1 Overall efficiency equations.

Aircraft type	Overall efficiency	
Conventional aircraft (AC)	$\eta_{oAC} = \eta_{propAC} \times \eta_{thermAC}$	(1.5)
Fully turboelectric aircraft (TE)	$\eta_{oTE} = \eta_{propTE} \times \eta_{thermTE} \times \eta_{elecTE}$	(1.6)
Partially turboelectric aircraft (PE)	$\eta_{oPE} = \dfrac{\eta_{propPE} \times \eta_{thermPE} \times \eta_{elecPE}}{(1-\xi)\eta_{elecPE} + \xi}$	(1.7)
Parallel hybrid electric aircraft (HE)	$\eta_{oHE} = \dfrac{\eta_{propHE} \times \eta_{thermHE} \times \eta_{elecHE}}{(1-\xi)\eta_{elecHE} + \xi\eta_{thermHE}}$	(1.8)

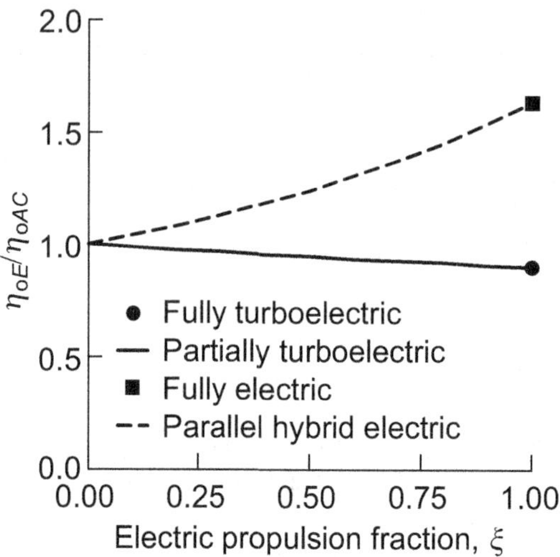

Figure 1.8 Ratio of electrified to conventional aircraft overall efficiency.

of the specific energy of the battery and the battery mass. Thus, the input energy equations are

$$E_{\text{fuel}} = \frac{Se_{\text{fuel}}}{g} W_{\text{fuel}} \tag{1.9}$$

and

$$E_{\text{batt}} = \frac{Se_{\text{batt}}}{g} W_{\text{batt}}. \tag{1.10}$$

1.2.5 Relationship among Aircraft Component Weights

The final part of the breakeven analysis relates the specific power of the EAP system to the other component weights. We know that the initial aircraft weight is defined as the sum of the OEW, payload weight, fuel weight, electric drive system weight (for electrified aircraft), and battery weight (for HE aircraft):

$$W_{\text{i}} = W_{\text{OEW}} + W_{\text{payload}} + W_{\text{fuel}} + W_{\text{elec}} + W_{\text{batt}}. \tag{1.11}$$

From Equation (1.11) we can see that

$$\frac{W_{\text{elec}}}{W_{\text{i}}} = 1 - \frac{W_{\text{OEW}}}{W_{\text{i}}} - \frac{W_{\text{fuel}}}{W_{\text{i}}} - \frac{W_{\text{batt}}}{W_{\text{i}}} - \frac{W_{\text{payload}}}{W_{\text{i}}}, \tag{1.12}$$

noting that the payload weight and the ratio of OEW to initial aircraft weight are constant among the aircraft.

For the TE aircraft, where all the power must pass through the electric drive system, Sp_{elec} will be defined based on takeoff power rather than cruise power.

If we denote the ratio of takeoff to cruise power as α, then the electric drive system weight ratio is [1]

$$\frac{W_{\mathrm{elec}TE}}{W_{\mathrm{i}TE}} = \frac{\alpha v_{\mathrm{cruise}}}{\left(\frac{L}{D}\eta_{\mathrm{prop}}\right)_{TE} Sp_{\mathrm{elec}}/g}. \tag{1.13}$$

Alternatively, it is assumed for the partially turboelectric and parallel hybrid electric cases that electric propulsion power, which is the product of ξ and propulsion power, is not required for takeoff, so the electric drive system weight ratio is defined as [2]

$$\frac{W_{\mathrm{elec}HE,PE}}{W_{\mathrm{i}HE,TE}} = \frac{\xi v_{\mathrm{cruise}}}{\left(\frac{L}{D}\eta_{\mathrm{prop}}\right)_{HE,PE} Sp_{\mathrm{elec}}/g}. \tag{1.14}$$

1.3 Breakeven Results

1.3.1 Fully Turboelectric Aircraft (TE)

Equations for the fully turboelectric aircraft for the range, input energy, and component weight equations, respectively, are as follows:

$$\ln\left(1 - \frac{W_{\mathrm{fuel}TE}}{W_{\mathrm{i}TE}}\right) = \frac{\left(\frac{L}{D}\eta_{\mathrm{prop}}\eta_{\mathrm{therm}}\right)_{AC}}{\left(\frac{L}{D}\eta_{\mathrm{prop}}\eta_{\mathrm{therm}}\eta_{\mathrm{elec}}\right)_{TE}} \ln\left(1 - \frac{W_{\mathrm{fuel}AC}}{W_{\mathrm{i}AC}}\right) \tag{1.15}$$

$$\frac{W_{\mathrm{i}AC}}{W_{\mathrm{i}TE}} = \frac{\left(\frac{W_{\mathrm{fuel}TE}}{W_{\mathrm{i}TE}}\right)}{\left(\frac{W_{\mathrm{fuel}AC}}{W_{\mathrm{i}AC}}\right)} \tag{1.16}$$

and

$$\frac{W_{\mathrm{elec}TE}}{W_{\mathrm{i}TE}} = \left(1 - \frac{W_{\mathrm{fuel}TE}}{W_{\mathrm{i}TE}} - \frac{W_{\mathrm{OEW}}}{W_{\mathrm{i}}}\right) - \frac{W_{\mathrm{i}AC}}{W_{\mathrm{i}TE}}\left(1 - \frac{W_{\mathrm{fuel}AC}}{W_{\mathrm{i}AC}} - \frac{W_{\mathrm{OEW}}}{W_{\mathrm{i}}}\right). \tag{1.17}$$

Several observations can be made from Equations (1.15) to (1.17). First, Equation (1.15) shows that the fuel fraction for the turboelectric aircraft will be reduced if the product of L/D and overall efficiency is increased compared to the baseline aircraft. Then Equation (1.16) shows that the aircraft weight will increase compared to the baseline, which is a result of the added electric drive system.

To solve this set of equations for Sp_{elec}, we first assume a value of η_{elec} (e.g., $\eta_{\mathrm{elec}} = 100$ percent). Equation (1.15) then yields the fuel fraction $W_{\mathrm{fuel}TE}/W_{\mathrm{i}TE}$, given the baseline fuel fraction and assumed values for L/D and η. From Equation (1.16), we find the ratio of conventional initial aircraft weight to turboelectric initial aircraft weight, which is then substituted Equation (1.17) to give the electric drive system weight ratio $W_{\mathrm{elec}TE}/W_{\mathrm{i}TE}$. Finally, Equation (1.13) is solved for Sp_{elec}. This is repeated for a range of values of η_{elec}, resulting in a curve of η_{elec} vs. Sp_{elec} for the turboelectric system. This procedure is used in a similar way for the partially turboelectric and parallel hybrid electric propulsion systems, using the appropriate equations for those aircraft.

Similar to the results given by Jansen et al. [1], the electric drive specific power and efficiency required to break even on range and input energy were determined, based on expected propulsive improvements. Again, the difference between this analysis and the previous analysis is that the breakeven is based on constant input energy and constant ratio of OEW to initial weight in this study, whereas it is based on constant initial weight and OEW in the previous study.

The turboelectric aircraft studied here is based on the NASA N3-X hybrid wing body fully turboelectric aircraft. In Felder et al. [5], the N3-X was compared to two different baseline aircraft configurations – a conventional tube-and-wing aircraft (a Boeing 777-200 LR) and an intermediate hybrid wing body aircraft with conventional propulsion (a NASA N3A concept). Table 1.2 details the parameters used in the analysis. For all the aircraft, it is assumed that the transfer efficiency is 80 percent (which is multiplied by the propulsive efficiency supplied in the study to give η_{prop}), and the thermal efficiency η_{therm} is assumed to be 55 percent.

First, we look at the effect of aero and propulsive benefits on the breakeven curves. Here the baseline parameters η_{prop} and L/D are based on the Boeing 777 aircraft, and the maximum benefits are those for the fully turboelectric N3-X aircraft. We look at three benefit levels between the baseline 777 and N3-X; these include combined aero and propulsive benefits of 7, 18, and 29 percent for minimum, medium, and maximum benefits, respectively. The 29 percent benefit is representative of the N3-X versus the 777 baseline with the L/D and η_{prop} improvements shown in Table 1.2.

The breakeven curves for the three levels of propulsive benefits are shown in Figure 1.9. Electric drive systems with performance above each curve should result in lower fuel burn. Clearly, improving L/D and η_{prop} leads to lower demands on the electric drive system. Table 1.2 includes the specific power and efficiency expected of a superconducting electric drive system, 7.1 kW/kg and 98.54 percent. With these values, only the medium and maximum benefits case would result in lower fuel burn.

Table 1.2 Fully turboelectric aircraft parameters.

Parameter	Baseline 777	Baseline N3A	Turboelectric N3-X
α	2.0	1.8	
v_{cruise} [m/s]	255	255	255
$W_{\text{fuel}AC}/W_{\text{i}AC}$	36%	24%	
$W_{\text{OEW}}/W_{\text{i}}$	48%	54%	48%/54%
L/D	19	22	22
η_{prop}	69.6%	72.2%	77.1%
Sp_{elec} [kW/kg]			7.1
η_{elec}			98.54%

Relaxing the efficiency to 90 percent, as for a non-superconducting electric drive system, only the maximum benefits case would result in lower fuel burn.

The ratio of electric drive weight to initial turboelectric aircraft weight as a function of specific power is illustrated in Figure 1.10. Clearly, the higher the specific power is, the lighter the electric drive system will be. For the minimum allowable specific power of 3.1 kW/kg for maximum benefits at 100 percent efficiency, the electric drive system comprises 9.6 percent of the aircraft weight. This number quickly falls with increasing specific power.

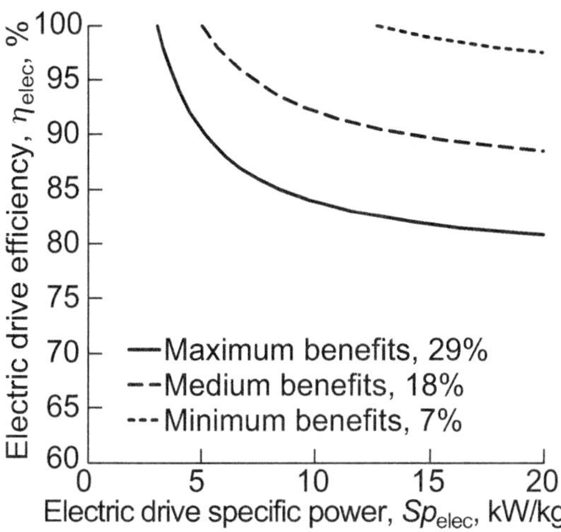

Figure 1.9 Breakeven curves for turboelectric propulsion.

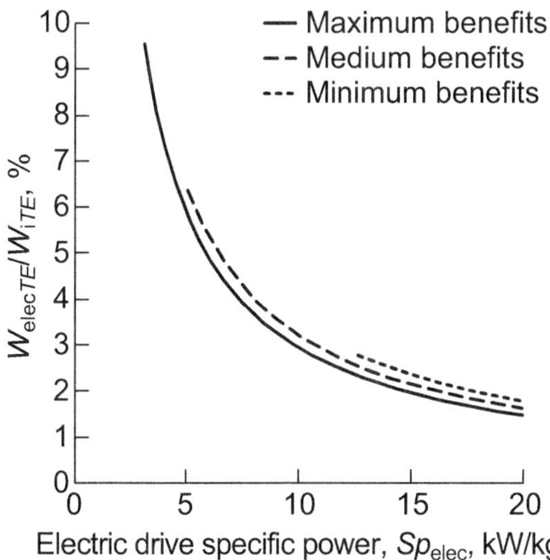

Figure 1.10 Electric drive weight ratio for turboelectric propulsion.

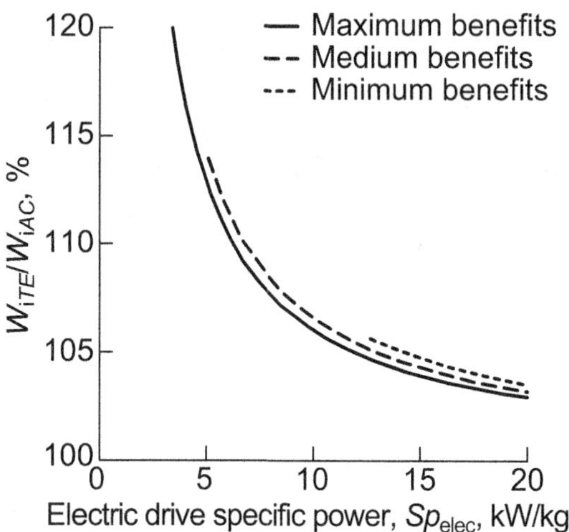

Figure 1.11 Ratio of turboelectric to baseline aircraft weight in breakeven analysis.

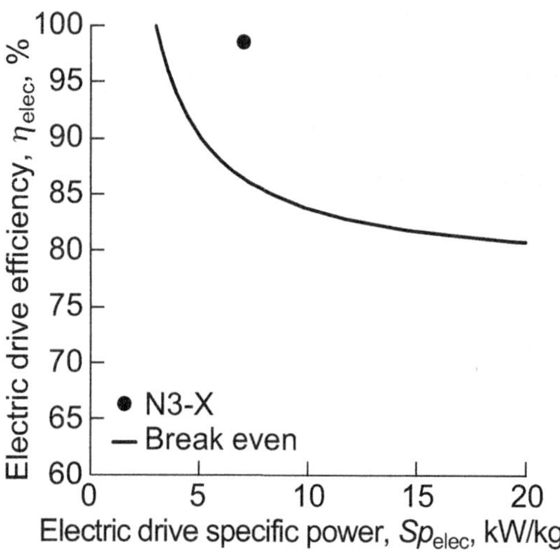

Figure 1.12 Breakeven for N3-X vs. 777.

Finally, Figure 1.11 shows the increase in the turboelectric aircraft weight as a function of electric drive specific power. This particular breakeven analysis results in heavier aircraft, but with the same fuel burn as the baseline aircraft.

The electric drive breakeven curves for the turboelectric N3-X versus the baseline 777 and the baseline N3A, respectively, are shown in Figures 1.12 and 1.13. The electric drive efficiency and power indicated by the orange symbols is for a superconducting system, which has very high performance. In Figure 1.13, we see that the electric drive system used in the N3-X analysis does not provide fuel burn benefits in

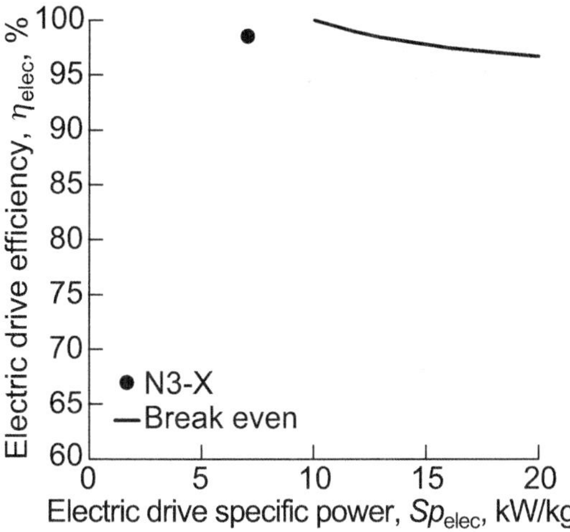

Figure 1.13 Breakeven for N3-X vs. N3A.

this breakeven analysis, even though results in [5] indicate reduced fuel burn. The discrepancy lies in the breakeven analysis assumptions. Here we are assuming equal input power, which in this case is equal fuel burn. This results in a larger aircraft compared to the baseline N3A. However, the N3-X aircraft actually had a 7 percent lower aircraft weight than the baseline N3A. This illustrates the sensitivity of this breakeven analysis to the key assumptions. However, Figure 1.13 does clearly indicate the necessity of choosing the high-performance superconducting electric drive.

1.3.2 Partially Turboelectric Aircraft (PE)

Equations for the partially turboelectric aircraft for the range, input energy, and component weight equations, respectively, are as follows:

$$\ln\left(1-\frac{W_{\text{fuel}PE}}{W_{\text{i}PE}}\right)=\frac{\left(\frac{L}{D}\eta_{\text{prop}}\eta_{\text{therm}}\right)_{AC}}{\left(\frac{L}{D}\frac{\eta_{\text{prop}}\eta_{\text{therm}}\eta_{\text{elec}}}{(1-\xi)\eta_{\text{elec}}+\xi}\right)_{PE}}\ln\left(1-\frac{W_{\text{fuel}AC}}{W_{\text{i}AC}}\right) \tag{1.18}$$

$$\frac{W_{\text{i}AC}}{W_{\text{i}PE}}=\frac{\left(\frac{W_{\text{fuel}PE}}{W_{\text{i}PE}}\right)}{\left(\frac{W_{\text{fuel}AC}}{W_{\text{i}AC}}\right)} \tag{1.19}$$

and

$$\frac{W_{\text{elec}PE}}{W_{\text{i}PE}}=\left(1-\frac{W_{\text{fuel}PE}}{W_{\text{i}PE}}-\frac{W_{\text{OEW}}}{W_{\text{i}}}\right)-\frac{W_{\text{i}AC}}{W_{\text{i}PE}}\left(1-\frac{W_{\text{fuel}AC}}{W_{\text{i}AC}}-\frac{W_{\text{OEW}}}{W_{\text{i}}}\right). \tag{1.20}$$

Table 1.3 Partially turboelectric aircraft parameters.

Parameter	Baseline N3CC	Partially turboelectric STARC-ABL
ξ		45%
v_{cruise} [m/s]	206	206
$W_{\text{fuel}AC}/W_{\text{i}AC}$	17%	
$W_{\text{OEW}}/W_{\text{i}}$	57%	57%
L/D	21.4	22.3
η_{prop}	64%	75.1%
Sp_{elec} [kW/kg]		2.0
η_{elec}		90%

These equations are similar to the fully turboelectric case, except in the definitions of overall efficiency (Eqs. (1.7) vs. (1.6)) and electric drive system weight (Eqs. (1.14) vs. (1.13)).

The effect of electric propulsion fraction on required electric drive system performance was examined for the case of the partially turboelectric STARC-ABL aircraft concept. In [9], Welstead and Felder performed a systems study of the STARC-ABL aircraft compared to an N+3 Conventional Configuration (N3CC) baseline conventional fuel-powered turbofan aircraft. Table 1.3 shows the baseline and partially turboelectric aircraft parameters used in the breakeven analysis. The propulsive efficiency for a CFM56 fan is assumed to be 80 percent, which is multiplied by the transfer efficiency of 80 percent to give 64 percent. Similarly, the propulsive efficiency of 93.9 percent for the GE hFan is used for the STARC-ABL analysis and is multiplied by 80 percent to give 75.1 percent.

If we assume that L/D and η_{prop} are constant with changing electric propulsion fraction, then the breakeven curves are as shown in Figure 1.14. The STARC-ABL aircraft has an electric propulsion fraction ξ of 45 percent at cruise, and if we assume that the aero and propulsive parameters L/D and η_{prop} for the STARC-ABL in Table 1.3 scale with ξ, then the breakeven curves are as shown in Figure 1.15. This shows the effect of the benefits versus the costs of the electric drive system and the importance of predicting those benefits in this type of analysis.

In Figure 1.16 the electric drive weight ratio is shown, assuming constant η_{prop} and L/D. The weights are lower for partially turboelectric compared to the fully turboelectric, since the electric drive system is sized based on cruise power rather than takeoff power. The ratio of initial weights, comparing partially turboelectric aircraft to conventional aircraft, is plotted in Figure 1.17.

The results of the breakeven analysis for the STARC-ABL concept at its design electric propulsion fraction of 45 percent are plotted in Figure 1.18. Here we see that the electric drive efficiency and specific power used in [9] does result in an aircraft with lower fuel burn. Unlike the N3-X example, the STARC-ABL aircraft actually has a 3 percent higher initial weight than the baseline, whereas the breakeven analysis shows a 7 percent higher initial weight at $Sp_{\text{elec}} = 2$ kW/kg.

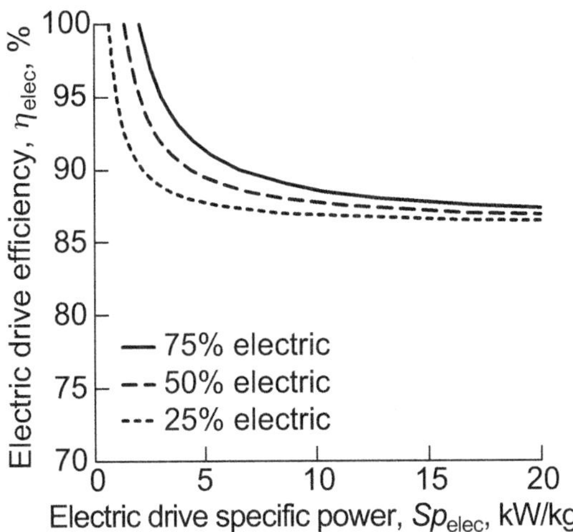

Figure 1.14 Breakeven curves for partially turboelectric aircraft with constant aero and propulsive benefits.

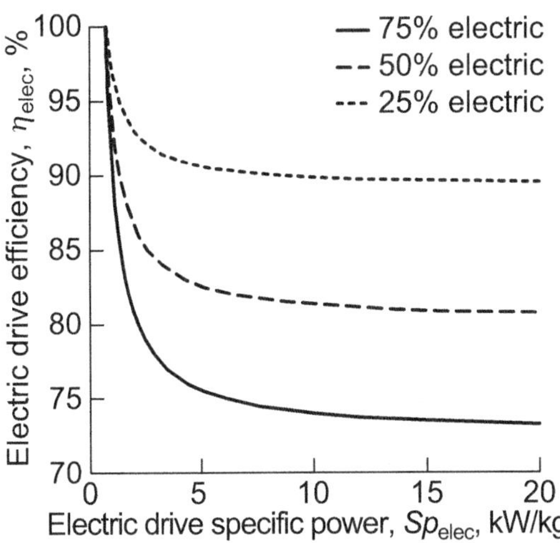

Figure 1.15 Breakeven curves for partially turboelectric aircraft with scaled aero and propulsive benefits.

In general, the breakeven analysis assumptions are similar to the systems study in [9]; therefore, the results are similar.

1.3.3 Parallel Hybrid Electric Aircraft (HE)

Equations for the parallel hybrid electric aircraft for the fuel range, electrical propulsion fraction, input energy, and component weight equations, respectively, are as follows:

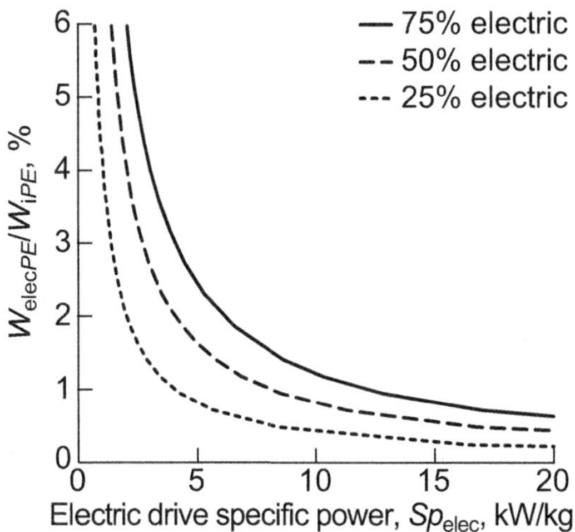

Figure 1.16 Electric drive weight ratio for partially turboelectric aircraft with constant aero and propulsive benefits.

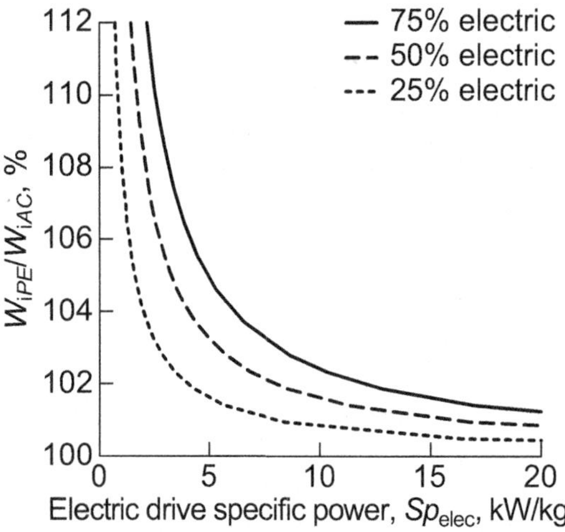

Figure 1.17 Ratio of partially turboelectric to baseline aircraft weight with constant aero and propulsive benefits.

$$\ln\left(1-\frac{W_{\text{fuel}HE}}{W_{\text{i}HE}}\right)=(1-\xi)\frac{\left(\frac{L}{D}\eta_{\text{prop}}\eta_{\text{therm}}\right)_{AC}}{\left(\frac{L}{D}\eta_{\text{prop}}\eta_{\text{therm}}\right)_{HE}}\ln\left(1-\frac{W_{\text{fuel}AC}}{W_{\text{i}AC}}\right) \tag{1.21}$$

$$\frac{W_{\text{batt}HE}}{W_{\text{i}HE}}=\left(\frac{\xi}{1-\xi}\right)\frac{Se_{\text{fuel}}}{Se_{\text{batt}}}\frac{\eta_{\text{therm}HE}}{\eta_{\text{elec}HE}}\frac{W_{\text{fuel}HE}}{W_{\text{i}HE}} \tag{1.22}$$

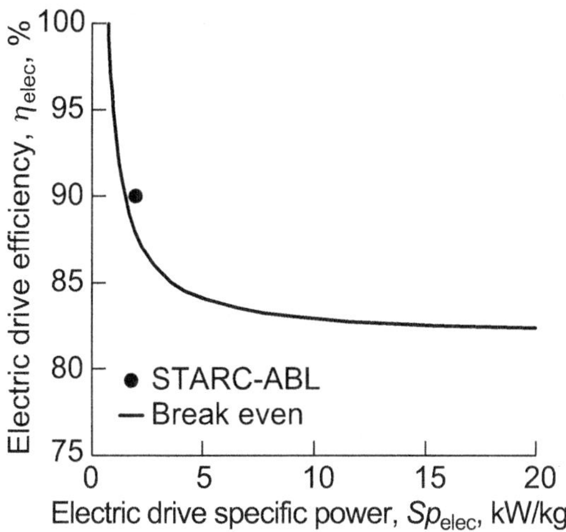

Figure 1.18 Breakeven for STARC-ABL vs. N3CC.

$$\frac{W_{iAC}}{W_{iHE}} = \frac{Se_{batt}\left(\frac{W_{battHE}}{W_{iHE}}\right) + Se_{fuel}\left(\frac{W_{fuelHE}}{W_{iHE}}\right)}{Se_{fuel}\left(\frac{W_{fuelAC}}{W_{iAC}}\right)} \tag{1.23}$$

And

$$\frac{W_{elecHE}}{W_{iHE}} = \left(1 - \frac{W_{fuelHE}}{W_{iHE}} - \frac{W_{OEW}}{W_i} - \frac{W_{battHE}}{W_{iHE}}\right) - \frac{W_{iAC}}{W_{iHE}}\left(1 - \frac{W_{fuelAC}}{W_{iAC}} - \frac{W_{OEW}}{W_i}\right). \tag{1.24}$$

The additional equation in this case, Equation (1.22), results from the assumption that the battery-powered portion of the thrust is defined by the electrical propulsion fraction, ξ. We can see from Equation (1.22) that the ratio of battery weight to initial aircraft weight is directly proportional to the ratio of fuel specific energy to battery specific energy. The fuel specific energy is approximately 12,000 Wh/kg, compared to projected battery specific energy of 500, 750, or 1,000 Wh/kg. It is easy to see that the battery weight can become quite large, making hybrid electric configurations more difficult to implement than partially turboelectric configurations, despite the better overall efficiency. However, there are some conditions under which the hybrid electric configuration is more successful. To that end, we investigate the effect of range, Se_{batt}, and electric propulsion fraction ξ on the breakeven curves.

A breakeven analysis was performed for the parallel hybrid electric aircraft described by Antcliff et al. [10, 11]. This is a short-range aircraft devised for 48 passengers; the shorter range makes it a better choice for hybrid electric. The baseline conventional aircraft is the ATR 42-500, which utilizes two turboprop engines. There is an intermediate parallel hybrid electric concept with a range of 600 NM and the parameters shown in Table 1.4. Here the propulsive efficiencies are calculated assuming a transfer efficiency of 80 percent and $\eta_{\text{therm}} = 55$ percent. The parallel hybrid electric PEGASUS concept has a 400 NM range, and a fully electric (at cruise) PEGASUS concept has a 200 NM range.

To start, the effect of aircraft range was examined. The aircraft range is approximately proportional to the baseline aircraft fuel fraction, $W_{\text{fuel}AC}/W_{\text{i}AC}$. Therefore, examining the effect of this fuel fraction in the breakeven analysis is essentially the same as examining the effect of the range. We looked at two values of baseline fuel fraction: 0.05 (shorter range) and 0.091 (baseline 600 NM). Compared to the aircraft in the turboelectric and partially turboelectric studies, this range is quite small. Figure 1.19 shows the electric drive performance required for the two ranges for $Se_{\text{batt}} = 750$ Wh/kg and $\xi = 25$ percent. Clearly, the parallel hybrid electric

Table 1.4 Parallel hybrid electric aircraft parameters.

Parameter	Baseline	Parallel hybrid electric
ξ		25%, 50%, 75%
v_{cruise} [m/s]	150	150
$W_{\text{fuel}AC}/W_{\text{i}AC}$	9.1%	
$W_{\text{OEW}}/W_{\text{i}}$	64%	64%
L/D	11	15
η_{prop}	60%	72%
Se_{batt} [Wh/kg]		500, 750, 1,000
Sp_{elec} [kW/kg]		7.3
η_{elec}		90%

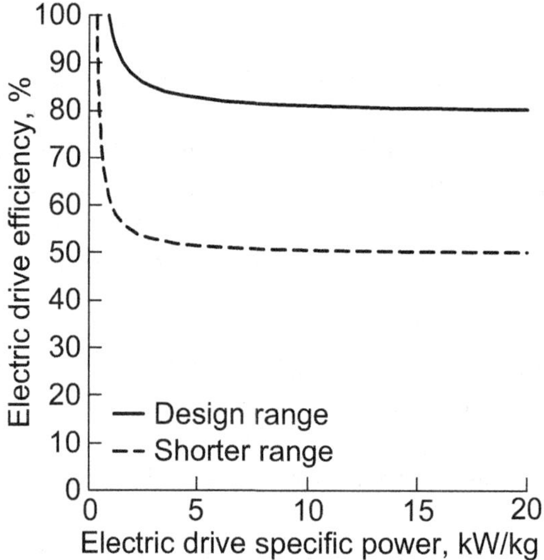

Figure 1.19 Breakeven curves based on aircraft range, $\xi = 25$ percent, $Se_{\text{batt}} = 750$ Wh/kg.

configuration is a better option for shorter range flights, which is expected. Note that the electrical efficiency required for the shorter-range flight is very low. This is a result of the parallel configuration. For constant η_{therm}, as long as $(\eta_{\text{elec}}\eta_{\text{prop}})_{HE} > (\eta_{\text{therm}}\eta_{\text{prop}})_{AC}$, the overall efficiency will be higher than the baseline. There are certainly weight penalties, especially for the battery weight, but these can be overcome depending on the aero and propulsive benefits, which are quite high for this case.

Next, the effect of battery specific energy was examined for the shorter range $W_{\text{fuel}AC}/W_{\text{i}AC} = 0.05$. The results for $\xi = 25$ percent for $Se_{\text{batt}} = 500, 750$, and 1,000 Wh/kg are given in Figure 1.20. As expected, carrying the heavier batteries increases the performance required of the electric drive system.

Figure 1.21 shows the breakeven curves for various values of electric propulsion fraction for $W_{\text{fuel}AC}/W_{\text{i}AC} = 0.05$ and $Se_{\text{batt}} = 750$ Wh/kg, assuming the aero and propulsive benefits are constant. If we assume that these η_{prop} and L/D change with ξ, normalizing the benefits to $\xi = 50$ percent, then the breakeven curves are as shown in Figure 1.22. There is a big difference between the two charts, and it clearly illustrates the balance between the aero and propulsive benefits and the costs of the battery and electric drive system.

Returning to the assumption that the aero and propulsive benefits remain constant, Figures 1.23–1.25, respectively, show the electric drive weight fraction, the battery weight fraction, and the ratio of hybrid electric aircraft weight to conventional aircraft initial weight. Compared to the fully and partially turboelectric aircraft, the hybrid electric aircraft requires significant added weight.

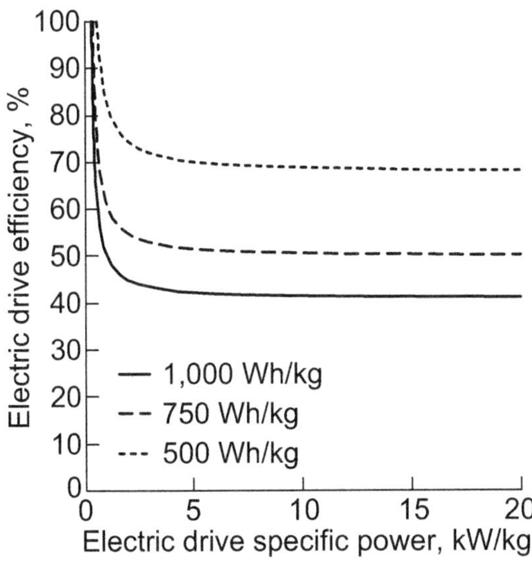

Figure 1.20 Breakeven curves based on battery specific energy, $\xi = 25$ percent, $W\text{fuel}_{AC}/W\text{i}_{AC} = 0.05$.

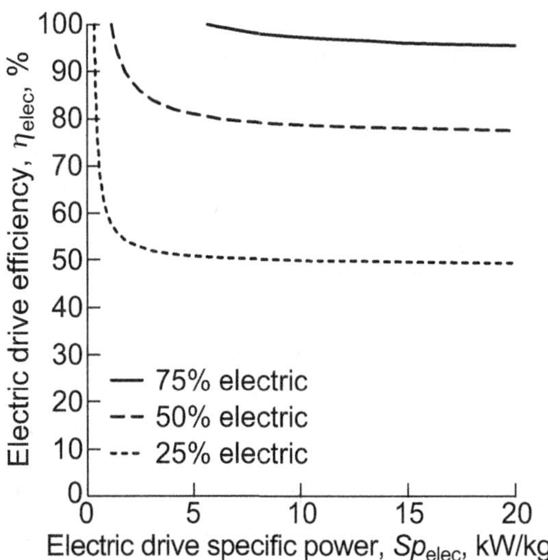

Figure 1.21 Breakeven curves based on electric propulsion fraction with aero and propulsive benefits constant, $Se_{\text{batt}} = 750$ Wh/kg, and $W\text{fuel}_{AC}/W\text{i}_{\text{AC}} = 0.05$.

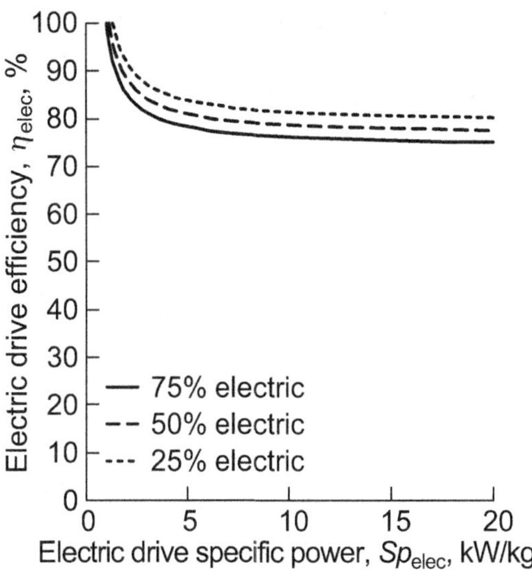

Figure 1.22 Breakeven curves based on electric propulsion fraction with aero and propulsive benefits scaling with ξ, $Se_{\text{batt}} = 750$ Wh/kg, and $W\text{fuel}_{AC}/W\text{i}_{\text{AC}} = 0.05$.

Now we look at the 600 NM range parallel hybrid electric aircraft described in Table 1.4, with a fuel fraction of 0.091. We compare breakeven results with those found in [11], which show that the 750 Wh/kg battery approximately breaks even on input power. This is one of our analysis assumptions, making it a good study for

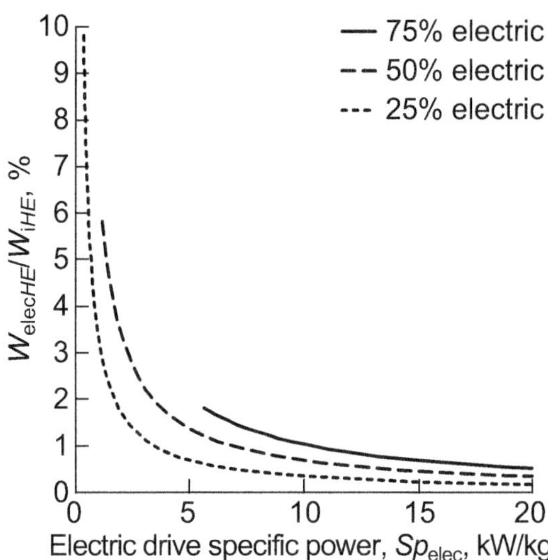

Figure 1.23 Electric drive weight ratio with equal benefits, $Se_{\text{batt}} = 750\ \text{Wh/kg}$, and $W\text{fuel}_{AC}/W\text{i}_{\text{AC}} = 0.05$.

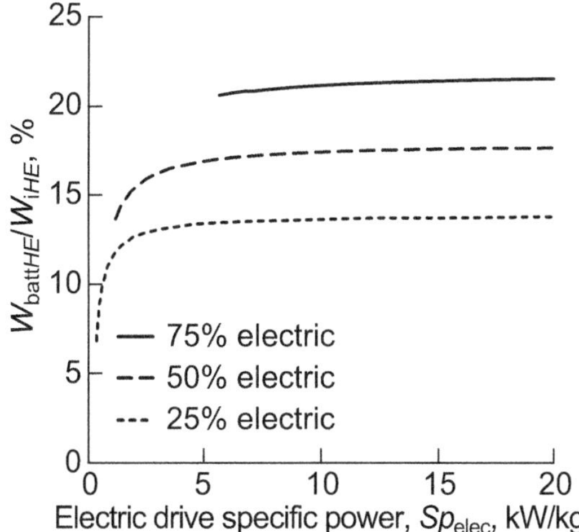

Figure 1.24 Battery weight ratio with equal benefits, $Se_{\text{batt}} = 750\ \text{Wh/kg}$, and $W\text{fuel}_{AC}/W\text{i}_{\text{AC}} = 0.05$.

comparison. The 500 Wh/kg battery increases total energy, and the 1,000 Wh/kg battery decreases total energy.

Figure 1.26 shows the results for the parallel hybrid electric concept in our breakeven analysis for an electric propulsion fraction of 25 percent. As expected, the 750 Wh/kg battery breakeven line was relatively close to the electric drive efficiency and specific power used in the systems study, which found nearly equal

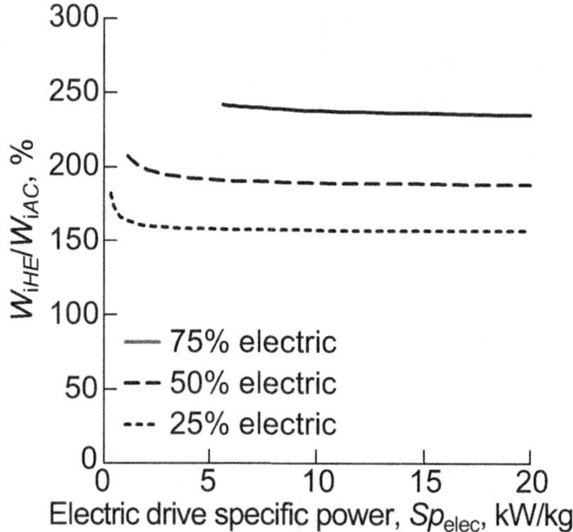

Figure 1.25 Ratio of hybrid electric aircraft weight to conventional aircraft weight with equal benefits, $Se_{\text{batt}} = 750$ Wh/kg, and $W\text{fuel}_{AC}/W\text{i}_{\text{AC}} = 0.05$.

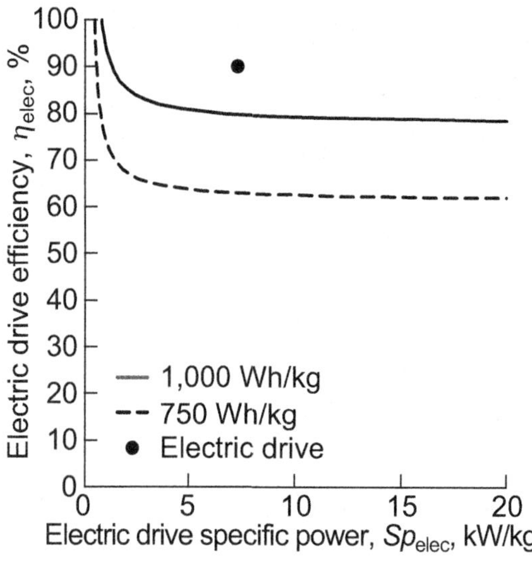

Figure 1.26 Breakeven for parallel hybrid electric aircraft example at $\xi = 25$ percent.

input power for that configuration. Improving Se_{batt} to 1,000 Wh/kg allows a relaxation in the electric drive performance. The breakeven analysis did not yield any viable electric drive performance for the 500 Wh/kg battery, as expected.

These results look good; however, increasing the electric propulsion fraction to 50 percent or higher does not yield feasible electric drive properties in this breakeven analysis, while in [11] the authors did find viable configurations. Further inspection of those results reveals that the assumption of W_{OEW}/W_i remaining constant is not true for that study. We made an assumption that the aircraft would need to be sized up to carry the weight of the added batteries. If the assumption is made that $W_{OEW}/(W_i - W_{batt})$ remains constant, which is similar to the Antcliff results, then viable electric drive configurations can be found for $\xi > 25$ percent.

1.4 Summary

The electrified aircraft propulsion concepts for commercial transport aircraft include a very wide range of propulsion airframe integration options as well as electric drive train options. Bounding analyses or parametric trade studies can be very useful to help narrow choices for detailed studies as well as guide technology development choices. Specific power, efficiency, and electric propulsion fraction have been proposed as KPPs for the electric drive system of an electrified aircraft. The boundary of the system is defined between the output shafts of the turbine to the input shaft of the propulsor and includes the electrical machines, power distribution, any other power components related to propulsion, as well as any thermal systems associated with the power system. Equations were developed that compare the benefits and costs of an electrified aircraft propulsion system compared to the baseline conventional aircraft. Some key conclusions include the following:

- Fully turboelectric aircraft
  - The electric drive system must provide power for takeoff results in tougher requirements on specific power than for partially turboelectric aircraft.
- Partially turboelectric aircraft
  - Assuming constant aero and propulsive benefits, a higher electric propulsion fraction requires a better performing electric drive system, due to the added weight of the electric drive system.
  - Assuming propulsive benefits that scale with electric propulsion fraction, a higher electric propulsion fraction relaxes the requirements of the electric drive system, since the higher aero and propulsive benefits cancel the costs of the electric drive system.
- Parallel hybrid electric aircraft
  - Parallel hybrid electric aircraft is better suited to shorter range.
  - Improving battery specific energy will make hybrid electric configurations more feasible.
  - Assuming constant aero and propulsive benefits, increasing the electric propulsion fraction increases the demands on the electric drive system to an even larger extent than the partially turboelectric system because of the added battery weight.
  - Assuming propulsive benefits that scale with electric propulsion fraction, a higher electric propulsion fraction relaxes the requirements of the electric drive system. Again, the higher aero and propulsive benefits cancel the costs of the

electric drive system. However, the added battery weight makes the benefits less dramatic compared to the partially turboelectric system.

- All electric aircraft
  - The breakeven curves are very sensitive to the propulsive benefit assumptions.
  - The breakeven analysis is sensitive to the component weight assumptions. Here it was assumed that the ratio of OEW to initial aircraft weight remains constant. It may be that other component assumptions are better for a given configuration, which could easily be incorporated into the breakeven analysis.
  - In general, at low specific power, the efficiency of the electric drive system dominates. But increasing specific power above a certain level yields diminishing returns.

Abbreviations

AC	conventional, fuel-based aircraft propulsion system (used as a subscript)
AFPM	axial flux permanent magnet
BLI	boundary layer ingestion
BPR	bypass ratio
EAP	electrified aircraft propulsion
HE	parallel hybrid electric propulsion system (used as a subscript)
HWB	hybrid wing body
KPP	key performance parameter
L/D	lift-to-drag ratio
LiAir	lithium-air
LiS	lithium-sulfur
N3CC	N+3 conventional configuration
NASA	National Aeronautics and Space Administration
NM	nautical mile
OEW	operating empty weight
PE	partially turboelectric aircraft propulsion system (used as a subscript)
PEGASUS	Parallel Electric-Gas Architecture with Synergistic Utilization Scheme
STARC-ABL	single-aisle turboelectric aircraft with aft boundary layer propulsor
TE	fully turboelectric aircraft propulsion system (used as a subscript)

Variables

D	drag, [N]
g	acceleration due to gravity on Earth, [9.81 m/s^2]
E_{batt}	input energy of battery-based aircraft, [J]

E_{fuel}	input energy of fuel-based aircraft, [J]
L	lift, [N]
P_{batt}	battery output power, [kW]
P_{elec}	electrical drive system output power, [kW]
P_{fuel}	fuel output power, [kW]
P_{prop}	propulsive output power, [kW]
P_{turb}	turbine engine output power, [kW]
R_{batt}	range of battery-based aircraft, [m]
R_{fuel}	range of fuel-based aircraft, [m]
Se_{batt}	battery specific energy, [Wh/kg] (in text) or [J/kg] (in equations)
Se_{fuel}	fuel specific energy, [Wh/kg] (in text) or [J/kg] (in equations)
Sp_{elec}	electric drive specific power, [kW/kg]
v_{cruise}	cruise velocity, [m/s]
W_{batt}	battery weight, [N]
W_{i}	cruise weight of aircraft, initial value, [N]
W_{f}	cruise weight of aircraft, final value, [N]
W_{elec}	electric drive weight, [N]
W_{fuel}	aircraft fuel weight, [N]
W_{payload}	payload weight, [N]
W_{OEW}	operating empty weight of aircraft, [N]
α	ratio of takeoff to cruise power
η_{elec}	efficiency of electric drive system
η_{o}	overall aircraft efficiency
η_{prop}	propulsive efficiency of aircraft
η_{therm}	thermal efficiency of turbine engine
ξ	electric propulsion fraction

References

[1] R. H. Jansen et al., "Turboelectric aircraft drive key performance parameters and functional requirements," presented at the 51st AIAA/SAE/ASEE Joint Propulsion Conf., Reston, VA, 2015, Paper AIAA 2015-3890.

[2] R. H. Jansen, K. P. Duffy, and G. V. Brown, "Partially turboelectric aircraft drive key performance parameters," presented at the 53rd AIAA/SAE/ASEE Joint Propulsion Conf., Atlanta, GA, 2017, Paper AIAA 2017-4702.

[3] J. L. Felder et al., "Turboelectric distributed propulsion engine cycle analysis for hybrid-wind-body aircraft," presented at the 47th AIAA Aerosp. Sci. Mtg., Orlando, FL, 2009, Paper AIAA 2009-1132.

[4] G. V. Brown, "Weights and efficiencies of electric components of a turboelectric aircraft propulsion system," presented at the 49th AIAA Aerosp. Sci. Mtg., Orlando, FL, 2011, Paper AIAA 2011-225.

[5] J. L. Felder et al., "Turboelectric distributed propulsion in a hybrid wind body aircraft," presented at the 20th Intl. Soc. for Airbreathing Engines Meeting, Gothenburg, Sweden, 2011, Paper ISABE-2011-1340.

[6] A. T. Wick et al., "Integrated aerodynamic benefits of distributed propulsion," presented at the 53rd AIAA Aerosp. Sci. Mtg., Kissimmee, FL, 2015, Paper AIAA 2015-1500.

[7] R. H. Jansen, C. Bowman, and A. Jankovsky, "Sizing power components of an electrically driven tail cone thruster range extender," presented at the 16th AIAA Aviation, Tech., Integration, and Ops. Conf., Washington, DC, 2016, Paper AIAA 2016-3766.

[8] T. P. Dever et al., "Assessment of technologies for noncryogenic hybrid electric propulsion," NASA, Cleveland, OH, Tech. Rep. GRC-E-DAA-TN10454, 2015.

[9] J. Welstead and J. L. Felder, "Conceptual design of a single-aisle turboelectric commercial transport with fuselage boundary layer ingestion," presented at the 54th AIAA Aerospace Sci. Meeting, San Diego, CA, 2016, Paper AIAA 2016-1027.

[10] K. R. Antcliff and F. M. Capristan, "Conceptual design of the parallel electric-gas architecture with synergistic utilization scheme (PEGASUS) concept," presented at the 18th AIAA/ISSMO Multidisciplinary Analysis and Optimization Conf., Denver, CO, 2017, Paper AIAA 2017-4001.

[11] K. R. Antcliff et al., "Mission analysis and aircraft sizing of a hybrid-electric regional aircraft," presented at the 54th AIAA Aerosp. Sci. Mtg., San Diego, CA, 2016, Paper AIAA 2016-1028.

[12] M. Hepperle, "Electric flight – potential and limitations," presented at the Energy Efficient Tech. and Concepts of Operation Workshop, Lisbon, Portugal, 2012, Paper STO-MP-AVT-209.

2 Aircraft Electric Power System Design, Control, and Protection

Robert Bayles

Introduction

Chapter 1 has described the substantial benefits of electrified aircraft propulsion (EAP). Clearly, achieving economical and safe EAP in transport aircraft would constitute an enormous leap forward in aviation. However, as with all engineering breakthroughs, "the devil is in the details." In this chapter, we begin to examine "the details" by introducing the electric power system (EPS), summarizing both its design and its control and protection functions. With the electrification of propulsion systems, EPS power levels (i.e., generation, distribution, and loads) are expected to increase by at least an order of magnitude, which will have far-reaching implications on the system design. All aspects of the EPS that are described in this chapter will be impacted by this, so a thorough understanding and appreciation of the EPS and its functions is necessary to fully comprehend the challenges ahead. In this chapter, several key EPS components and functions are described in order to give context to the rest of the book. A thorough understanding of the material herein prepares the reader for what follows in subsequent chapters: focused discussions of individual system components.

2.1 Certification Authority

Power distribution and protection is the term that describes the safe and efficient delivery of power to the airplane loads. There are multiple approaches to and configurations for delivering power, but the most important objective is to provide power safely. Aircraft safety requirements are driven by the United States (US) Federal Aviation Administration (FAA) and other governing bodies. Federal aviation regulations establish certain requirements that drive the overall system reliability. These, combined with component reliability requirements, limit the architecture design space.

Besides safe delivery, providing well-controlled power to the loads while also providing fault protection in the event of failures is critical. Well-controlled power is defined by *power quality* (PQ) requirements, which characterize power features such as voltage level, frequency, voltage ripple, and waveform distortion. There are both normal and abnormal conditions (for example, overvoltage and undervoltage) that must be considered in the distribution and protection system.

To certify an airplane for flight, the FAA establishes requirements – often called Federal Aviation Rules, or FARs – for minimum functionality and capability. For transport category airplanes, the rules are established in Title 14 of the *Code of Federal Regulations* (CFR), *Part 25*, entitled, *14 CFR Part 25 – Airworthiness Standards: Transport Category Airplanes* [1]. The following sections discuss some of the key FARs in [1] that regulate the EPS design and operation.

2.2 Power Generation and Distribution

Regarding airplane power generation and distribution, the FARs highlighted in the following sections have the greatest impact on architectural trades and the ultimate design of the airplane system. The following is not a complete duplication of the regulations but, rather, highlights of the key elements that impact the electrical power distribution and protection system. Specific sections are cited, but other sections also apply (e.g., *14 CFR Part 25, Subpart F – Equipment*).

2.2.1 Equipment, Systems, and Installations (*14 CFR Part 25.1309*)

These FARs require the EPS to perform its intended function for all foreseeable operating conditions, including failures. This drives a fault-tolerant design, usually achieved via redundant and backup systems. The EPS must be designed so that the probability that any one failure can disrupt the safe flight and landing of the airplane is extremely low (less than 10^{-9} failures per hour (FPH) probability).

These FARs also require provisions for alerting the crew in the case of a fault or failure and for identifying essential (i.e., flight-critical) loads. Additional information can be found in advisory circular *AC 1309-1A*, which provides guidance in system design and analysis to show compliance with *14 CFR Part 25.1309.*

2.2.2 General (*14 CFR Part 25.1351*)

This section requires that electrical power sources, main power buses, transmission cables, and associated control, regulation, and protective devices be designed for independence and proper operation during failures. It defines PQ, identifies essential loads, and requires protection from faults and operation without normal electrical power for at least five minutes while only receiving power from the system battery.

2.2.3 Electrical Equipment and Installations (*14 CFR Part 25.1353*)

These FARs focuses on battery installation and charging. Until now, the battery has been used for backup and auxiliary power only. However, if and when the battery becomes the primary source of power, these FARs will likely be brought into sharper focus as EAP and all-electric aircraft become more prevalent.

2.2.4 Distribution System (*14 CFR Part 25.1355*)

The distribution system – which includes the distribution buses, their associated feeders, and each control and protective device – is the focus of these FARs. It also requires that backup power sources be automatically switched in or manually selectable to maintain equipment or system operation. All modern airplanes are automatically reconfigured in the event of an EPS failure.

2.2.5 Circuit Protective Devices (*14 CFR Part 25.1357*)

In these FARs, automatic protective devices are required to protect for wiring faults or serious load failures. The generating system must also be designed with protective functions for overvoltage and other malfunctions. If the ability to reset a circuit breaker is essential to safe flight, pilots must have access to this functionality while in flight.

2.2.6 Electrical Wiring Interconnection System (EWIS) (*14 CFR Parts 25.1701–1733*)

The interconnecting wiring of the airplane must comply with the EWIS FARs. One key element is system separation requirements in *14 CFR Part 25.1707*. The wiring system must be designed and installed to provide adequate physical separation from other EWIS and airplane systems so that EWIS component failures will not create a hazardous condition. Separation can include distance, barriers, or other means.

2.3 Reliability

The EPS provides power to essential airplane systems and functions. For these, the reliability must be very high with the probability of failure (POF) less than 10^{-9} FPH, or one failure or less per one billion hours of operation. Most often, to achieve this level of reliability, an architecture with at least three channels or "lanes" of power is utilized. Each channel consists of a generating source and a distribution path. In the most common architecture – a three-channel system – the channels are electrically and physically isolated so that a single failure within one channel does not impact the others. Failures include all forms of electrical faults as well as fire, explosions, and bird strikes, to name a few. Each channel must be protected from causing loss or failure of other channels in the case of a failure.

An example aircraft EPS is shown in Figure 2.1. This figure contains a top-down view of an airplane, where the nose is on the left and the tail is on the right. When looking from tail to nose, the left and right side of the plane are on the bottom and top of the figure, respectively. This EPS shows three electrical channels, each indicated by a large rectangle containing one or more power panels. The left and right channels are powered by the electrical generators on the left and right engines, respectively.

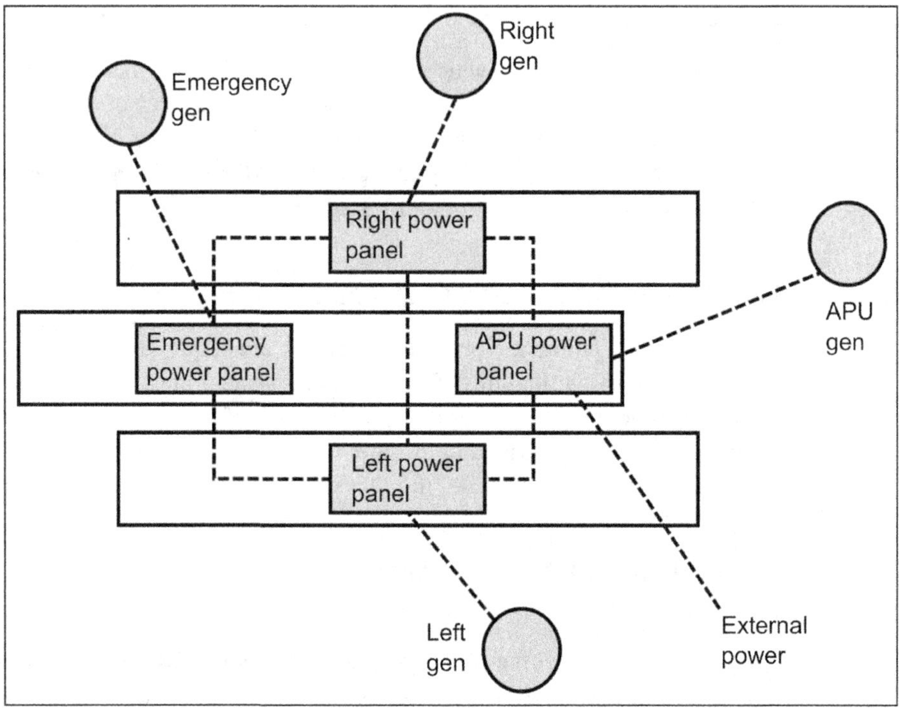

Figure 2.1 A notional three-channel EPS.

These generators are denoted "left gen" and "right gen." Under normal conditions either the left or right channel also powers the center channel, which often contains the essential power bus, the bus which supplies power to flight-critical loads. Because of their flight criticality, the essential loads should also be powered from an independent, emergency source, such as a ram air turbine (RAT) or auxiliary power unit (APU), as denoted by "emergency gen" and "APU gen," respectively, in Figure 2.1. It is also common for external power, often supplied by a ground cart during maintenance, to connect to the EPS through the center channel.

There is a good reason why the three-channel architecture is so prevalent. It turns out that, at a minimum, three channels are necessary to reduce the EPS POF to less than 10^{-9} FPH with today's standard EPS components. To see this, let us assume that the reliabilities (mean time between failures (MTBF), in hours) and corresponding failure rates for a single EPS channel are as given in Table 2.1. With one generator, one control unit, and one contactor in series to supply a channel, a cascade of the failure rates shown in Table 2.1 results in the POF of one channel to be 5.5×10^{-5} FPH. With two independent channels instead of one, the POF of the combined system (which now requires both channels to fail concurrently before the EPS fails) drops to 3×10^{-9} FPH. This is still three times too high to meet the certification reliability requirement; thus, unless the component failure rates are reduced substantially from

Table 2.1 Example reliability and failure rate of EPS components.

Component	MTBF [h]	Failure rate [FPH]
Generator	30,000	3.33×10^{-5}
Control Unit	60,000	1.67×10^{-5}
Contactor	200,000	5×10^{-6}

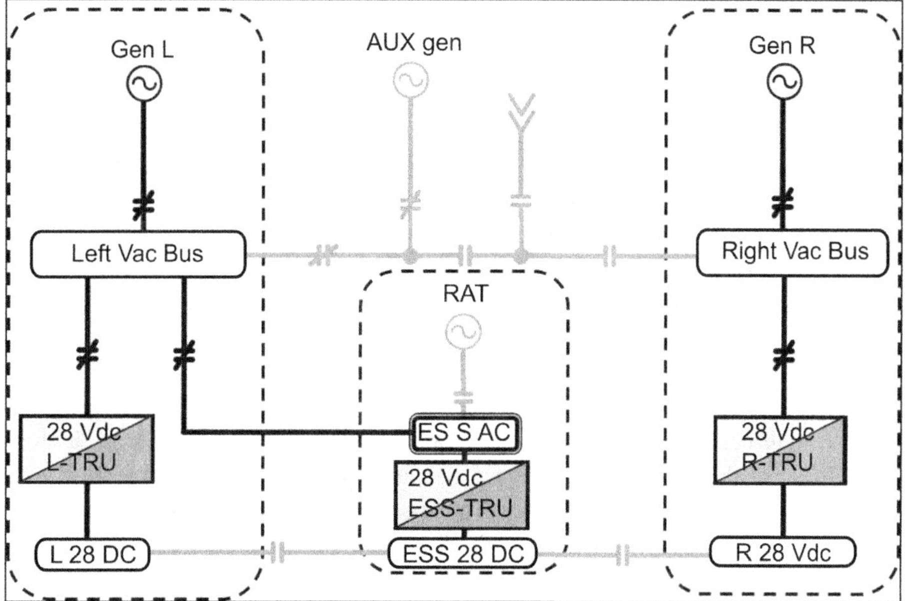

Figure 2.2 Single-line diagram of example aircraft EPS with three lanes of power.

the typical numbers given in Table 2.1, a third channel is necessary to reduce the EPS POF to an acceptable level.

A common means of presenting an electrical generation and distribution system is through the use of a single-line, or one-line, diagram, an example of which is shown in Figure 2.2 for a three-channel arrangement. The single-line diagram represents the power flows from sources to buses without regard to the number of wires or phases. For alternating current (ac) sources and buses, there are typically three phases with three or four associated wires. The phases are referenced to a common ground. For direct current (dc) sources, a single power feeder wire is utilized, and it, too, is referenced to the common ground.

In our example EPS, each channel is powered from an ac source. The left and right channels are powered by the associated electrical generator on the left and right engines, and transformer rectifier units (TRUs) provide power to the dc buses. As shown in Figure 2.2, in normal operation one of these channels also provides power to the center channel, which contains the essential power bus. The buses, such as the left vac bus and the L 28 Vdc bus, provide power to their respective loads.

The emergency source is often an independent source. The main electrical generating channels usually use a common generator design and are the same configuration. In nearly all transport aircraft, the generators are wound-field synchronous machines powered by the main engines, which share a common fuel source. The emergency channel utilizes a different energy source, architecture, and components in order to avoid common mode failures, which are discussed in Section 2.5.

2.4 Controllability

FAR *14 CFR Part 25.671 General* requires that an airplane be designed so that it is controllable if all engines fail. Examples of alternative power sources with all engines failed are given in Table 2.2. Of these, we will discuss the RAT in more detail to highlight some of the important considerations for controllability.

A RAT is a wind turbine-powered system, which is deployed automatically into the airstream to provide emergency electrical or hydraulic power (or both). Its power comes from the kinetic energy of the wind created by the aircraft motion. In the example EPS shown in Figure 2.1, a RAT would represent the emergency gen supplying the emergency power panel. Electrical power can be directly provided from a generator driven by the RAT or from hydraulic power via the RAT pump, which drives a hydraulic motor that operates an emergency generator. When the aircraft nears an emergency landing with no engine power, the airspeed is too low for the RAT to continue sourcing power, so an alternative power source, such as a battery, must automatically connect into the system to complete a safe landing.

The RAT is independent from the main electrical generating system. It has different sources of power and cooling, being air driven rather than engine driven and air cooled rather than oil cooled. Its speed is controlled via a governor. Care is taken to ensure that there are no common complex electronic parts in the RAT and main generator controllers, and the controllers that supply power to the essential load buses are

Table 2.2 Examples of emergency power sources that are available with all engines failed.

Emergency power source	Comments
Battery	Limited by its state of charge (SOC).
RAT	Powered by kinetic energy in the wind. Inoperable at low airspeed.
Hydraulic motor generator	Often powered by RAT hydraulic pump.
Permanent magnet alternator	Powered by engine windmilling.
APU	Possible common fuel source with generator and often not certified for in-air start at altitude.

separate from the main bus power control units. This separation is an example of the results associated with the *common cause analysis*, which is discussed in the next section.

2.5 Common Cause Analysis

The use of failure probabilities for assessment of acceptable failure conditions is based on the assumption that failures between systems are independent. To ensure that this is the case, a common cause analysis is required. There are three areas of evaluation required for a common cause analysis of the distribution system: *zonal safety analysis*, *particular risk analysis*, and *common mode analysis*.

Zonal safety analysis ensures that the installation of equipment within different zones meets an adequate safety standard regarding installation standards, interfaces between systems, and maintenance errors. If adequate standards are not demonstrated, then changes to the system are required.

For the *particular risk analysis*, risks are defined as events or influences that are external to the system but have the possibility of impacting the system performance. These can include rotor bursts, bomb blasts, thermal stress (extreme heat or cold), excessive moisture, tire bursts, and bird strikes. The goal of the analysis is to show that the systems are independent from simultaneous or cascading effects based on the risk event.

Common mode analysis is used to confirm that independence is considered in combination with some failure condition. This includes common elements of the specification, manufacturing, installation, maintenance errors, and design. Some common distribution system characteristics associated with common cause analysis include the location of power panels on the left and right sides of the airplane; wire runs separated in isolation channels on the left, right, top, or bottom of the airplane; and location of hardware to avoid rotor burst or bird strike areas.

2.6 Power Quality

As described in the FAR, the PQ requirement is intended to ensure proper load operation. PQ specifications define the electrical characteristics at both the power sources and the loads. Figure 2.3 shows a simple power flow diagram with power flowing from a source (e.g., electrical generator) to an electrical load via a distribution system. PQ generation specifications are most often defined at a generator's point of regulation (POR), which is typically at the input of the generator line contactor (GLC). The GLC controls power flow from the generator to the main power bus. The loads have also been designed to operate based on a required input PQ. Both source POR and load input PQ requirements must be consistent and account for the impedance of the distribution system. They must also account for the impact the load may have on the distribution system – e.g., via harmonics.

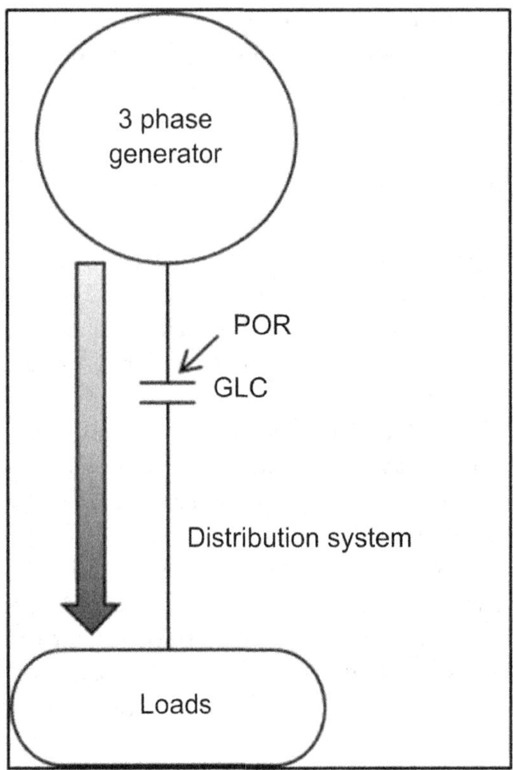

Figure 2.3 Power flow diagram for a three-phase circuit.

PQ typically degrades the farther it is measured from the regulated source. The distribution system has impedance from feeder wires and contactors, resulting in a slightly lower voltage at the loads than at the source. The generation and load PQ specifications must account for this voltage drop, and the distribution system must be designed to ensure a certain maximum allowable voltage drop. The distribution system is designed such that if the source is providing its required PQ, then the required load PQ is ensured. Typical PQ electrical characteristics for sources and loads are listed in Table 2.3.

A widely used PQ specification is MIL-STD-704F [2], which specifies PQ at electrical loads. Another well-known specification, which gives PQ requirements for generator systems, is MIL-PRF-21480B (formerly MIL-G-21480A) [3]. Numerous other PQ specifications exist, some of which are proprietary and not publicly available.

2.7 Voltage and Frequency

The two basic elements addressed by PQ are voltage and frequency. Most commercial transport airplanes utilize three-phase, 115 Vac (line-to-neutral, root-mean-square

Table 2.3 Typical PQ electrical characteristics.

PQ characteristics
Voltage levels
Unbalanced voltage
Frequency and phase
Power factor
Voltage ripple
Voltage modulation
Distortion factor
Crest factor
dc content within ac system
Abnormal conditions (e.g., voltage, frequency)

(RMS)), 400 Hz power, which is produced by engine-driven generators via accessory drive gearboxes. Power generation sources for commercial transport aircraft also include 115 Vac, 400–800 Hz variable frequency power, and 230 Vac at constant or variable frequency. On the ground and in certain emergency situations, an APU is operated to provide mechanical shaft power to a generator, which can provide power independently to one or more buses. A generator's output frequency is directly related to its rotor speed and inversely related to its pole count.

All transport airplanes also utilize 28 Vdc, which powers most of the critical loads. As is the case in the example power system shown in Figure 2.2, this voltage is primarily generated with TRUs, which transform the 115 Vac to 28 Vdc through the use of a transformer to step down the voltage and a rectifier to convert it from ac to dc. The flight-critical 28 Vdc power is also backed up with a battery in case of the loss of one or more generating sources. A few flight-critical 28 Vdc loads, such as flight control computers, are powered independently by a small permanent magnet generator (PMG) or alternator (PMA) in conjunction with a power supply to condition the power to 28 Vdc. These sources operate independently of the main power electrical system.

Smaller airplanes with low electrical power needs, such as those of the general aviation or small commuter classes, may only require 28 Vdc. In these cases, the main generators will produce dc directly at the required voltage.

Over the years, airplanes have used a variety of voltages for electrical power. This has been driven by the need to develop systems that are as light as possible. As electrical loads have increased over time, voltage levels have followed to maintain low equipment weight and, very importantly, lighter wires. Because power is the product of voltage and current, higher voltage requires less current than lower voltage for the same power, which results in smaller-gauge, lighter wires that are more pliable and easier to install. Unfortunately, higher voltage also comes with a higher risk of corona discharge and arcs, especially at high altitudes where the breakdown voltage of the atmosphere is much lower than on the ground.

Early aircraft used low-voltage dc systems, most often at 28 V. As power needs increased, the resulting currents became very large, requiring large wires and heavy systems. Around 1943 the Army Air Corps commissioned a study to recommend the future electric standards for aircraft. For several reasons, the study's authors recommended 115 Vac at 400 Hz. The US electrical grid was at the time (and still is today) approximately 115 Vac, 60 Hz, which meant that significant engineering knowledge and associated technology for this level was readily available. Despite this, the 60 Hz frequency was not an ideal choice for aircraft. Higher frequencies allow smaller transformers and smaller, lighter, higher-speed motors. Specifically, the 400 Hz frequency was selected to maximize the performance of motors and transformers with minimum weight while also avoiding significant issues with skin effect on the distribution system. The use of 115 Vac was selected as a compromise for a lighter weight distribution system while avoiding high risks of arcing or corona discharge. It also reaped the ancillary benefits of being the common distribution voltage in the United States.

Although 115 Vac became widespread after the Army Air Corps study, there are examples of airplanes that use 230 Vac, often called *double voltage systems*. These are usually large transport aircraft with very large electrical loads, and the higher voltage is required to keep the wiring weight manageable. In double voltage systems, additional care must be taken to control the insulation system to avoid issues with faults and the resulting arcs.

More recently, many systems have evolved to utilize power electronics or motors to replace traditional mechanical systems and/or mechanical control systems. This has resulted in loads with built-in power conditioning units that locally rectify the ac voltage into dc power for use by their associated power electronic controllers. Source and load power specifications have essentially become decoupled, and the use of 400 Hz sources has become less important. Variable-frequency ac and high-voltage dc systems are thus becoming more common and this is expected to continue into the future.

2.7.1 Variable Frequency ac

The use of variable frequency ac has evolved and today is used on several aircraft, including the Airbus A350 and Boeing 787. Historically, the electrical generation system has been powered from an auxiliary gearbox that is connected to the high spool shaft of the engine. The generator speed range of this spool is usually approximated to vary 100 percent (e.g., from x to $2x$ rpm). The output frequency of a generator is directly proportional to its shaft speed. In the traditional 400 Hz system, the input speed of the generator is controlled by a hydraulic unit called a *constant speed drive* that conditions the variable input speed from the engine (idle to maximum thrust speeds) into a constant speed that is supplied to the generator. The generator is designed to produce a constant frequency 400 Hz output at this shaft speed.

Alternatively, in a variable frequency system, the constant speed drive is eliminated, and the variable engine speed directly drives the generator, resulting in an output frequency that varies accordingly. Usually, the system is designed to output a frequency that ranges from approximately 400 to 800 Hz. This is desirable because it keeps the minimum frequency within the historic range of experience and the top end frequency reasonably low to avoid excessive issues with wire skin effect. The distribution system is not significantly impacted by this frequency variation. Both constant and variable frequency systems distribute three phase power with a common ground network.

2.7.2 High-Voltage dc

The use of high-voltage dc (HVDC) electric power has been a significant trend in military applications and is a potential future trend for transport airplanes. Numerous trade studies have advocated the use of HVDC, and an HVDC electrical system standard has been established for 270 Vdc in MIL-STD-704F [2].

The reason why certain voltages are selected as standards is interesting. Often, the selection is based on leveraging engineering experience and existing standards. Case in point: when three-phase 115 Vac – a long-existing standard – is rectified using a full wave bridge rectifier, the resultant voltage is approximately 270 Vdc. By utilizing this voltage directly, dc load equipment can function without the previously required 115 Vac rectifiers. The use of 270 Vdc impacts the distribution system significantly since a 270 Vdc feeder and dedicated return wire are needed for each load. This is in contrast to a three-phase ac system, which uses a common ground.

The adoption of 270 Vdc leads to another future trend for high-power aircraft systems: the use of ±270 Vdc (with respect to the system ground). In this system, no feeder differs in potential from ground by more than 270 V, which is the same as the 270 Vdc system. The advantage is that now the load terminals can be connected to the positive and negative 270 Vdc feeders, respectively, which provides current to the load at 540 Vdc. Provided the positive and negative feeders are run far apart until they are brought together at the load, the current is halved without increasing the risk of arcing or corona discharge as compared to a 270 Vdc system. In principle, this is very promising; however, at present, there is very little experience with this higher voltage in airplane applications at altitude. Because the distributed voltage on each wire is still 270 Vdc when compared to ground, the ±270 Vdc system allows for the use of existing technologies and experience. Although many airplanes benefit from a HVDC primary power system, most of the performance gains come from careful system integration (e.g., weight and cost reductions) of loads, wiring, and source designs.

2.7.3 Future Trends

The next generation of HVDC standards must address the very large power needs of future vehicles, such as electrically propelled aircraft. There are environmental and regulatory requirements and objectives – similar to those for electric automobiles – which are driving aircraft toward EAP. The power required for the electric propulsion of even the smallest aircraft will be at least two to four times that of most airplane systems today. Larger airplanes will require 10 times more power than these small aircraft. Supplying this power without significant losses and feeder weight will demand higher distribution voltages. Projections start at approximately 1,000 Vdc to meet near-term needs. Demonstrations are being planned for uses up to 3,000 Vdc. As with the approach for 540 Vdc, it is expected that these voltages will be centered about ground or neutral such that the bus voltage will be ±500 Vdc or ±1500 Vdc, respectively. Projections with longer time horizons are for even higher voltages as power needs evolve for larger airplanes. This continues the historic, long-term trend in aviation that has seen larger and larger loads requiring higher voltages to minimize the size and weight of the distribution system.

The high-power needs of EAP will require large currents at today's standard distribution voltage levels. Because ohmic losses increase with the square of the current, feeder losses can quickly escalate to unmanageable levels if not carefully dealt with. These losses are critical, as they require power sources, such as batteries, to be larger (to supply the losses) and additional thermal management systems to help deal with the heat.

Research into how best to deal with the issues associated with high-voltage aircraft power systems is ongoing. These issues include corona, arcing, insulation breakdown, life, switch gear design, and fault detection and isolation. This is a dynamic time in the history of aircraft power distribution systems, and much is yet to be learned.

2.8 Power Transfer Approaches

A key consideration of aircraft power distribution is how best to transfer power under abnormal conditions. The most common approach is to provide a system with isolated channels – i.e., an *isolated system* – that can be cross-connected as needed. As shown in Figure 2.2, the generators (power sources) operate in a single-channel configuration: the associated buses – the left and right vac buses, respectively – are powered by one source at any one time. Focusing on the left side for this discussion, the left vac bus normally is powered exclusively by the generator on the left engine (gen L), but it can also be powered by the APU (aux gen), an external power unit (via the ext pwr plug), or the generator on the right engine (gen R) in case of a failure or if the left engine is not rotating (e.g., on the ground). This approach is used almost universally for recent commercial transport aircraft in either two-channel (two generator sources) or four-channel configurations. A common term used for this configuration is a *break power transfer* system, meaning that the loads on the associated buses see a short power interruption while one contactor opens to remove power and another closes to apply power from another source.

There is one exception to this operation for 400 Hz systems. While on the ground, a *no-break power transfer* can occur by momentarily paralleling two sources. As demonstrated in Figure 2.4, an example of this is the aux gen powering the left vac bus while on the ground before the engines have started. After the left engine starts, gen L becomes powered and able to take over as the source to the left vac bus. By controlling the phase angle between the two sources and ensuring they are both within normal PQ requirements, the two sources are temporarily paralleled while the gen L power is applied and the aux gen power is removed. This avoids any power

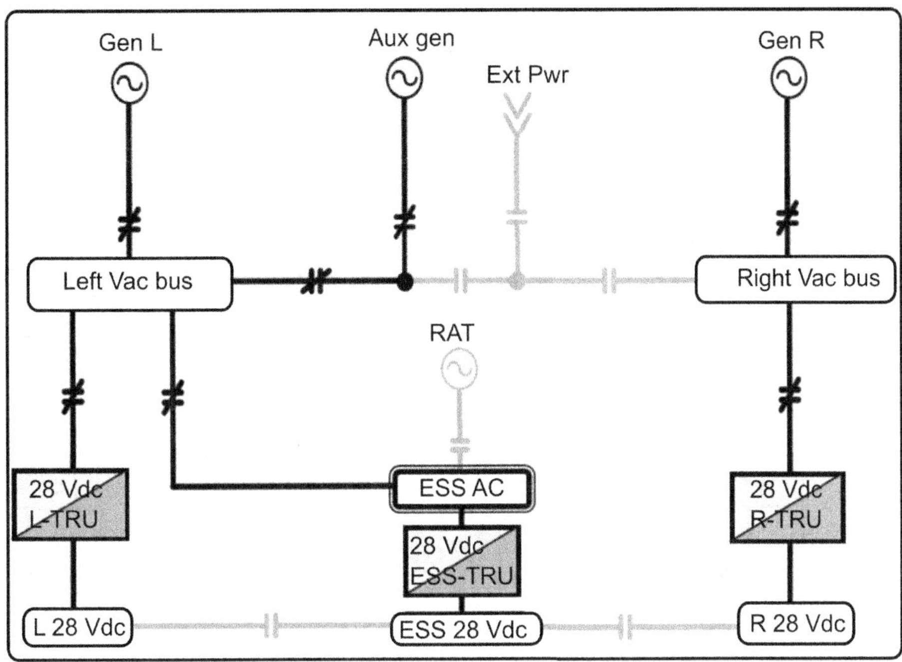

Figure 2.4 Example of no-break power transfer.

interruption to the loads. This option is only available on systems where the source frequency and phase angle can be controlled, such as in systems that use constant speed drives. Typical uses for no-break power transfer are for transferring power from external power to/from APU power, from APU power to/from an integrated drive generator (IDG), and from the left channel IDG to/from the right channel IDG.

In many older, multiengine transport airplanes – such as the Boeing 707, McDonnell-Douglas DC-10, or Lockheed L1011 – the electrical systems operate in parallel. An example of a parallel distribution system is shown in Figure 2.5 for a two-channel system, although this approach is more common in three- and four-channel systems. This is still used on some large military airplanes with large electrical needs. In a parallel system, the generator outputs are connected, and the voltage, frequency, and phase angle differences of each generator must be controlled within predefined limits. If these limits are not met for any reason, the distribution system will disconnect the generators, allowing them to operate in isolation.

There are various benefits and overhead to consider when choosing between an isolated (ac or dc) or parallel ac system. The isolated system has simpler controls, because a single generator regulates the bus voltage, and smaller load disturbances, because faults only impact one channel. Fault currents are limited to the capacity a single source, allowing smaller, lighter-weight switching and protective devices. Isolated channels can utilize either constant or variable frequency, but loads must be spread among the various buses and no one load can be larger than the source that

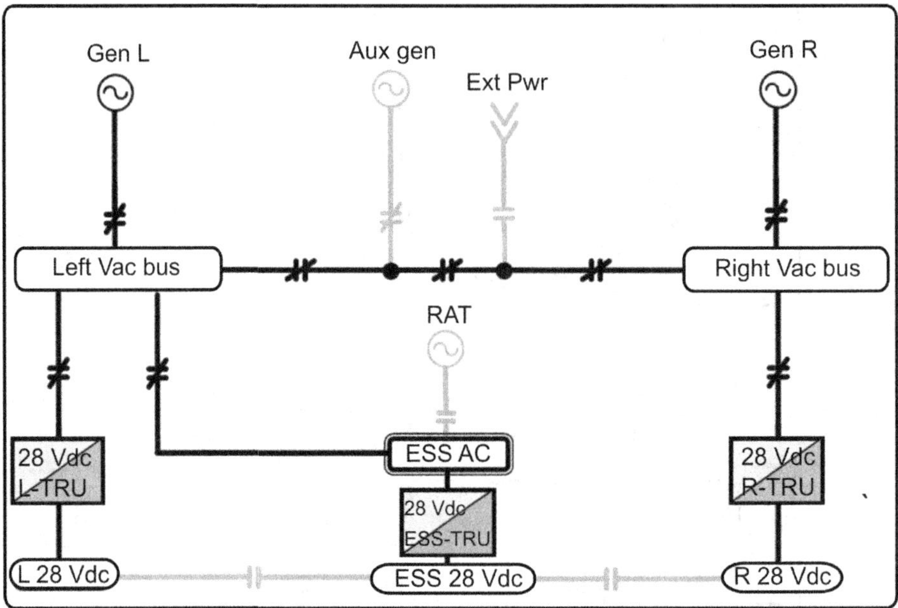

Figure 2.5 Example of paralleled distribution system.

feeds it. A parallel ac system requires the use of a constant frequency. It provides greater fault-clearing capabilities and a lower-impedance (stiffer) source capability, resulting in lower-voltage transients and larger interconnected load capability. But this too comes with the need to control the frequency and phase angle of the two interconnected systems and to provide additional protections to monitor for failures of the control system. Because faults can be fed by multiple sources, large fault currents are possible if protective devices are not designed appropriately.

For safe paralleling, certain limits for voltage, frequency, and phase angle must be met and maintained, as shown in Figure 2.6. Real load current is maintained within specific limits by controlling the frequency and phase angle differences between the sources (two sources are shown in Figure 2.6). Reactive load current is maintained within specific limits by controlling the voltage differences between the sources. The machine voltages are controlled via the exciter field of the wound field synchronous generators. The frequency and phase are controlled by the rotor speed and position, which are typically controlled via the constant speed drive, as discussed earlier. There are built-in protective functions to isolate the system if the real or reactive current flow exceeds predetermined limits.

2.9 Power Switching Devices

Power switching devices play a key role in the distribution system. Traditionally, these are electromechanical or electromagnetic devices referred to as *contactors*,

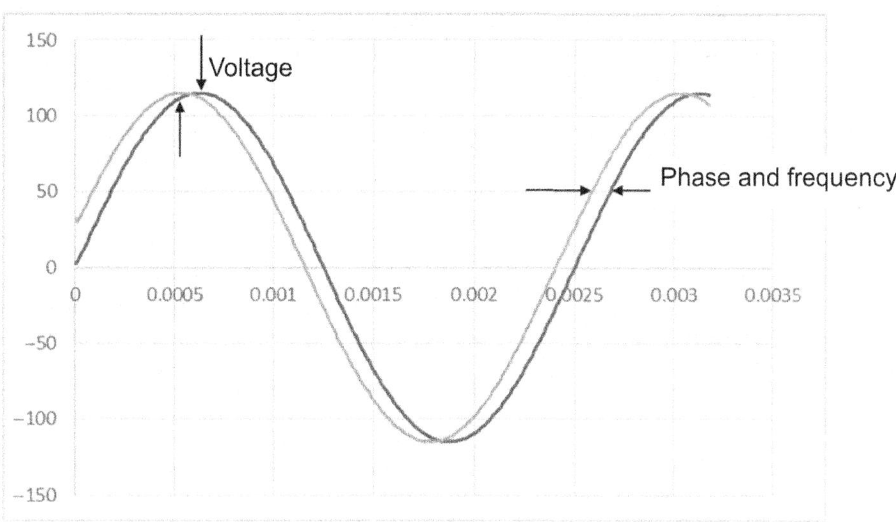

Figure 2.6 Key parameters controlled for parallel systems.

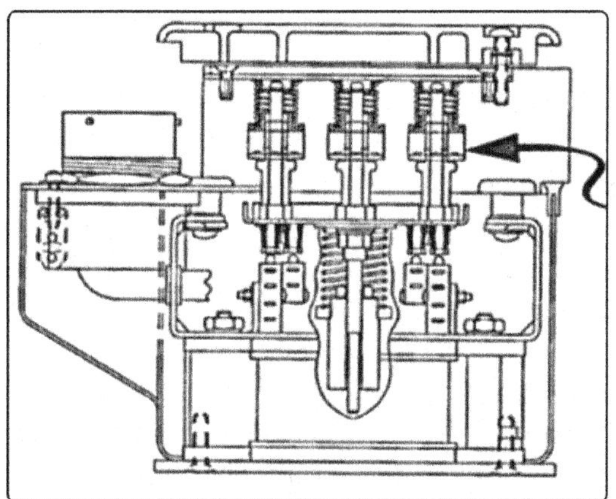

Figure 2.7 A typical electromechanical contactor.

relays, or *breakers*. Most often, these devices (herein called contactors), are controlled via a 28 Vdc signal to the devices' solenoid. The solenoid forms the coil of an electromagnet, which moves a plunger with contacts that makes (closes) or breaks (opens) the electrical connection. When the solenoid releases, the plunger returns to its previous position, and the electrical connection is then reversed (open to closed or closed to open). Auxiliary switches are often included and used for contactor position sensing and control functions. A diagram of a typical contactor is shown in Figure 2.7.

There are many contactor configurations. Several key characteristics that determine the configuration are listed in Table 2.4. While not an exhaustive list, this table

Table 2.4 Contactor characteristics and configurations.

Characteristic/feature	Example/typical options
Nameplate rating	kVA or A
Contact voltage levels	230 or 115 Vac, 270 or 28 Vdc
Contact phases	Single or three phases
Solenoid/coil control voltage	28 Vdc or 115 Vac
Solenoid latch type	Magnetic or electric
Normal (de-energized) state	Closed or open
Solenoid economizer (to reduce holding current when energized)	Mechanical or electrical
Contact type	Single or double throw

provides some insight into the various options designers have in the development of the distribution system.

A new type of power switching device that is becoming common in distribution systems is the *solid-state power controller* (SSPC). The SSPC is usually composed of a field-effect transistor (FET) for the switching element, circuitry to control the on/off function of the FET (a gate drive), and protections for voltage and/or current monitoring. SSPCs can be used for direct replacement of lower-current contactors but are most often used to replace relays and circuit breakers. Historically, a cockpit contained many circuit breakers and relays to control power to individual loads. Recent aircraft have replaced many of these devices with SSPCs that can be controlled through software that interfaces with cockpit displays and input devices.

SSPCs enjoy many advantages over traditional electromechanical contactors. They are more reliable, while providing more capabilities (e.g., current limiting and arc fault mitigation). They provide flexibility in wiring and system layout with their remote monitor function, allowing them to be located outside of the cockpit. They are also fully programmable for automated, fast power transfer and system reconfiguration.

Despite their many advantages, SSPCs are limited by their relatively modest power ratings and the thermal challenges they present. Electromechanical contactors have very low forward voltage drop, and SSPCs dissipate more power in their closed state. These are important considerations as EPS power levels increase drastically, as is expected with the adoption of EAP.

2.10 Control and Protection Methods

The distribution system, in conjunction with the power generation system, must provide protections to ensure proper PQ at the loads. Table 2.5 lists some of the most common protections, as well as how their limits are set. Many other protections are available to the EPS designer, but this is a sample of key ones that will be included at minimum.

The locations of these protections can vary depending on the system design. Figure 2.8 shows some of the typical units that are used to perform these functions,

Table 2.5 Common distribution system protections and limits.

Protection	How limits are set
Steady-state overvoltage (OV)	• Slightly below MIL-STD-704F max limits
Transient OV	• OV slightly below MIL-STD-704F max limits • Inverse time delay for faster trip at higher voltages
Undervoltage (UV)	• Fixed threshold level for a fixed time delay
Ground fault (GF)/Overcurrent (OC)	• GF: Fixed threshold level for a fixed time delay • OC: Inverse time delay for faster trip at higher currents
Over-/Underfrequency	• Fixed threshold level for a fixed time delay
Phase sequence	• Ensures a-b-c positive sequence, esp. for ground power

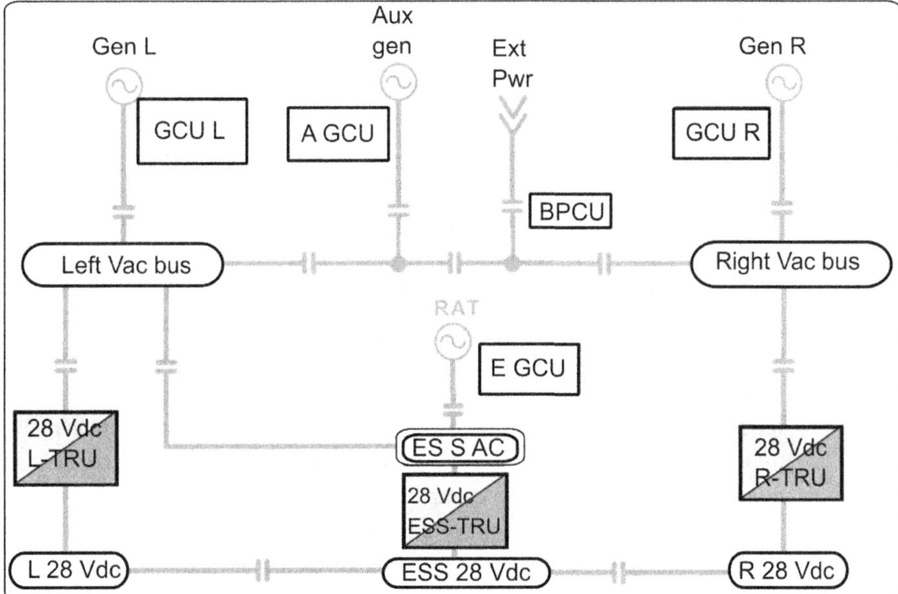

Figure 2.8 Control units that implement EPS protection functions.

including the *generator control units* (GCUs), *bus power control units* (BPCUs), and the *emergency control unit*. Other units, such as *ground power control units,* can also be utilized, or additional functions can be integrated into some of these other units such as the BPCU.

2.11 Load Management

The total electrical loads on a multiengine airplane will exceed the capacity of one generator. The loads are typically split evenly between the generators (two for a twin-engine airplane). The steady-state loads on each side of the aircraft might be 60 percent or more of a single generator's capacity. Transient loads such as motor starts will exceed this level. Generators are rated for steady-state performance with allowances for temporary overloads. While this varies, most generators will have one overload rating for five minutes and another, higher, overload rating for five seconds.

Aircraft EPSs are designed to support all essential (i.e., flight-critical) functions with only one generator. To accomplish this, certain "less important loads" can be turned off, or *shed*, as part of a load management system. Essential loads are never shed; they receive power in all situations when power is available. Shedding of nonessential loads historically has been accomplished through the process of shedding buses, whereby each load is powered by a specific bus. Each bus can be de-energized by opening its feeding contactor if the loss of a generator were to occur and create an overload condition. The generator overload rating (e.g., 5 seconds or 5 minutes) provides time for the protection system to sense this condition and take appropriate action.

As outlined in Section 2.10, SSPCs provide an alternative means to accomplish this. SSPCs receive commands through a data bus that can provide the load management function through software control. The load management software can command a set of SSPCs to shed/de-energize a set of predefined loads as problems arise. This allows for the creation of "virtual" buses – as opposed to physical buses – that can be easily spread across the airplane, and it prevents the need for the extra wiring of centralizing a physical bus.

Another means of load management is to allocate a power allowance to specific loads, again through software control. This can take different forms. One form is the simple example of providing power to a galley. A power allocation (e.g., 20 kVA) can be assigned, through software, to the galley control unit, which ensures that the galley loads do not exceed their allowance. Another form of load management is the allocation of power to a motor controller that controls the speed of a motor. The controller limits the speed to ensure the motor power does not exceed its allowance. If either the galley or motor control unit fails to limit its respective load's power usage, the load management control software may shed that load.

2.12 Summary

The generation and distribution system is used to safely and efficiently deliver electric power to the airplane loads. The EPS configurations used in practice – especially in commercial transport aircraft – have been designed to meet stringent certification and safety requirements. However, even under these strict guidelines, the design space is very broad, and general configurations can vary greatly in their specific details, reflecting a wide range of airframer experience and preferences across the industry.

Producing and delivering well-controlled power to the loads while also providing fault and failure protection are the key functions the system. Well-controlled power is defined by the PQ requirements for both normal and abnormal conditions. In the event of faults and failures, a well-designed system provides for appropriate power transfers, system reconfiguration, and power management to assure the availability of power to the as many loads as possible but always to the critical loads.

This chapter has laid the groundwork for understanding the basic layout and operation of an aircraft EPS. It has explained why certain standard features are necessary and how the overall system must operate to ensure safe flight. All aspects of the system described herein will be profoundly impacted by the order-of-magnitude increase in power level expected with EAP. The next four chapters will cover some of the critical components of the EPS and how they will be impacted by and/or enable propulsion electrification. Chapters 3 and 4 deal with generators, Chapters 5 and 6 focus on the power electronic devices (converters, inverters, etc.), and Chapter 7 addresses the thermal challenges.

Abbreviations

ac alternating current
APU auxiliary power unit
BPCU bus power control unit
CFR Code of Federal Regulations
dc direct current
EAP electrified aircraft propulsion
EPS electric power system
EWIS electrical wiring interconnection system
FAA Federal Aviation Administration
FAR Federal Aviation Rule
FET field-effect transistor
FPH failures per hour
GCU generator control unit
GF ground fault
GLC generator line contactor
HVDC high-voltage dc
IDG integrated drive generator
MTBF mean time before failure
OC overcurrent
OV overvoltage
PMA permanent magnet alternator
PMG permanent magnet generator
POF probability of failure
POR point of regulation
PQ power quality
RAT ram air turbine
RMS root mean square
SOC state of charge
SSPC solid-state power controller
TRU transformer rectifier unit
UV undervoltage

References

[1] Code of Federal Regulations, Title 14 – Aeronautics and Space, Chapter 1 – Federal Aviation Administration, Department of Transportation, Subchapter C – Aircraft, Part 25 – Airworthiness Standards: Transport Category Airplanes.
[2] Department of Defense Interface Standard: Aircraft Electric Power Characteristics, MIL-STD-704F, 2004.
[3] Performance Specification: Aircraft, Generator System, Electric Power, 400 Hertz Alternating Current, Aircraft, General Specification for, MIL-PRF-21480B, 2010.

3 Megawatt-Scale Electric Machines for Electrified Aircraft Propulsion

Tim C. O'Connell and Xiaolong Zhang

Introduction

This text has thus far made a strong case for the electrification of aircraft, through electrifying either (1) the traditionally non-electric (non-propulsion) subsystems – the "more-electric aircraft" (MEA) concept – or, more germane to the focus of this text, (2) the propulsion system, via hybrid electric and turboelectric concepts. Over the past several decades, power electronics, batteries and energy storage devices, electric machines (EMs), and other enabling technologies have gotten smaller, lighter, and cheaper while simultaneously increasing power capabilities. Because of this, aircraft electrification is no longer the stuff of science fiction; it has become reality. The numerous benefits promised by electrification have led many to envision an aviation future that is quieter, greener, and safer than today. No doubt, it is an exciting time to be in the aerospace industry.

And yet, despite the many technological leaps that have brought the industry to its current state, numerous challenges remain. Identifying and addressing these challenges are the goals of this and the following chapters. This chapter focuses on the large, conventional (i.e., non-cryogenic), megawatt (MW)-scale EMs that are required to facilitate electrified aircraft propulsion (EAP) on a large scale. Crucially, the *specific power* (SP; *power per unit mass*, or *power-to-mass ratio*) of these machines must be very high; in fact, we will see that the SP of today's state-of-the-art (SOTA) high-specific-power (HSP) machines must increase by roughly an order of magnitude in order to facilitate the goals of electric propulsion. This chapter gives a comprehensive overview of conventional large EMs for EAP applications, with a focus on methods for mass reduction and SP improvement. Superconducting machines (SCMs) with cryogenic systems are very promising candidates for HSP EAP, but they are not likely to achieve the necessary technology maturity for at least 20–30 years [2]. For this reason, and because they represent an entirely new class of machines, in this text, SCMs are deemed "unconventional" machines and are covered separately in Chapter 4.

This chapter is based on the authors' review paper [1], but the material has been reorganized, expanded, and updated for this text. Section 3.1 identifies the major

This chapter draws on material contained within a previous paper by the authors.

design challenges that must be overcome in order to realize HSP MW-scale machines. Section 3.2 provides a review of EAP powertrain architectures and how HSP EMs enable them. In Section 3.3, motor and generator systems are introduced, with a focus on propulsion motors, high-speed generators, and system considerations unique to the EAP powertrain. Design principles that guide high-power, lightweight, MW-scale EM development are presented in Section 3.4 – including sizing equations, key physical variables and their constraints, thermal limitations, and power and speed scaling and their thermal ramifications. Section 3.5 reviews the major machine topologies commonly used (or considered) for aircraft generators and propulsion motors, and Section 3.6 presents a comprehensive survey of more than 50 SOTA HSP machines that are either commercial off-the-shelf (COTS) designs or laboratory prototypes. Common features of HSP machines are identified and promising options for further weight reduction are discussed. Despite their SOTA classification, the surveyed machines fall well short of the lofty requirements necessary for next generation devices, and Section 3.7 discusses some emerging technologies that may help to remedy this. This segues into the detailed coverage of SCMs found in Chapter 4.

3.1 The Electrified Propulsion Machine Design Challenge

3.1.1 Discussion

As detailed in Chapter 1, EAP has been proposed for large commercial aircraft to improve fuel economy and to reduce emissions and audible noise. To fully transition to EAP, the necessary electrical components must be carefully designed. Their masses must be minimized so that their benefits – which include improved turbine efficiency, distributed propulsion, and propulsion-airframe integration – are not canceled out by their weight penalty. This puts stringent requirements on the large EMs used in the system, both those that generate electric power from the turbine shaft and those that drive propellers or ducted fans, because they are among the heaviest of the added electric components. A key machine design metric in this application is the SP, typically expressed in kW/kg.

Generators and motors are key EAP system components. More stringent performance requirements are imposed on these EMs than on those used in terrestrial (industrial, rail, automotive) or marine applications. An aircraft's aerodynamic performance and fuel efficiency are very sensitive to its components' masses; thus, HSP machines are a must. For quantitative context, consider that the SP of general-purpose industrial motors, (for fans, pumps, compressors, milling machines, etc.) is in the range of 0.1–0.5 kW/kg. With their advanced materials and designs, automotive electric vehicle (EV) drive motors have SP values in the range of 1–3 kW/kg. However, projections have calculated that the SP required for passenger-class aircraft propulsion machines will be at least 10 kW/kg [3]!

Highlighted in Table 3.1 is the large jump in SP from today's SOTA machines for various applications to University of Illinois (15 kW/kg) and Ohio State University

Table 3.1 SP of various SOTA EMs versus the NASA target for a future aircraft electric propulsion motor.

Application	Manufacturer	SP [kW/kg]
General purpose	Marathon [4]	0.2
EV drive	Remy [5]	2
Aircraft propulsion	Siemens [6]	5
Aircraft propulsion, MW class (research prototype)	U. of Illinois [7, 8]; Ohio State U. [8]	>13
Aircraft propulsion, MW class (NASA Glenn Research Center Target [8])	N/A	16

(13 kW/kg) MW-class machine research prototypes and the U.S. National Aeronautics and Space Administration (NASA) Glenn Research Center target (16 kW/kg).

A combination of lighter materials and size reduction techniques are necessary to increase the SP of EMs if they are to achieve the lofty targets required by EAP [7]. To address this, sizing tools that optimize material usage have been developed for various machine types [9–11]. SP does not tell the whole story, however. The maximum power ratings of SOTA automotive EV motors are only a few hundred kilowatts, while the required power ratings for aircraft propulsion machines will be in the megawatts. A 20-year performance projection shows that the power capability of aircraft EMs is expected to reach 1–3 MW, and the SP is expected to reach 9 kW/kg (as with all projections, assuming sufficient budgetary and human resource inputs) [3].

In turboelectric powertrains (reviewed in Section 3.2.2), generators that enable EAP are likely either to be mounted on, or in parallel to, the turbine engine shaft or inside the engine compressor region. These locations are in environments with extreme vibrations and temperatures, which pose complex mechanical and thermal design challenges.

3.1.2 Design Drivers

The facts reveal that every EAP system-based EM design must address four main challenges. (1) The aerospace environment is much more restrictive than those a machine designer may previously have encountered. Admittedly, terrestrial (automotive, industrial, rail) and marine environments pose unique challenges, but none come with the combination of size and weight restrictions and mechanical and environmental challenges of aerospace applications. After all, airplanes fly. Unlike in a factory, where space and mass are virtually un-restricted, in an airplane weight and volume fetch a premium because everything onboard must be lifted off the ground. Not only that – once airborne, airplanes can move/rotate on up to six axes, posing extreme mechanical design challenges not present in (stationary) industrial or (one- or two-axis motion) automotive/rail applications. Throw in the extreme temperature and pressure swings experienced by aircraft, and the scope of the challenges posed by the aerospace design environment begins to reveal itself. (2) Aerospace EMs, especially those enabling EAP, must have HSP. Because size and weight are at a premium, EMs must squeeze every last watt out of each kilogram of mass and cubic centimeter of volume if they are to "buy their way" onto the plane. In fact, SP in the 13–16 kW/kg range is required, far exceeding today's SOTA machines. (3) The power rating must be very high – i.e., in the megawatt range. Today's 100–200 kW SOTA machines cannot supply enough power to propel large aircraft. However, as will become apparent in Section 3.4, scaling up machine power via traditional methods is not a straightforward process. Finally, (4) machine efficiency must be extremely high because as machine power goes up, so does the thermal impact of its waste heat. At 1 MW of power, even a 95 percent efficient machine generates 50 kW of relatively low-quality heat, enough to overwhelm most SOTA aircraft thermal management systems. For this reason, it is likely that 98 percent efficiency is the minimum acceptable value for MW-scale machines.

To summarize:

(1) The aerospace environment is uniquely challenging, requiring machines with the following attributes:
(2) HSP (>13 kW/kg),
(3) High power (>1 MW), and
(4) High efficiency (>98 percent).

3.2 Powertrains for Electrified Propulsion

EAP requires a significant redesign of conventional aircraft powertrain architectures in order to support the EMs and fans that will propel the aircraft. Although covered in an earlier chapter, EAP powertrains are briefly revisited in this section in order to better preface the material that follows.

A conventional commercial aircraft powertrain consists of a gas turbine engine (the "turbo" part of a turbofan) powering a rotating shaft. The shaft drives either a large bypass fan (the "fan" part of a turbofan) or, in smaller planes, a propeller. The plane's fans/propellers provide its thrust. Thus, all propulsive power is supplied by the plane's gas turbines mechanically coupled to its propulsive devices. Conversely, in an aircraft with electrified propulsion, some fraction (potentially all) of the thrust is produced by electrically driven fans. Thus, some or all the propulsive power must either be stored and supplied electrically or converted from some other form (chemical, mechanical, etc.) to electricity.

Although there exist numerous powertrain architectures for EAP, all can generally be classified according to the six categories shown in Figure 3.1: parallel hybrid electric, series hybrid electric, series/parallel partial hybrid electric, all turboelectric, partially turboelectric, and all electric [12]. Broadly speaking, the three hybrid architectures use a combination of fuel and energy storage to power some combination of electric and turboelectric propulsion fans; the two turboelectric architectures use fuel exclusively to power electric propulsion fans and possibly turbofans; and the all-electric architecture uses electric energy storage exclusively to power electric

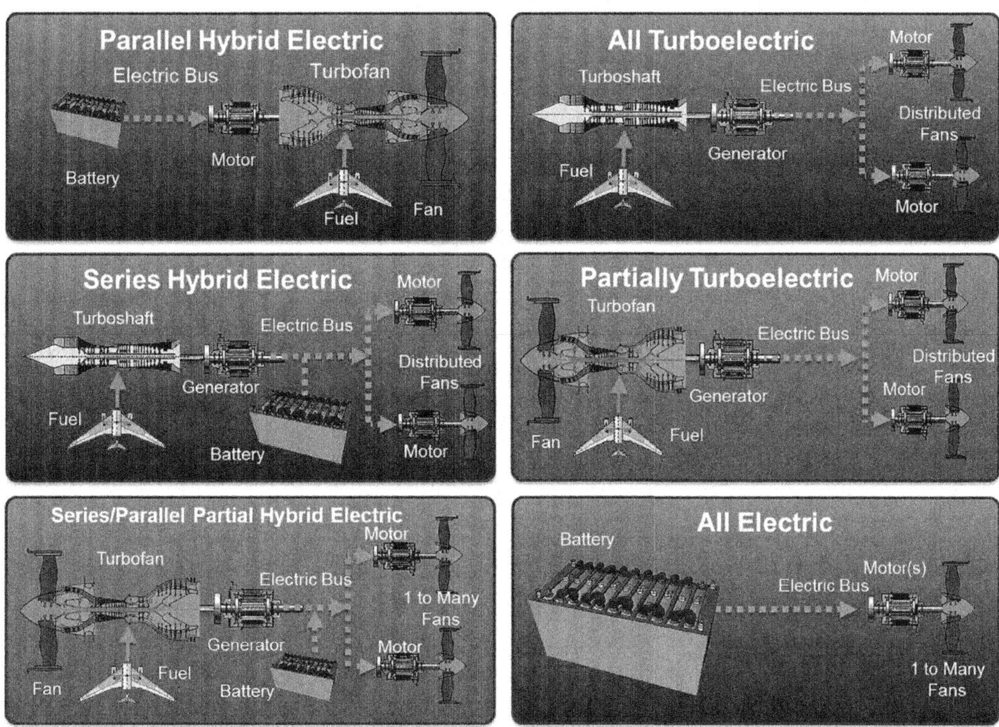

Figure 3.1 Felder's classification of power system architectures for electrified aircraft propulsion [12].

propulsion fans. Note that Figure 3.1 is only notional, and, in practice, the "battery" shown would include a power electronics interface to the power system.

The EAP powertrains shown in Figure 3.1 are briefly reviewed in the following sections.

3.2.1 Hybrid Powertrains

The three hybrid systems, which occupy the left column of Figure 3.1, have one common trait: their primary propulsion power is supplied by a combination of batteries (electrical energy storage) and fuel (chemical energy storage). They are differentiated by their method of using that energy to produce thrust. Parallel hybrids contain a conventional fuel-powered turbofan whose shaft can also be driven by a battery-powered electric motor. The turbine engine and motor are mounted on the same propulsion fan shaft so that the fan can be driven by either of the two energy sources independently. Converting a conventional turboelectric powertrain to a parallel hybrid powertrain is probably the most straightforward of all the options in Figure 3.1 because it only requires adding a battery and coupling a motor to the main turboshaft.

Unlike the parallel hybrid, the series and series/parallel partial hybrids utilize electric motor-driven propulsion fans, which are powered by a combination of battery and gas turbine–generated electricity. In the series hybrid, all the power in the gas turbine shaft is converted by a generator to electricity, which then powers the propulsion motors and charges the batteries. In such a system, only the electric motors are mechanically connected to the fans. Series/parallel partial hybrid systems, as indicated by their name, combine features of series hybrid and parallel hybrid electric powertrains. The main difference between the series and series/parallel partial hybrid architectures is that the series hybrid has only a turboshaft (it is propelled exclusively by motor-driven fans), whereas the series/parallel partial hybrid has a turbofan (it is propelled by both a turbo-driven fan and motor-driven fans).

Although useful for categorizing powertrain architectures, the classification diagram shown in Figure 3.1 leaves out some important details, some of which are the size/rating ratio of the various powertrain elements with respect to each other, the optimum mix of battery and turbo power for a given application, and the timing of *when* to use the various energy sources and propulsion methods. It is the engineer's job to make these decisions based on the aircraft and its intended performance. Generally speaking, the most efficient combination of fuel and propulsion at any given time should be used, but, ultimately, these decisions are made on a design-by-design and a mission-by-mission basis. As an example, [13, 14] describe the process of determining when is the optimal time to use a hybrid electric propulsion system's batteries during a flight. The authors in [13] analyze the benefits of battery-augmented power in takeoff and climb, while the authors in [14] postulate that the optimal mission segments for battery use are at takeoff and at the end of the mission.

3.2.2 Turboelectric Powertrains

The turboelectric architectures are the top two shown in the right column of Figure 3.1. These convert the chemical energy in jet fuel to mechanical and electrical energy via a gas turbine generator. The only difference between the all- and partially turboelectric powertrains is that the partial system uses a fraction of the turbine shaft power to drive a turbofan, whereas the all-turboelectric powertrain converts all of the turboshaft power into electricity. The electric power is distributed to other parts of the vehicle, where it drives distributed propulsion fans. The electrical components for a partially turboelectric system do not have to be sized to handle the full propulsive power of the aircraft and can therefore be developed with smaller advances beyond the SOTA than are required for an all turboelectric system. Note that commercial aircraft already generate both mechanical shaft power and electrical power; however, currently the electricity is used exclusively for non-propulsive power requirements. It is this new propulsion element that drives the massive increases in EM power and SP requirements.

It is interesting to note that the only difference between the series hybrid and the all-turboelectric powertrains is the addition of a battery in the series hybrid. A similar observation applies to the comparison between series/parallel partial hybrid and partial turboelectric powertrains. Because of this, some authors believe there are only four main powertrain categories rather than the six shown in Figure 3.1. Regardless of which classification is used, however, the EM design implications are the same.

3.2.3 All-Electric Powertrains

All-electric powertrains use electrochemical energy storage, typically batteries, exclusively to supply propulsive power. Essentially, in all but the parallel hybrid electric architecture, converting to an all-electric powertrain requires the removal and/or replacement of the turbine generator with a battery. To convert the parallel hybrid electric architecture to an all-electric one, the turbofan is replaced with electric motor-driven fans. Because all-electric powertrains have no chemical energy storage (fuel), the size of aircraft that can use an all-electric solution depends entirely on the availability of batteries with sufficient energy density.

3.2.4 Summary

As emphasized in earlier chapters, several aircraft system studies have indicated that there are significant benefits to using electrified propulsion to either replace or augment traditional fuel-based propulsion. The electric powertrains that will enable this upgrade – whether they be hybrid, turboelectric, or all electric – require EMs with much higher power, SP, and efficiency than are available today. Significant research and development is required across the breadth of the supporting technologies to achieve these gains.

3.3 Motor and Generator Systems

Numerous EAP studies have identified EMs, power conversion electronics, and circuit protection devices as key components requiring significant technological advancements. This chapter focuses on one of these three components – EMs – and, specifically, how to reduce the weight of large EMs and make them compatible with the overall aircraft system.

In this section, EMs are classified and organized by topology and then ranked according to a series of key characteristics specific to high-performance aerospace applications. Large EMs for EAP fall into two broad categories: propulsion motors and high-speed generators. Each of these is discussed in turn. Finally, some important considerations that apply when designing an EAP powertrain using large EMs are presented.

3.3.1 Aerospace Electric Machine Classification

Aerospace EMs can be classified according to the diagram in Figure 3.2, which was originally proposed by Ganev in [15] as a way to organize and rank high-performance

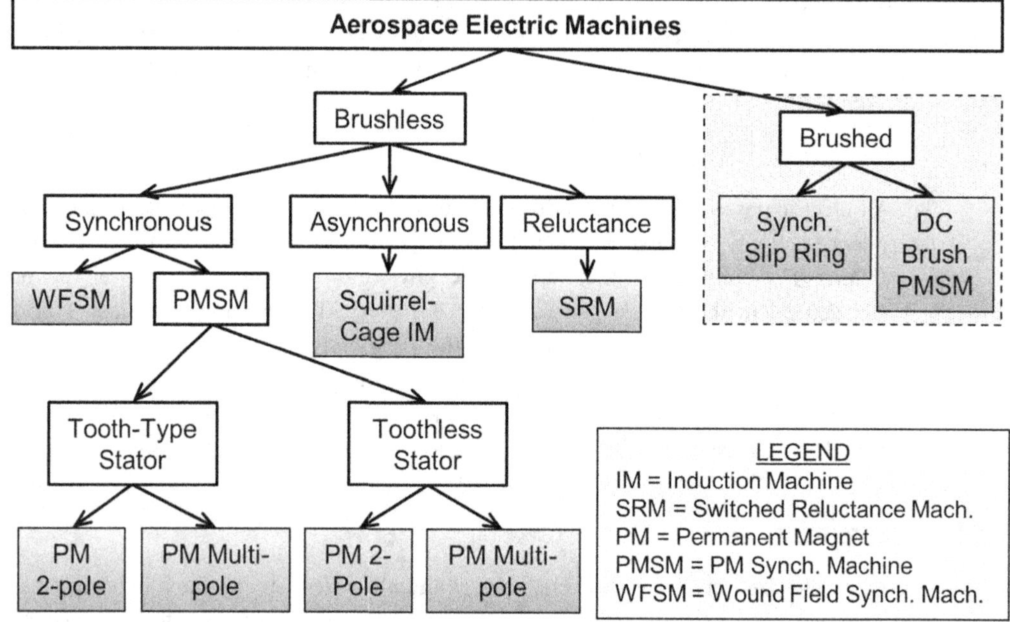

Figure 3.2 Ganev's classification of aerospace EMs [15].

EMs for aerospace high-power generation systems. Although his original intent was to classify generators, Ganev's diagram can be used to more generally classify EMs for EAP powertrains. In it, the white boxes represent EM categories, which become more specific when moving from top to bottom. Each category "pathway" ends on a gray box, which represents a specific EM topology. For example, the permanent magnet (PM) multipole machine indicated in the lower left is a PMSM with a tooth-type stator, which is a synchronous, brushless machine.

Ganev ranked the brushless machine types classified in Figure 3.2 according to several factors important to aerospace applications. Each factor was scored on a scale of 1–10, with 1 being the lowest and 10 being the highest. The brushed machines, due to low reliability, high maintenance costs, and arcing hazard, were not ranked. Ganev also chose not to score the wound-field synchronous machine (WFSM), citing its speed limitations and complex construction. However, it is included and scored here for comparison. The rankings are shown in Table 3.2, where it is revealed that PM machines as a whole score higher than WFSMs, induction machines (IMs), and switched reluctance machines (SRMs). Amongst the PM machines, the PM toothless, multipole design scored the highest, with its only noted limitations being rotor thermal limitations and short circuit behavior. Validating Ganev's assumption, the WFSM ranked the lowest. For more details on the scoring system, the reader is referred to [15].

The EM classification and associated rankings, which are based on the current SOTA, provide some guidance for designing HSP machines for EAP. However,

Table 3.2 Ranking of EMs for aerospace applications from [15], with the WFSM scored and ranked for comparison.

Key characteristic	WFSM	IM	SRM	PM 2-pole	PM multipole	PM toothless, 2-pole	PM toothless, multipole
Rotor losses	6	6	6	10	10	10	10
Stator losses	8	8	8	9	10	8	9
Windage losses	4	5	1	9	9	10	10
Rotor thermal limitations	7	8	10	4	4	4	4
Cooling options	5	5	5	9	9	10	10
Rotor mechanical limitations	4	5	7	9	9	10	10
Torque-to-inertia ratio	6	5	7	9	9	10	10
Torque pulsation	6	9	3	6	6	10	10
Compatibility with bearings	9	5	5	9	9	10	10
High-speed capability	3	5	7	9	9	10	10
Short circuit behavior	9	10	10	4	4	3	3
Machine complexity	6	7	10	9	9	8	8
Current density	10	7	7	10	10	8	8
Power density	7	7	8	10	10	8	8
TOTAL	**90**	**92**	**94**	**116**	**117**	**119**	**120**

technical advancements can occur, significantly changing the scores of certain key characteristics in Table 3.2. If this happens, the relative scoring may change and low-scoring topologies, such as the SRM or WFSM, may become more competitive. New topologies that outperform the members of this list may also be invented and developed. Thus, this ranking, or any ranking based on the current SOTA, should be periodically updated to reflect technological advancements. Further, it is not the intent here to judge ranking systems, and other scoring systems may be devised that lead to alternative rankings. A particular ranking has been included here to highlight the various key attributes that must be considered when comparing different machines.

3.3.2 Propulsion Motors

Propulsion motor systems consist of the propulsion motor, power electronics, and possibly a gear reduction unit. For distributed EAP, the output power required of each

EM is 1–2 MW. The propeller or ducted fan speed will be 2.5–4 krpm and likely will have blade pitch control to limit excessive speed excursions. Due to noise-reduction goals, the lower end of this speed range will likely be targeted. Unless a gear reduction unit, with its associated weight and efficiency penalties, is employed, this combination of speed and power serves to define the motor torque. It also hampers the ability to use high-speed machines as a means to reduce size and mass. The input to the propulsion motor system is the power bus, which may be a variable-voltage dc bus (typically hundreds or thousands of volts), fixed frequency ac bus (typically hundreds or thousands of hertz, hundreds or thousands of volts), or variable-frequency ac bus (typically hundreds or thousands of hertz, hundreds or thousands of volts).

3.3.3 High-Speed Generators

A high-speed generator system consists of a generator mounted on the high-speed turbine shaft, which supplies power to the power electronics. The power electronics convert the electricity supplied by the generator to a form that can be used by propulsion fans. The aircraft propulsion power structure requires redundancy; thus, one high-speed generator system will drive multiple, but not all, propulsion fans. The result is that the required generator output power is 5–10 MW. The turbine speed determines the generator speed. High-speed turbines with speeds of 20–40 krpm are common. The output from the generator system is to the power bus, which may be a variable voltage dc bus, fixed frequency ac bus, or variable frequency ac bus.

3.3.4 Electrified Aircraft Propulsion Powertrain Design Considerations

When an aircraft propulsion powertrain is electrified, two conditions must be satisfied under both steady-state and transient conditions to ensure the normal operation of the power system: power flow balance and speed and frequency matching.

Steady-state power flow balance is accomplished by appropriately sizing each component in the system – turbine engine, propulsion/ducted fans, converters, generators, and motors, etc. Transient power flow balance is accomplished by active power control algorithms, which are implemented in every energy-conversion stage and possibly through the use of energy storage devices. Control issues that may be encountered are similar to those present in other electric power system applications such as (hybrid) EV systems and distributed-generation microgrids. Effective power control methods have been developed in these other areas and can be readily adapted and used in the EAP system.

EAP system design optimization often focuses on addressing the mismatch between the various frequencies and rotational speeds at which different mechanical and electrical components operate. Variables include the gas turbine rotational speed (20–40 krpm is typical), generator speed and frequency, ac grid frequency (variable or fixed, typically up to a few kHz), propulsion motor speed and frequency, and the propulsion fan/propeller speeds (around 2.5–5 krpm). Any mismatches must be addressed through optimum selection of the number of poles on the EMs and the use of intermediate energy conversion stages, such as gearboxes at mechanical interfaces, and power electronics converters at electrical interfaces.

For mechanical speed mismatches (e.g., in the turbine-generator, or motor-fan subsystems), a direct drive topology is generally preferred because it lacks a heavy, high-maintenance gearbox. However, in the case of the motor-fan subsystem, the fan/propeller speed is typically much slower than that of the motor, especially when the motor speed has been maximized to give HSP. In this case, a step-up gearbox is probably unavoidable, but the impact of the gearbox weight and reliability on the system performance must be considered.

Several options for the EAP motor-fan subsystem exist. A few are discussed here for comparison purposes. In one option, the EM interfaces directly between the fan (a variable-speed load) and the ac source/grid. There is no power electronics interface, so the EM must operate at the grid voltage and frequency. The machine speed cannot be increased independently from the fan speed or grid frequency; thus, most HSP machine designs are not compatible, rendering this a poor/infeasible choice for EAP. Alternatively, a full-power back-to-back power electronics converter can be inserted at the electrical interface between the motor and the grid so that the amplitude and frequency of the operating voltages at the motor terminals are decoupled from the grid. In this architecture, the EM is free to operate at whatever voltage and frequency the power converter can supply, but there still must be a gearbox to interface with the fan if the machine is to operate at high speed. This architecture is a good candidate for EAP, but it relies on power electronic converters that are rated for full EM power (MW scale). A third option is a doubly fed IM (DFIM)-based architecture, which is common in the wind power industry. In this arrangement, a fraction of the machine power – typically about one-third – is pulled from/pushed to the ac bus and used to excite the rotor windings. The advantage is that the power electronics circuit, which is now in parallel with the main power circuit, only processes this excitation power; thus, it is a fraction of the size of the in-line option. If voltage and frequency ratings are closely matched between the motor and the grid, a DFIM may be used. Although the DFIM requires a relatively small power converter for field control, it comes with its own penalties, several of which are indicated in Table 3.2. For example, the power density of a DFIM is significantly lower than that of a multiple-pole PMSM. Also, the speed regulation capability is limited by the power rating of the field control converter. The relatively poor fault ride-through capability of the DFIM compared to other machine types is also a disadvantage.

Regardless of the architecture chosen, power flow balance and speed and frequency matching are key design constraints. Any EAP system must be designed with these in mind. These system considerations are important to remember while studying the fundamental EM design principles described in the next section.

3.4 Design Principles

The design of large EMs is guided by a few key principles, which are consequences of the way the governing physical principles of small machines are affected by scaling.

These principles are introduced here. This section is not intended to replace a comprehensive machine design text – e.g., Lipo's *Introduction to AC Machine Design* [16], Adkins' *General Theory of Alternating Current Machines* [17], or Alger's *Induction Machines* [18]. Rather, it is included to introduce only the most important concepts governing large EM design. These principles help to emphasize the greatest technological challenges that must be overcome in order for EAP to be feasible on a large scale.

3.4.1 Electromagnetic Design and the Sizing Equation

There are a few basic physical principles of power and force production in EMs that impact the machine size/mass, summarized as follows:

- For a rotating EM operating as a motor, the mechanical output power P_{out}, in W, is the electromagnetic torque T_e, in N-m, multiplied by the rotational speed ω_m, in rad/s:

$$P_{\text{out}} = T_e \omega_m \tag{3.1}$$

- Electromagnetic force, necessary for torque, is created by the interaction of the *electric loading*, K_s, in A-turns/m, defined as the number of stator A-turns per unit airgap surface circumference length, and the *magnetic loading*, B_{g1}, in T, defined as the peak of the fundamental magnetic flux waveform at the airgap. The combined effect of magnetic and electric loading is often referred to as the *electromagnetic loading*. The expression for the *shear stress* or *surface force density*, σ_m, in N/m^2, is

$$\sigma_m = \frac{\sqrt{2}k_1}{2} K_s B_{g1}, \tag{3.2}$$

where k_1 is the *winding factor*, a fraction that represents the aggregated effect of the stator winding (pitch, skew, etc.) on the net mmf waveform [16].

- Torque is calculated by multiplying the shear stress by the rotor surface area, in m^2, and the airgap radius R_{ag}, in m, i.e.,

$$T_e = \sigma_m \times (2\pi R_{ag} l_a) \times R_{ag}, \tag{3.3}$$

where l_a is the axial length, in m. The rotor volume V_R, in m^3, is expressed as

$$V_R = \pi R_{ag}^2 l_a = \frac{\pi}{4} D_{ag}^2 l_a, \tag{3.4}$$

where D_{ag} is the diameter of a cylinder whose outer surface lies in the center of the airgap. Combining (3.3) and (3.4) gives

$$T_e = 2\sigma_m V_R. \tag{3.5}$$

Finally, we define the *torque per unit rotor volume* (TRV), in N/m^2, as

$$\text{TRV} \triangleq \frac{T_e}{V_R} = 2\sigma_m; \tag{3.6}$$

I.e., the TRV is equal to twice the shear stress.

- Combining (3.1), (3.2), and (3.5) results in (3.7), the so-called *sizing equation*, which expresses EM volume as a function of output power, speed, and other major design parameters:

$$V_R = \frac{P_{\text{out}}}{\sqrt{2}k_1 K_s B_{g1} \omega_m}. \tag{3.7}$$

Equation (3.7) has two major implications for machine sizing. First, size can be reduced and SP increased by designing a machine with high magnetic and electric loading. Because both B_{g1} and K_s are factors in the expression of shear stress, increasing either or both of them increases TRV, allowing the machine size needed for a given power rating to decrease. Of course, increasing them is not a trivial task, because both B_{g1} and K_s are constrained by thermal system limits, and B_{g1} is constrained by the properties of the magnetic material. Clearly, these are important areas to explore if large increases in SP are to be made.

Second, the sizing equation makes clear the benefits of high-speed operation. For a fixed electromagnetic loading and machine volume, higher speed generates higher power output. This partially explains the importance of electrical frequency and tip speed in achieving HSP because, as we will see, expressions of each parameter contain the speed term ω_m. However, as explained in Section 3.3, in propulsion motor applications, the rotating speed of the EM is restricted by the propeller or ducted fan speed if no gearbox is used, and it may not be freely increased. There are also mechanical and thermal machine speed limitations, which are discussed in Sections 3.4.5 and 3.4.6.

3.4.2 Frequency and Pole Count

The electrical frequency f, in Hz, of the machine input excitation is proportional to the rotational speed and the number of poles, P:

$$f = \frac{P}{2}\frac{\omega_m}{2\pi}. \tag{3.8}$$

Thus, because (3.1) dictates that higher speeds lead to higher power, (3.8) means that many SOTA HSP machines have high electrical frequencies.

Equation (3.8) also indicates that, in addition to high rotational speed, a high pole count will increase machine frequency. What are the implications of this? For a given magnetic loading (flux density), a high pole count reduces the total flux per pole so that thinner ferromagnetic yokes can be used, significantly reducing the amount of iron material. Thus, a machine of a given size can be made lighter, which increases its SP. General-purpose COTS machines typically have between two and eight poles; much larger numbers can be chosen to increase the SP for HSP machines. However, this does not come without its challenges. The higher frequencies that come with high pole counts are limited by the speed of power electronics drives and by frequency-dependent losses, copper ac losses, eddy current, and hysteresis losses in ferromagnetic materials, etc. All of these issues must be carefully considered when designing a machine with a high pole count.

3.4.3 Machine Diameter and Tip Speed

SP tends to be higher in large-diameter machine designs. There are several reason for this. First, a larger diameter is required to accommodate a high pole count because a certain per-pole airgap circumference-to-depth ratio needs to be maintained to prevent large pole-to-pole leakage flux. The larger the diameter, the larger will be the per-pole airgap circumference. Second, rotordynamics considerations (i.e., keeping the machine mechanically balanced) favor large-diameter designs (more pancake-shaped) over long-axial-length (more pencil-shaped) at high rotational speeds beyond 10 krpm. Third, based on scaling laws that assume constant electric loading and current density and no end effects [16], the copper loss is proportional to the rotor surface area ($P_{\text{loss, Cu}} \propto D_{ag} l_a$), whereas the output power is proportional to the rotor volume ($P_{\text{out}} \propto D_{ag}^2 l_a$). These two expressions differ from each other by an order of D_{ag}, suggesting that a larger-diameter, shorter-axial-length design would be more efficient in the utilization of copper material than a smaller-diameter, longer-axial-length design.

The design parameter most closely related to the diameter is the rotor *tip speed* v, in m/s, where

$$v = \frac{D_{ag}}{2}\omega_m. \tag{3.9}$$

In inner-rotor topologies, the tip speed is proportional to the airgap diameter if airgap depth is small enough to be neglected. In outer-rotor topologies, the thickness of the outer rotor ring needs to be added. High tip speeds induce high mechanical hoop stresses in the rotor, as well as high windage losses on the rotor surface. These stresses and losses must be kept within design limits when expanding the rotor diameter. More details on this are given in Section 3.4.6.

3.4.4 Current Density

Another important design parameter that affects SP is the excitation *current density* J, in A/m^2. In conventional machines (as opposed to SCMs), the total mass of components that are iron/magnetic is much higher than that of those that are copper/conducting. However, if the iron yoke thickness is significantly reduced by a high-pole-count design, the copper material mass begins to take a larger share of the total. Current density and electric loading are directly related by

$$K_s = d_s J k_{\text{Cu}} \frac{w_s}{w_s + w_t}, \tag{3.10}$$

where d_s is the *slot depth*, k_{Cu} is the *slot copper packing factor*, w_s is the *slot width*, and w_t is the *tooth width*. The depth and width parameters are expressed in m.

It is apparent that increasing J can reduce the slot depth for a given electric loading. In slotted machines, this means the total mass of the windings and iron teeth is reduced. In slotless machines, large current density is even more advantageous; without it, the only way to achieve high electromagnetic loading is through the magnetic field. But the large magnetizing reluctance that is caused by the lack of teeth would likely force the magnetic loading to be increased past its limits, or more magnets on the rotor would need to be used.

Current density is limited primarily by the temperature limits of the conductor and its insulation. These limits are directly related to the ability of the cooling system to remove heat. This is detailed in the following section.

3.4.5 Thermal Limitations and Cooling Systems

Any increase in magnetic or electric loading, electrical frequency, rotor tip speed, or current density leads to higher losses, which inevitably raise the temperatures of various portions of the machine. However, temperature limitations must be respected to avoid machine breakdowns caused by melted electric conductors and/or insulation, permanent magnet demagnetization, bearing and lubrication failure, and other factors. It is the cooling system's job to remove excess heat and limit temperature increases. An HSP machine is most likely to operate at or near its thermal limits with armature current magnitudes pushed to their full capacity. Therefore, a very effective and efficient cooling system is essential. For this, there are three heat transfer modes of interest: conduction, convection, and radiation. In [19–21], modern cooling techniques based on these heat transfer modes are reviewed.

The heat conduction rate depends on the thermal conductivity of the machine materials. Metals – e.g., copper, aluminum, and steel – have high thermal conductivities in the range of 15–400 W/(m-K), while the thermal conductivity of insulators such as resins and paper is much lower, typically in the range of 0.1–1 W/(m-K) [19]. For comparison, air has an average thermal conductivity of 0.026 W/(m-K). The large differences between the thermal conductivities of these materials lead to efforts to limit the amount of insulating air by increasing slot fill factor and filling voids in motor windings. These can be achieved by using preformed tooth-wound coils with fractional slot winding configurations [22], encapsulation, and impregnation with epoxy resin [20]. The idea behind all of these methods is to reduce the amount of air insulating the coils and iron parts while increasing the contact surface area between the hot machine parts and high-conductivity cooling materials.

Convection is heat transfer by mass motion of a fluid such as air or water when the heated fluid is caused to move away from the source of heat, carrying energy with it. In the design of motor cooling systems, a distinction is made between natural and forced convection techniques. In natural convection, air extracts heat from hot surfaces of motor parts, becomes less dense, and then rises. In forced convection, the fluid motion is due to an external force created by a special device, such as a fan or pump, which means that the flow rate will be much higher than that of a natural convection process. The heat transfer coefficients for different convection techniques are as follows: air natural convection – 5–10 W/(m^2-K); air forced convection – 10–300 W/(m^2-K), and liquid forced convection – 50–20,000 W/(m^2-K) [19].

Radiation is the transfer of heat energy by electromagnetic waves. The radiation resistance between a motor and its surrounding environment is usually much higher than the forced convection resistance and thus is often neglected [20]. However, in

Table 3.3 Forced convection cooling techniques for HSP machines [19–21].

Method	Description	Max. J [A/mm^2]
Forced air	Forced air through radial/axial channels/ducts by fans	5–12
Indirect water	Water pumped into cooling jacket around stator frame/ yoke	10–15
Indirect oil	Oil or lubricant circulation for cooling on stator frame/yoke	10–15
Liquid bathed	Whole machine bathed in dielectric liquids such as mineral oils	up to 25
Direct liquid	Coolant in direct contact with conductors to take heat away	up to 30
Oil spray	Oil sprayed onto end turns and heat extraction by vaporization	Greater than 28
Hollow shaft	Oil through hollow shaft to cool otherwise inaccessible rotor parts	-

aircraft and aerospace applications, where the air density is low, standard convection-based cooling schemes are less effective. In these cases, radiative heat transfer can play a significant role. Radiative cooling can be enhanced by increasing housing emissivity and environment system absorptivity.

Most HSP machines rely on aggressive forced convection cooling techniques to extract and remove the heat from a highly constrained space. These cooling techniques and their achievable continuous current densities are summarized in Table 3.3. In general, liquid convection is more efficient in extracting heat than air convection because liquid has higher density and specific heat than air. Liquid coolants frequently used in motor cooling systems are water, oil, and their combinations. Structural optimization of fins and heat sinks is also useful to increase the heat transfer surface area and enhance the cooling performance.

3.4.6 High-Speed Operation

According to (3.1), all else being equal, increasing rotational speed increases output power and SP. However, as speed is increased ever higher in pursuit of HSP, thermal and mechanical design constraints begin to rear their ugly heads. At high enough speeds, these limits often dominate the design. Some considerations in this regard are discussed in this section.

Extensive surveys on high-speed EMs have been made in [23–28]. These report numerous high-speed machine prototypes for applications as varied as gas turbines, compressors, flywheel energy storage systems, turbo-molecular pumps, machine tool spindle drives, etc. The authors in [23] summarize the technical aspects and features of each of the machine prototypes. In [24], such issues as performance requirements dictated by various applications, high-standard materials adopted for overcoming physical limits, and different features originating from varied topologies are discussed. Electrical and mechanical design considerations and trade-offs are presented in [25].

The mechanical aspects include the shaft dimensions, bearings and lubrication, and cooling system. Electromagnetic utilization, loss components, types of motors, high-speed bearings, inverter–motor interaction, etc., are addressed in [26]. All of these papers plot power-speed distribution diagrams for prototypes found in literature, and a common finding is that the maximum power *decreases* as the nominal speed increases. This is the opposite of what is suggested by (3.1). So, what is going on here?

In [26, 27], the power-speed profile correlation is obtained empirically as

$$P_{\text{nom}} \propto \frac{1}{f_{\text{nom}}^{3+k}}, \qquad 0 \le k < 1, \tag{3.11}$$

where P_{nom}, in W, is the nominal output mechanical power and f_{nom}, in Hz, is the nominal synchronous frequency. It is suggested that this relation arises from thermal constraints, or, more specifically, that the power output is limited by the maximum frequency-dependent losses that must be dissipated through the cooling surface area. Based on this assumption, both [26, 27] have developed formulas of similar forms to (3.11). Besides the thermal constraints, the authors in [27] also consider mechanical speed constraints. Analytical equations or numerical methods are introduced to calculate radial and tangential centrifugal stress distribution within the rotor. Methods to find the flexural critical speed and stability threshold speed in terms of rotordynamics are also reported.

The authors in [28] suggest that the limitations of high-speed EMs can be classified by introducing a so-called high-speed index, or HS index for short, which is the product of nominal speed and nominal power $(n_{\text{nom}}P_{\text{nom}})$. For high-speed machines with a relatively low nominal speed (approximately 10–20 krpm), the HS-index is mainly limited by the rotor elastic limits (i.e., the maximum rotor tip speed, v_{max}), as indicated by (3.12):

$$\text{HS-index:} \quad n_{\text{nom}}P_{\text{nom}} \propto v_{\text{max}}^2, \qquad n_{\text{nom}} < \sim 20 \text{ krpm}. \tag{3.12}$$

For high-speed machines with higher nominal speeds, the HS index is mainly limited by the additional electromagnetic losses, and the following relationship is derived:

$$\text{HS-index:} \quad n_{\text{nom}}P_{\text{nom}} \propto \frac{v_{\text{max}}^3}{n_{\text{nom}}^2}, \qquad n_{\text{nom}} > \sim 20 \text{ krpm}. \tag{3.13}$$

The study in [28] also discusses other limitations – bearings, cooling techniques, power electronics, rotor losses, etc.

Alternatively, the authors in [24] suggest that the maximum power capability in high-speed machines is limited by mechanical dynamic issues such as critical speeds, high values of bearing $D_b n_b$ (the product $D_b n_b$ is a bearing performance indicator, where D_b – in m – is the mean diameter of the bearing, and n_b – in rpm – is its rotational speed), peripheral speeds and stresses, and sensitivity to good balancing. The concept of $n_{\text{nom}}\sqrt{P_{\text{nom}}}$, where n_{nom} is the nominal machine speed in rpm and P_{nom} is the nominal output mechanical power in W, is an empirical parameter originally introduced and described in [29] in a broad rotating machinery context.

This parameter is used as a "guide number" to assess the likely severity of those dynamic problems. The paper comments that dynamic problems are moderate for rotating machinery that operate between 500 and 1,000 krpm$\sqrt{\text{kW}}$. Greater than this, mechanical problems become more and more difficult to manage. In Table 3.5, this krpm$\sqrt{\text{kW}}$ figure of merit is calculated for the numerous surveyed HSP machines. All but one of the numerous high-speed HSP machines surveyed lie within or below the $500-1{,}000$ krpm$\sqrt{\text{kW}}$ range.

Current studies of EAP systems indicate that power and speed ratings of high-speed machines lie predominantly in the range of 1–10 MW and 10–40 krpm, respectively. Correspondingly, the aforementioned figure of merit varies between 300 and 4,000 krpm$\sqrt{\text{kW}}$. The lower end is apparently achievable through the current technology, while the upper end will require major technological breakthroughs, for example, through the maturing of SC technologies.

3.4.7 Power Scaling

Simply taking the design equations and scaling up for speed and power is not as easy as it sounds. For example, (3.1) indicates that if torque can be held constant, machine power can be scaled up by simply increasing the rotational speed. This is true. However, as highlighted in Section 3.4.6, there are physical limitations that prevent this strategy from being successful at high speeds. This section provides additional insights on this important topic of scaling.

In a typical rotating machine with a cylindrical geometry and radial flux, the output power is proportional to speed, electrical and magnetic loading, and volume,

$$P_{\text{out}} \propto \frac{\pi^2}{120} n(D_{ag}^2 l_a) B_{g1} K_s k_1 = \frac{\pi}{30} n V_R B_{g1} K_s k_1, \tag{3.14}$$

where n, in rpm, is the rotational speed [16]. Dividing power by volume in (3.14) gives the *power density,* or *power per unit volume* (not SP) in W/m^3, ρ_P as

$$\rho_P = \frac{P_{\text{out}}}{V_R} \propto \frac{\pi}{30} n B_{g1} K_s k_1. \tag{3.15}$$

Assuming similar *mass density,* or *mass per unit volume* ρ_m, in kg/m^3, across similar machine designs, ρ_P and SP are directly related: maximizing one will necessarily maximize the other. According to (3.15), ρ_P is maximized by maximizing speed, magnetic loading, and electrical loading (the winding coefficient k_1 is considered to be fixed for this analysis; there are ways to increase its value, but these do not affect the power density nearly as much as increases in the other variables do). Also, recall that the TRV is twice the shear stress σ_m (Equation (3.6)), and $\sigma_m \propto k_1 K_s B_{g1}$. Thus, ρ_P is directly proportional not only to speed n but also to TRV:

$$\rho_P \propto n \times \text{TRV}. \tag{3.16}$$

According to (3.16), for a fixed torque density the power density can be increased by speeding up the machine. However, this only works for so long; the rotor will eventually reach its mechanical tip speed limit. Typically, v is limited to roughly 200 m/s before centrifugal forces begin to tear the rotor apart. The power equation (3.14) can be expressed in terms of tip speed. First,

$$P_{\text{out}} \propto \frac{\pi^2}{60}\left(\frac{60}{2\pi}\omega_m\right)\left(\frac{D_{ag}}{2}\right)D_{ag}l_aB_{g1}K_sk_1. \quad (3.17)$$

Then, simplifying and writing in terms of v using (3.9) gives

$$P_{\text{out}} \propto \frac{\pi}{2}vD_{ag}l_aB_{g1}K_sk_1. \quad (3.18)$$

Expressing the power density using (3.18) and assuming the machine is operating at its maximum tip speed gives

$$\rho_P = \frac{P_{\text{out}}}{\frac{\pi}{4}D_{ag}^2l_a} \propto \frac{2v_{\max}B_{g1}K_sk_1}{D_{ag}} \propto D_{ag}^{-1}. \quad (3.19)$$

This reveals that the power density is inversely proportional to the machine diameter when the machine is tip speed limited. In other words, as the machine gets larger, its power density decreases. This doesn't bode well for scaling up in power.

How does this result manifest itself in machine SP, the key performance parameter of interest when designing MW-scale machines for EAP? If we assume that mass density ρ_m is roughly constant, mass scales directly with volume, which itself scales directly with $D_{ag}^2l_a$ according to (3.4). Next, from (3.5), assuming constant shear stress (the result of thermal, material property, and excitation requirement limits on electrical and magnetic loading in (3.2)), the torque T_e also scales as $D_{ag}^2l_a$. Therefore, mass, volume, and torque all scale directly with $D_{ag}^2l_a$.

Now, let's assume a fixed aspect ratio $\alpha \triangleq D_{ag}/l_a$, which results from maintaining relative machine dimensions as the machine size is scaled up. In other words, the machine is not allowed to become more "pancake-shaped" or "pencil-shaped" as it grows. With this assumption, the mass, volume, and torque all scale with D_{ag}^3/α or simply with D_{ag}^3.

Combining (3.1) and (3.9), recognizing that T_e scales with D_{ag}^3, and assuming max tip speed operation gives the following scaling for output power:

$$P_{\text{out}} = T_e\omega_m \propto D_{ag}^3\left(\frac{2v_{\max}}{D_{ag}}\right) \propto D_{ag}^2. \quad (3.20)$$

The machine SP now scales as

$$\text{SP} = \frac{P_{\text{out}}}{\text{mass}} \propto \frac{D_{ag}^2}{D_{ag}^3} = D_{ag}^{-1}. \quad (3.21)$$

This is the same scaling relationship that was revealed by (3.19) for the power density ρ_P. Mass and volume are directly related with the assumption of constant mass density ρ_m; thus, this is the expected result.

Finally, because mass $\propto D_{ag}^3$ but $P_{\text{out}} \propto D_{ag}^2$, we can write

$$\text{mass} \propto D_{ag}^3 = (D_{ag}^2)^{3/2} \propto P_{\text{out}}^{3/2}. \tag{3.22}$$

Writing in terms of SP gives

$$\text{SP} = \frac{P_{\text{out}}}{\text{mass}} \propto \frac{P_{\text{out}}}{P_{\text{out}}^{3/2}} = P_{\text{out}}^{-1/2} = \frac{1}{\sqrt{P_{\text{out}}}}.$$

Thus, if the machine is tip speed limited and its aspect ratio is fixed as it scales,

$$\text{SP} \propto \frac{1}{\sqrt{P_{\text{out}}}}. \tag{3.23}$$

We have arrived at the key scaling relationship. Clearly, (3.23) reveals a negative economy of scale for SP: when the machine is tip speed limited and its aspect ratio is constrained, scaling up in power reduces the SP. For example, if a 200 kW machine with SP of 6 kW/kg is scaled up to 1.8 MW (power multiplied by a factor of nine), its SP will be reduced to 2 kW/kg. Thus, when scaling conventional, existing machines up to MW power levels using standard techniques, HSP designs do not result.

3.4.8 Discussion

A machine's SP is directly and significantly impacted by its electric and magnetic loading, electrical frequency, rotor tip speed, and current density. All else being equal, increasing any of these (not necessarily independent) parameters will increase SP. However, all are constrained by some underlying physical factors such as material magnetic properties, joule losses, cooling capability, or material elastic properties. Generally, one parameter cannot be increased without some effect (positive or negative) on one or more of the others. For an optimal HSP machine design, trade studies should be conducted to find the best overall combination of values for all these parameters, which presents real challenges to design engineers [30]. For example, both f and B_{g1} are crucial parameters for iron losses, which are limited by the capacity of the cooling system. If f is increased, B_{g1} should be reduced to prevent excessive iron losses. Otherwise, the additional losses may lead to unacceptable temperature rise at some hot spots with limited cooling capacity.

For demanding high-performance EAP applications, EM sizing and optimization present both significant challenges and opportunities for innovation.

3.5 Review of High-Specific-Power Machine Topologies

According to the ranking in Section 3.3.1, PMSMs are superior for aerospace applications; however, technology advancements may impact those rankings significantly, potentially elevating another topology to the top of the list. It is therefore important to understand the key features, similarities, and differences of all the various candidates.

In this section, the four main conventional (i.e., non-SC) machine topologies that are candidates for achieving HSP designs are described. These are the permanent magnet synchronous machine (PMSM), induction machine (IM), switched reluctance machine (SRM), and wound-field synchronous machine (WFSM). This section is not an introduction to these machine types; rather, it assumes the reader has a solid foundation in machine design principles and is familiar with the basic operation of these devices. The four machine topologies are described in the context of MW-scale machine design, highlighting their various features and discussing their potential for use in EAP powertrains.

3.5.1 Permanent Magnet Synchronous Machines

HSP PMSMs use high-energy-density magnets, such as NdFeB and SmCo, to produce a strong magnetic field. These rare-earth PMs bring such benefits as high airgap flux density, high power factor, and low copper loss. PMSMs can make better use of high-pole-count short-pole-pitch designs than machines with rotor windings such as WFSMs and IMs to save iron yoke material [31]. For a permanent magnet (PM) pole, the magneto-motive force (mmf) is proportional to the length of the magnet in the magnetization direction and does not depend on the width of the magnet. While for a wound field pole, the mmf is determined by both length and width of the pole for a given current density in the coil. This difference indicates that PM excitation is comparatively more efficient when pole width is small. Recent commercial HSP PMSM designs with high pole counts and very thin yokes include the ThinGap 7140 [32] and Joby Motors JM1 [33].

Most high-speed HSP PM machines adopt a surface PM structure. Outside the magnets, a metallic or nonmetallic retaining ring is often needed to hold the rotor structure in the case of very high tip speed [34, 35]. The metallic ring and magnets are often segmented to reduce the eddy currents and associated joule losses. Carbon fiber is preferred to construct the retaining ring due to its low mass density and low conductivity.

In high-speed machines with conventional teethed stator structures, such as that shown in Figure 3.3a, iron losses often constitute a substantial portion of the total loss. In some high-speed PM machines, the iron teeth structures are removed to reduce mmf harmonics and iron losses. The saved space can then accommodate more conductors. This topology, shown in Figure 3.3b, is called an airgap winding. The airgap winding is often made using Litz wire (a bundle of very small strands twisted together) to reduce the high-frequency ac copper losses. In [36, 37], a 200 krpm HSP PMSM that utilizes a Litz wire airgap winding is presented. The lower losses and higher electric loading help to improve the SP.

Another PMSM option utilizes a Halbach magnet array, a configuration in which some magnets are not only radially but also tangentially magnetized. This configuration increases the equivalent magnetization length per pole and the total mmf in the flux loop. A Halbach array can be used to maintain the magnetic loading at a reasonable magnitude, even as the equivalent airgap is enlarged. With magnets

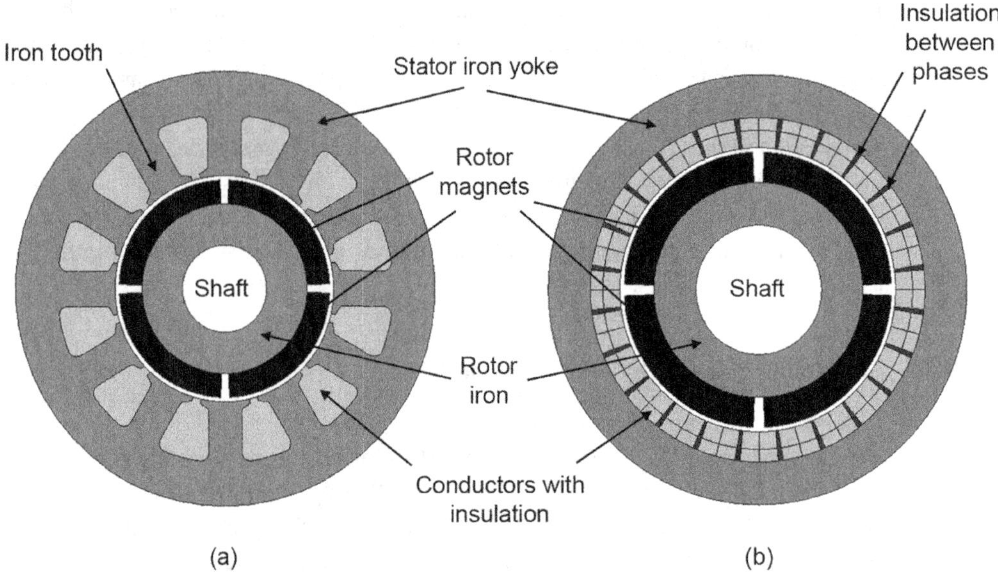

Figure 3.3 Stator core topologies with and without iron teeth structures. (a) Stator with conventional teethed iron corn. (b) Stator with airgap winding and no teeth.

magnetized at varied directions, the ferromagnetic yoke in the rotor could be minimized as well, which saves weight. In [38, 39], the design and testing of two high-speed aircraft PM motors that utilize the Halbach array structure to maximize airgap flux density are presented. Their SPs are estimated to be approximately 5 kW/kg.

Axial flux PMSM topologies share similar pros and cons to radial flux permanent-magnet (RFPM) machines. Compared to the conventional radial flux permanent-magnet RFPM topology, it is difficult to design and fabricate very high-speed axial flux permanent-magnet (AFPM) machines due to the large rotor diameter and complex bearings needed. However, AFPM also has its own advantages over RFPM such as better utilization of core material and being suitable for high-pole-count designs [40]. Several AFPM motor products have been able to achieve HSP values [41–43]. Similar to RFPM machines, AFPM machines can also use airgap windings and Halbach arrays to improve SP. Launchpoint has built an HSP AFPM motor with SP = 8 kW/kg using those structures [44].

The PMSM has some promising technology options for SP improvement. More than that, many of these technologies can be fit into one design to achieve superior performance. Figure 3.4 shows such an example [30, 45, 46]. This high-speed motor has a radial-flux, "iron-less," multiple-pole, outer-rotor architecture. The basic idea is to utilize a high pole count, an airgap winding, and Halbach magnet arrays to reduce the iron usage. The unconventional topology of the rotor creates mechanical challenges including high centrifugal stress and thin radial builds that affect rotordynamics. These issues are addressed in [45]. This 1 MW motor is expected to have SP that exceeds 10 kW/kg.

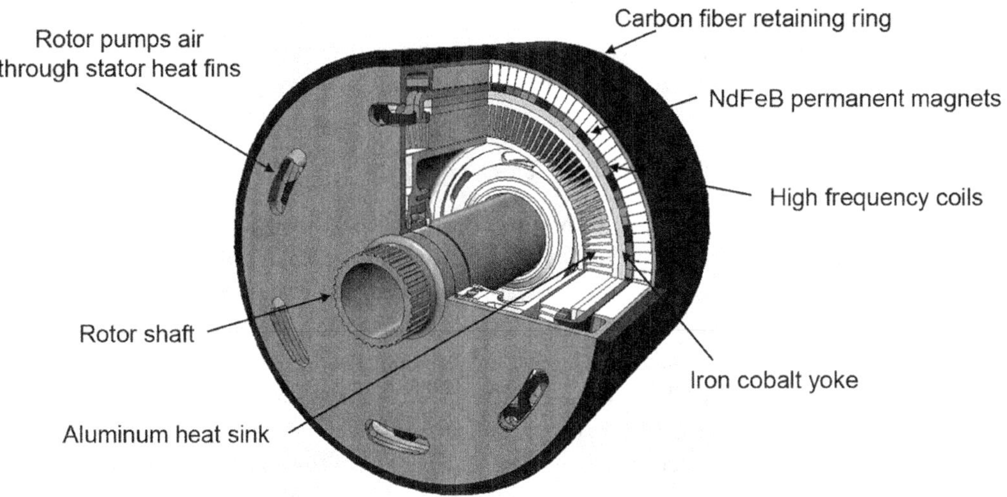

Figure 3.4 A radial-flux, "iron-less," very high frequency, outer-rotor PMSM for electrified aircraft propulsion [30, 45, 46].

3.5.2 Induction Machines

HSP IMs are mostly high-speed machines. Two kinds of rotor structures are often used: squirrel cage and solid. An IM with an aluminum or copper squirrel cage is the most common machine type for general-purpose industrial motors. The robust structure, mature fabrication technologies, and high reliability are the advantages for building a high-speed squirrel-cage IM. Its disadvantage is the additional magnetizing component needed in stator current, which makes its efficiency and power factor worse than PM machines. High pole count and airgap winding are not suitable for IM topology because they will further reduce the magnetizing inductance and worsen the machine performance.

The key aspect in the design and fabrication of a high-speed squirrel-cage IM is to choose appropriate materials and manufacturing techniques to fabricate a rotor that has sufficient electromagnetic performance while still maintaining structural integrity during high-speed rotation. The authors in [47] present an HSP IM that uses a high-strength copper composite called Glidcop to construct rotor bars. Stainless steel or metal matrix composite may also be used to retain the rotor end rings when necessary [34].

In a solid rotor IM, there is no cage or lamination structure on the rotor; the rotor is just a solid metal cylinder that carries the induced slip currents. This simplest rotor structure is very robust and can achieve the highest tip speed among various machine topologies. The maximum tip speed found in literature is 400 m/s [48], which is twice that of a conventional cage rotor or PM rotor. Solid rotor IMs can have large diameters and are suitable for high-speed, high-power applications. The authors in [49, 50] present an 8 MW, 6.6 kV, 12 krpm solid rotor IM for natural-gas-compression applications. The machine uses high-pressure natural gas as coolant and has a maximum rotor tip speed of 210 m/s. Its power density is 21.7 kW/l (although the weight is not provided, the SP is estimated to be 2–3 kW/kg).

3.5.3 Switched Reluctance Machines

Unlike PMSMs and IMs, SRMs have no magnets or windings on the rotor, which necessitates high current and mmf from the stator side to build the magnetic field. SRMs therefore tend to have lower power factors and torque densities than the other two machine types. Further, because large reluctance torque requires high rotor saliency, SRMs are prone to unbalanced radial forces, which can increase bearing wear and cause acoustic noise [51]. There are some other SRM features that help compensate for these disadvantages. First, SRMs have a robust rotor structure for high-speed operation. Second, the concentrated windings have short end parts, a feature that improves the thermal dissipation. Third, the absence of rare-earth magnets facilitates SRM operation in the harsh environment of an aircraft propulsion system. Finally, the simplicity of construction means that SRMs are relatively cheap compared to other machines of equivalent rating.

Most HSP SRMs found are high-speed prototypes for aerospace applications. These machines are typically cooled by engine lubricating or fuel oil and high magnetic and mechanical performance iron cobalt materials are used for stator and rotor cores to reduce machine volume and weight.

3.5.4 Wound-Field Synchronous Machines

WFSMs generally have high magnetic loading, high power factor, and a robust rotor forge suitable for high-speed rotation. All these features help to achieve HSP. One drawback of WFSM is the need for accessory parts to excite the rotating field windings such as brushes and slip rings, exciter machines, etc., which will add weight to the overall system.

In [52], two high-speed, high-power, HSP WFSM prototypes are presented and the importance of applying intensive cooling strategies is highlighted. The first machine is a 1 MW airborne synchronous generator, which utilizes cold plates in the rotor to cool its field windings. The current density in the field winding is increased to about 22 A/mm^2, and the SP is 4.8 kW/kg. The other machine is a 2.5 MW, 15 krpm, 1,500 Hz, oil-cooled synchronous generator developed by Electrodynamics Associates, Inc. The rotor windings are cooled by oil spray, and a current density of 30 A/mm^2 is attained. The machine runs at a 33 percent duty cycle, and the peak SP under this operating mode is 14 kW/kg. Note, however, that the *average* SP (which is reported later in Table 3.5) is only 9.5 kW/kg.

3.6 Survey of State-of-the-Art Machines

This section presents the results of a comprehensive survey of over 50 SOTA HSP machines. The surveyed machines are a mixture of COTS designs and laboratory prototypes. Before presenting the survey results, some general statements about the mass distribution in a SOTA EV motor are given for comparison.

Based on the survey results, common features of HSP machines are identified, and promising methods of further weight reduction and SP increase are discussed. Despite their SOTA status, none of the surveyed machines exceed an SP of 10 kW/kg. Thus, all fall well short of the lofty requirements necessary for EAP, leaving much room for innovation in the upcoming decades.

3.6.1 Subsystem Masses of Electric Vehicle Motors

In 2010, the U.S. Department of Energy published a detailed analysis of a 2010 Prius powertrain [53]. Comprehensive studies such as this are hard to come by. Thus, despite the fact that this motor is now close to a decade old, for comparison purposes the Prius motor is considered representative of a good SOTA baseline for automotive EVs; therefore, observations that are reasonably likely to apply to motors in general can be made.

The Prius motor has a peak SP of 1.6 kW/kg and power density ρ_P of 4,800 kW/m^3 (4.8 kW/l). Shown in Table 3.4 are the motor's component masses and their respective percentage of the total. An examination of other general-purpose industrial motors reveals similar subsystem mass distributions [54]. Thus, although not based on observations of MW-class motors, the following generalizations may provide some insights into system mass reduction opportunities for HSP machines:

- Casing is the heaviest single element in all motors (~40 percent).
- Stator laminations are second (~30 percent).
- Rotor and shaft are third (~20 percent).
- Neither the windings (~10 percent) nor the magnets (~2 percent) are major contributors to mass.

3.6.2 High-Specific-Power Machines

A survey of commercial and scientific literature for HSP EMs has been made. It was originally reported in [55] (partial results) and [1] (full results). It only includes existing machines that have been built and tested. Some of the machines are COTS products, and others are prototypes for scientific research or product development.

Table 3.4 2010 Prius motor mass distribution [53].

Item	Mass [kg]	% of total
Casing (stator, fan, front and back plates)	14.1	38.3
Stator laminations	10.36	28.1
Rotor and shaft	6.7	18.1
Windings	4.93	13.4
Magnets	0.77	2.1
Total	**36.86**	**100**

Thus, there are varied levels of technological maturity represented. Surveyed machines are used in military and civil aircraft, small unmanned aerial vehicles (UAVs), EV/HEVs, spindle drives, and centrifugal compressors. The topologies used are the following: PMSMs, IMs, SRMs, WFSMs, and synchronous reluctance machines (SynRMs). The PMSMs include both RFPM and AFPM variants.

Table 3.5 lists the surveyed machines and their performance traits, including nominal speed n_{nom}, nominal power P_{nom}, HS Index (HSI), frequency f, SP, and cooling method. Although these machines do not all qualify as MW-class large EMs (in fact, only six have nominal power ratings of 1 MW or greater), their design data can help us better understand the potential areas for mass reduction from a machine designer's perspective.

To better visualize the data, distribution diagrams plotting SP versus various parameters of the surveyed HSP machines are given in Figures 3.5–3.9. Each data point in the distributions represents one test-verified machine design, with its color and symbol indicating the machine topology used for that particular design. SC machines (SCMs) are included for comparison purposes, although they are not likely to evolve into being a mature technology for aircraft application in the next 20–30 years. Due to varied levels of product maturity and emphasis on different performance metrics based on their particular applications, it may not be easy to identify clear trends in these diagrams. However, these scattering patterns reveal some important differences in the key design parameters between HSP and non-HSP machines.

Figure 3.5a shows the SP versus nominal rotor speed distribution. HSP machines operating at speeds lower than 10 krpm are either PMSMs with high pole counts or SCMs with high electromagnetic loading. Other machines are designed to rotate at speeds greater than 10 krpm to reduce weight. Figure 3.5b plots SP versus rotor speed for PM machines only. In general, for a given machine type, SP tends to increase as the rotational speed increases.

SP versus rated power is plotted in Figure 3.6. Although specific trends are hard to identify quickly in the plot, in general SP tends to decrease as the power rating increases. An adverse scaling effect due to tip speed limitations was identified in Sections 3.4.6 and 3.4.7, and agrees with a similar conclusion in [87]. There are many possible explanations. First, machine losses scale with machine volume while the surface area available for heat dissipation does not increase proportionately. As the power rating and thus the machine size increase, the electromagnetic loading is reduced to maintain the same loss density in the cooling surface. Second, because the electromagnetic forces are larger in higher-power machines, the mass of inactive supporting structures is significantly higher. Third, at the MW scale, reducing machine material and manufacturing cost has been a main design objective. As a consequence, achieving high SP has not been a major design goal. Even so, a few MW-rated machines with impressive SP values have been found. Most of these machines adopt intensive cooling techniques to achieve very high electric loading.

The SP versus rotor tip speed distribution for the surveyed machines is plotted in Figure 3.7. Most of the HSP machines have a relatively high tip speed ($v > 50$ m/s),

Table 3.5 Survey of HSP EMs.

Topology	n_{nom} [krpm]	P_{nom} [kW]	HSI krpm$\sqrt{\text{kW}}$	f [Hz]	SP^ [kW/kg]	Cooling method	Ref. no.
RFPM	8.1	3.85	16	–	3.3	Forced air	[32]
RFPM	15	1	15	–	1.5	Forced air	[32]
RFPM	6.0	8.2	17	1,100	3.5†	Forced air	[33]
RFPM	6.0	13.2	22	1,100	3.9†	Forced air	[33]
RFPM	5.3	7.5	15	–	5.1	Forced air	[56]
RFPM	1.9	42	12.3	–	1.3	Forced air	[57]
RFPM	1.8	30	10	–	1.9	Forced air	[58]
RFPM	3.0	200	42	–	2.0	Indirect water (45 L/min)	[5]
RFPM	4.0	120	44	–	2.4	Indirect water	[59]
RFPM	10	65	81	–	1.2	Indirect water	[60]
RFPM	47	22.37	222	1,567	5.3*	Bathed in refrigerant	[34]
RFPM	8.0	90	76	400	4.1	Indirect water	[61]
RFPM	5.0	98	49	417	2.3	Indirect water	[62]
RFPM	28	25	140	1,400	2.3*	Spray oil	[63]
RFPM	200	2	283	3,333	8.4*	77 K environment	[36, 37]
RFPM	40	40	253	1,333	3.0*	Indirect water	[35]
RFPM	40	8	113	667	3.3*	–	[64]
RFPM	6.0	3,750	367	200	3.2*	Direct water	[65]
RFPM	15	16	60	750	5.2*	Bathed in oil	[38]
RFPM	10	1.7	13	–	3.1*	Forced air	[66]
RFPM	13	16	52	867	5.0	Indirect water	[67]
AFPM	4.0	42	26	917	3.4	Indirect water (8 L/min)	[41]
AFPM	4.0	100	40	667	4.9	Indirect water (8 L/min)	[41]
AFPM	3.2	75	28	–	1.3†	Indirect water (12 L/min)	[42]
AFPM	7.5	85	69	–	2.1†	Indirect water (12 L/min)	[42]
AFPM	5.0	150	61	–	1.9	Indirect liquid (12 L/min)	[43]
AFPM	6.0	15	23	1,200	3.1	–	[68]
AFPM	8.4	5.25	19	—	7.3	–	[44]
AFPM	4.5	15	17	600	1.1	Indirect water	[69]
AFPM	1.846	110	19	400	1.1†	Air	[70]
AFPM	1.5	50	11	125	0.9	Indirect water	[71, 72]
IM	50	21	229	833	1.7*	Indirect refrigerant	[47]
IM	45	45	302	1542	4.0*	Indirect liquid & forced air	[34]
IM	15	2,000	671	250	2.0*	Indirect oil & forced air	[73]
IM	56.5	12	196	942	1.6*	–	[74]
IM	12	8,000	1073	200	2.4*	High-pressure air	[49, 50]
IM	60	300	1039	1,000	3.8*	Water & forced air	[48]
IM	29	30	159	967	2.2*	Indirect water	[75, 76]
SynRM	48	60	372	800	2.7*	–	[77]
SRM	48	32	272	3,200	2.8*	Direct and indirect oil	[78]
SRM	25	90	237	1,666	6.9†	Direct and indirect oil	[79]
SRM	12.5	3.7	24	1,250	1.7†	Indirect oil	[80]
SRM	47	30	257	3,123	2.9†	Oil	[81, 82]
SRM	22	250	348	2,963	4.5	Direct oil, hollow shaft	[83, 84]
SRM	18	250	285	3,600	3.3*	–	[85]
SRM	28	25	140	2,800	1.1*	Spray oil	[63]

Table 3.5 *(cont.)*

Topology	n_{nom} [krpm]	P_{nom} [kW]	HSI krpm$\sqrt{kW}$	*f* [Hz]	SP^ [kW/kg]	Cooling method	Ref. no.
WFSM	13.6	2,500	680	680	3.7	Forced air	[86]
WFSM	10	1,000	316	–	4.8	Indirect liquid	[52]
WFSM	10	2,500	500	1,000	9.5	Indirect oil & oil spray	[52]

^ SP is the rated power over total machine mass, including both active and structural parts. In cases where total mass information is unpublished, it is estimated from the mass of active materials or machine geometrical dimensions.

† Estimated SP from the mass of only active materials.

* Estimated SP from the machine geometrical dimensions.

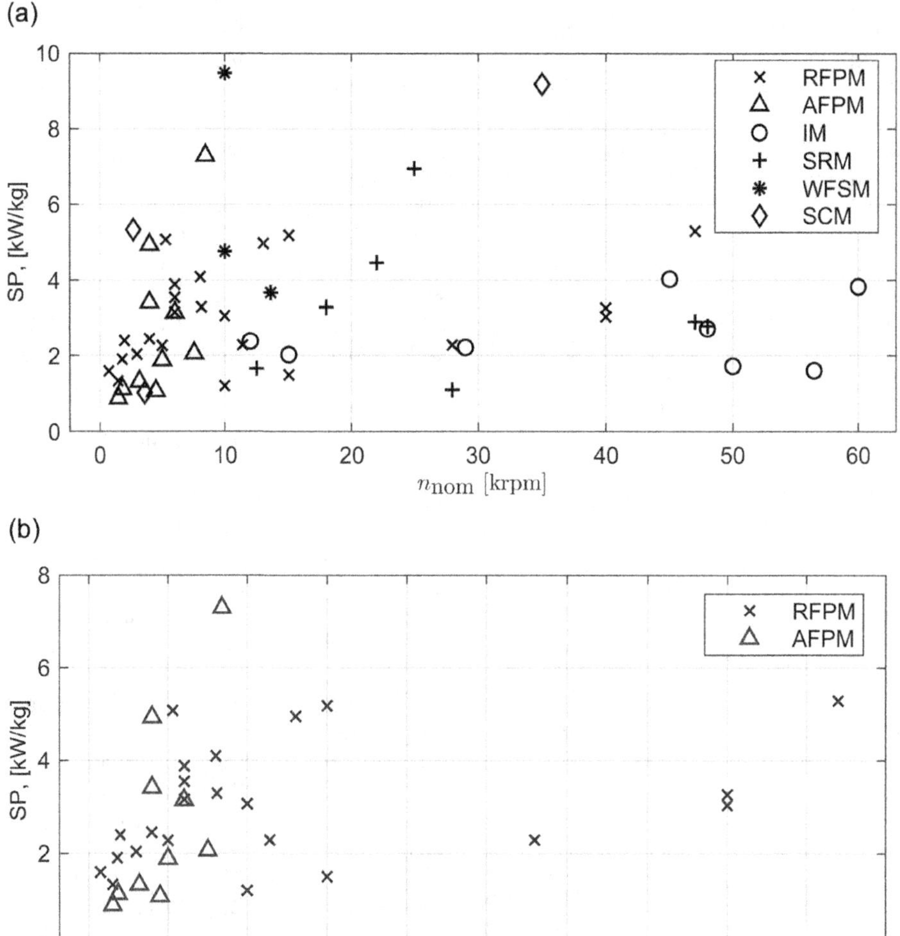

Figure 3.5 Specific power (SP) versus nominal rotor speed (n_{nom}) for (a) all surveyed machines tabulated in Table 3.5 and (b) PM machines only.

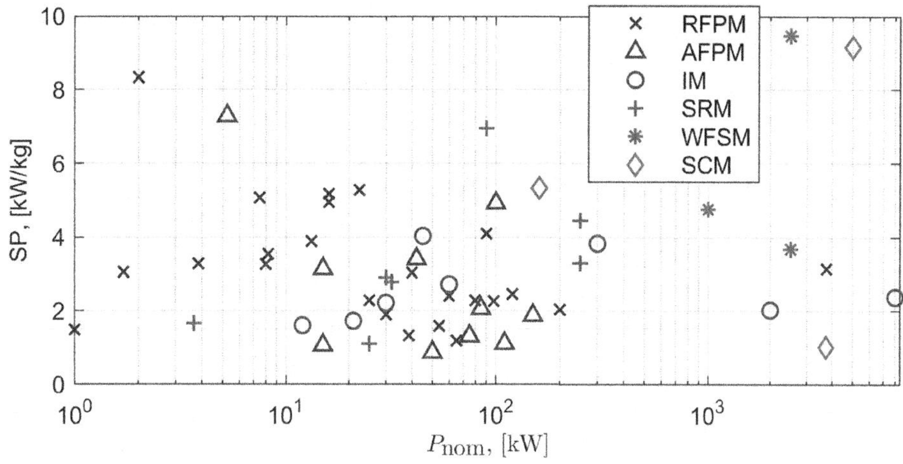

Figure 3.6 SP versus rated power for all surveyed machines tabulated in Table 3.5.

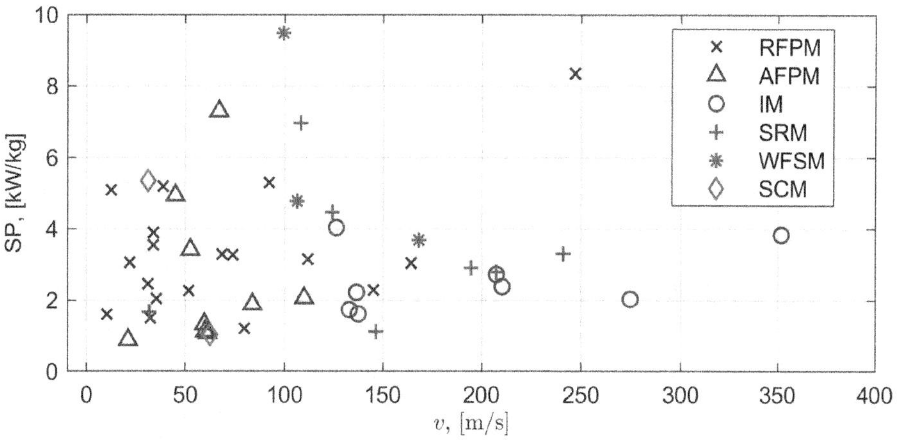

Figure 3.7 SP versus rotor tip speed for all surveyed machines tabulated in Table 3.5.

compared to general-purpose motors that have lower SP. It has been found in [87] that the benefits of high tip speed are best gained with intensive cooling systems because of the need for dissipating excessive mechanical losses. At the high end, rotor tip speed is generally limited to 250 m/s due to losses, material capability, and manufacturability limits.

Illustrated in Figure 3.8 is the SP versus frequency distribution. The frequency of HSP machines ranges from hundreds of hertz to a few kilohertz, significantly higher than the traditional 50/60 Hz motors. It is also indicated in the figure that SP generally increases as frequency increases for machines with the same topology. As expressed in (3.8), the frequency is determined by the rotational speed and the number of poles. Besides taking advantage of high rotational speed, many HSP machines also have remarkably high pole counts that are larger than 10. This leads to the high frequencies ranging from hundreds to thousands of hertz seen in the surveyed HSP machines.

The distribution of SP versus airgap shear stress is shown in Figure 3.9. According to (3.2), the shear stress is proportional to the product of magnetic and electric loading; thus, it can serve as a proxy for thermal stress. The shear stress of HSP machines ranges from a few kN/m^2 to 80 kN/m^2. High-specific-torque machines tend to have high shear stress. However, HSP machines appear to have relatively modest shear stress combined with high tip speed.

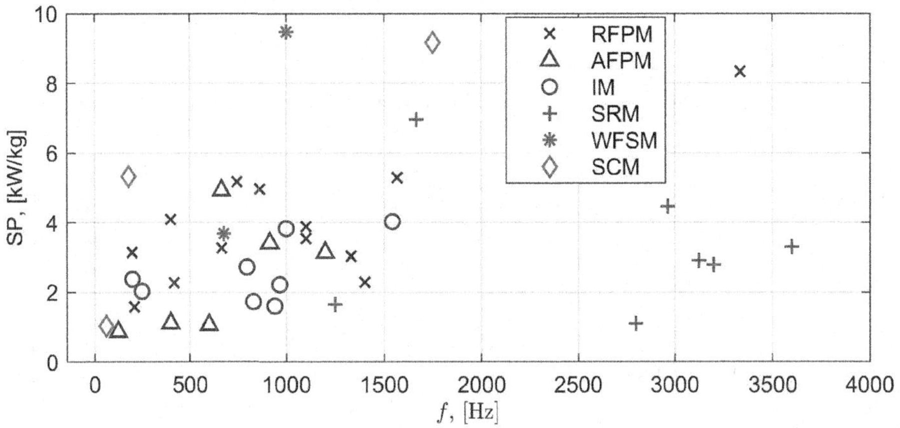

Figure 3.8 SP versus electrical frequency f for all surveyed machines tabulated in Table 3.5.

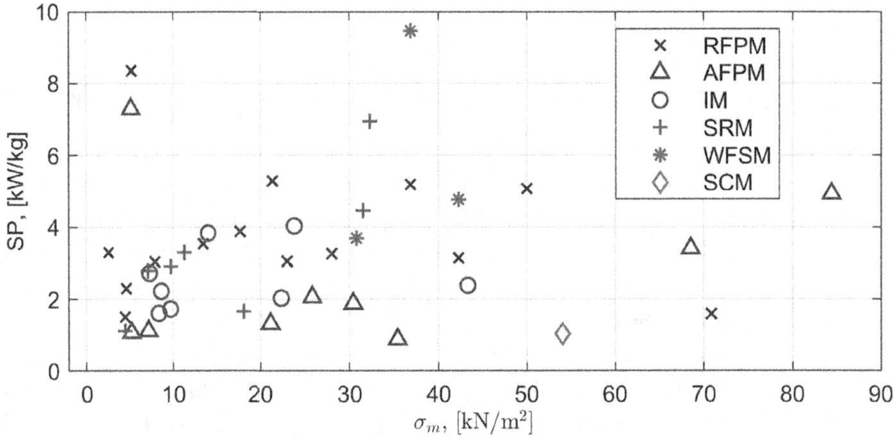

Figure 3.9 SP versus airgap shear stress σ_m for all surveyed machines tabulated in Table 3.5.

3.7 Emerging Technologies

This chapter has made clear that HSP machines are needed to enable EAP powertrains. However, today's SOTA HSP machines fall well short of the SP levels needed. In order to achieve the order-of-magnitude increase in SP that is required over the next one or two decades, NASA has proposed the 20-year roadmap shown in Figure 3.10 [88]. This aggregates projected motor technology contributions to SP increases for a theoretical 1 MW machine. The key technologies identified are *topology* (ring vs. disk designs, radial vs. axial field, inner vs. outer rotor, etc.); *advanced structures* to enable higher tip speeds; *quality manufacturing* to tighten tolerances (i.e., to reduce the airgap reliably and safely) and reduce material imperfections; *soft and hard magnetic materials* to increase the magnetic loading capability; and *advanced conductors* – such as nanoconductors – to improve fill factor, enable advanced cooling techniques, and increase electric loading. Starting from the baseline 2015 SOTA machine on the left, which has SP approximately equal to 3.3 kW/kg, each successive bar represents the projected SOTA after one additional motor technological advancement has been made. For example, moving to the 13.2 kW/kg design (the third bar) requires topology advancements that provide roughly a 5–6 kW/kg SP increase and advanced structures improvements that add roughly a 3–4 kW/kg SP increase. Additional improvements

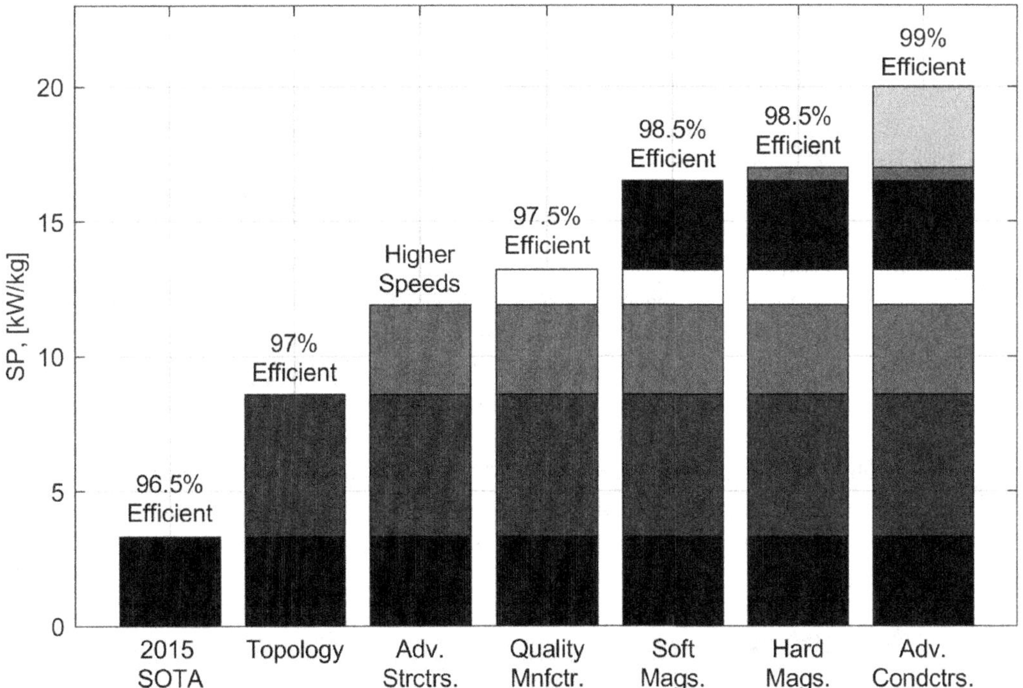

Figure 3.10 20-year roadmap of motor technology contributions to projected increases in machine SP for a 1 MW machine [88].

in the other key categories are projected to provide an additional 6–7 kW/kg SP increase, achieving SP equal to 20 kW/kg at the end of the 20-year time frame. The machine efficiency expected at each milestone is noted on the top of each bar. Although efficiency is only projected to improve from 96.5 percent to 99 percent, this translates to 25 kW of heat savings on a 1 MW machine. That large amount of heat has major implications for the size, weight, and efficiency of aircraft thermal management systems.

Advancements in materials science can play an important role, too. With the development of new materials, especially nanomaterials, many exotic cooling schemes such as thermoelectric, microchannel, nanofluid, and carbon nanotube become feasible. Because many of these are in early stages of research and development, they likely will not achieve the necessary technology maturity to make an impact in the 20-year time frame considered before. Thus, exotic cooling schemes are not explicitly included in Figure 3.10. However, it is reasonable to expect that some of these may play a role in the large improvements projected for advanced conductors, shown in the last bar on the right.

It is interesting to note that all of the projected improvements in Figure 3.10 are due to non-cryogenic technologies. Additional enormous leaps in SP are projected from advancements in cryogenically cooled SCMs. These are the topic of the next chapter and will not be discussed further here.

3.8 Summary

This chapter has described the challenges that exist for developing the MW-scale, HSP EMs necessary to enable EAP. For EAP to be practical in passenger transport aircraft vehicles, the SP of SOTA large EMs must increase significantly. Powertrain architectures for EAP have been reviewed in the context of highlighting their required high-power, HSP EMs. Motor and generator systems and how they fit into the larger EAP framework have been discussed, followed by an introduction to the important guiding design principles of MW-scale machines. These design equations have revealed that simply scaling existing SOTA machine designs by traditional methods results in a negative economy of scale with respect to SP: the higher the machine power, the worse is its SP. To overcome this scaling challenge, a comprehensive HSP machine survey has been conducted and has revealed that various machine technology options have the potential to substantially increase the SP of large machines. Finally, a brief discussion of emerging technologies, leading to the next chapter on SCMs, has been given.

Abbreviations

AFPM	axial flux permanent magnet
COTS	commercial off-the-shelf

DFIM	doubly fed induction machine
EAP	electrified aircraft propulsion
EM	electric machine
EV	electric vehicle
HEV	hybrid electric vehicle
HS index, HSI	high-speed index
HSP	high specific power
IM	induction machine
MEA	more-electric aircraft
MW	megawatt
NASA	National Aeronautics and Space Administration
PM	permanent magnet
PMSM	permanent magnet synchronous machine
RFPM	radial flux permanent magnet
SC, SCM	superconduct(or/ing), superconducting machine
SOTA	state-of-the-art
SP	specific power
SRM	switched reluctance machine
SynRM	synchronous reluctance machine
TRV	torque per unit rotor volume
WFSM	wound-field synchronous machine

Variables

B_{g1}	magnetic loading, [T]
d_s	slot depth, [m]
D_{ag}	airgap diameter, [m]
D_b	bearing mean diameter, [m]
f	electrical frequency, [Hz]
f_{nom}	nominal synchronous frequency, [Hz]
J	current density, [A/m^2]
k_{Cu}	copper slot packing factor
k_1	winding factor
K_s	electric loading, [A-turns/m]
l_a	rotor axial length, [m]
mmf	magneto-motive force, [A-turns]
n_b	bearing rotational speed, [rpm]
n_{nom}	machine nominal rotational speed, [rpm] or [krpm]
P	number of poles
$P_{\text{loss, Cu}}$	power loss in copper conductors, [W] or [kW]
P_{nom}	nominal machine output mechanical power, [W] or [kW]
P_{out}	mechanical output power, [W]

ρ_m	mass density (mass per unit volume), [kg/m^3]
ρ_P	power density (power per unit volume), [W/m^3]
R_{ag}	airgap radius, [m]
σ_m	shear stress (surface force density), [N/m^2] or [kN/m^2]
T_e	electromagnetic torque, [N-m]
v, $v_{\max}$	rotor (maximum) tip speed, [m/s]
V_R	rotor volume, [m^3]
w_s	slot width, [m]
w_t	tooth width, [m]
ω_m	rotor rotational speed, [rad/s]

References

[1] X. Zhang, C. L. Bowman, T. C. O'Connell, and K. Haran, "Large electric machines for aircraft electric propulsion," *IET Electr. Power Appl.*, vol. 12, pp. 767–779, 2018.

[2] K. Haran et al., "High power density superconducting rotating machines–development status and technology roadmap," *Supercond. Sci. Technol.*, vol. 30, pp. 1–41, 2017.

[3] National Academies of Sciences, Engineering, and Medicine, *Commercial Aircraft Propulsion and Energy Systems Research: Reducing Global Carbon Emissions*, Washington, DC: National Academy Press, 2016.

[4] Marathon Motors, Wausau, WI, USA. "Product Catalog: Commercial and Industrial SB300." (2014). Accessed: December 20, 2018. [Online]. Available: www.regalbeloit.com/-/media/Files/Literature/Marathon-Motors-Literature/SB300MarathonCatalog2014.pdf.

[5] Remy Electric Motors, Pendleton, IN, USA. "REMY HVH410–075-DOM Electric Motor Brochure." (2011). Accessed: December 20, 2018. [Online]. Available: https://cdn.borgwarner.com/docs/default-source/default-document-library/remy-pds—hvh410–075-sheet-euro-pr-3-16.pdf?sfvrsn=a742cd3c_13.

[6] Siemens AG, Munich, Germany. "Siemens Develops World-Record Electric Motor for Aircraft." (2015). Accessed: December 20, 2018. [Online]. Available: www.siemens.com/press/PR2015030156COEN.

[7] A. Yoon et al., "A high-speed, high-frequency, air-core PM machine for aircraft application," presented at the IEEE Power and Energy Conf. at Illinois, Urbana, IL, 2016.

[8] R. Jansen et al., "Overview of NASA electrified aircraft propulsion research for large subsonic transports," presented at the 53rd AIAA/SAE/ASEE Joint Propulsion Conf., Atlanta, GA, 2017, Paper AIAA 2017-4701.

[9] S. Huang et al., "A general approach to sizing and power density equations for comparison of electric machines," *IEEE Trans. Ind. Appl.*, vol. 34, pp. 92–97, 1998.

[10] S. Huang et al., "A comparison of power density for axial flux machines based on general purpose sizing equations," *IEEE Trans. Energy Convers.*, vol. 14 pp. 185–192, 1999.

[11] R. Qu and T. A. Lipo, "Sizing equations and power density evaluation of dual-rotor, radial flux, toroidally-wound, permanent-magnet machines," presented at the IEEE 16th Int. Conf. on Electr. Mach., Cracow, Poland, 2004.

[12] J. L. Felder, "NASA electric propulsion system studies," NASA, Cleveland, OH, Tech. Rep. GRC-EDAA-TN28410, 2015.
[13] C. E. Lents et al., "Parallel hybrid gas-electric geared turbofan engine conceptual design and benefits analysis," presented at the 52nd AIAA/SAE/ASEE Joint Propulsion Conf., Salt Lake City, UT, 2016, Paper AIAA 2016-4610.
[14] D. Trawick et al., "Development and application of GT-HEAT for the electrically variable engine(TM) design," presented at the 55th AIAA Aerosp. Sci. Mtg., Grapevine, TX, 2017, Paper AIAA 2017-1922.
[15] E. Ganev, "Selecting the best electric machines for electrical power-generation systems: High-performance solutions for aerospace more electric architectures," *IEEE Electrif. Mag.*, pp. 13–22, December 2014.
[16] T. A. Lipo, *Introduction to AC Machine Design*. Hoboken, NJ: IEEE Press/Wiley, 2017.
[17] B. Adkins and R. G. Harley, *The General Theory of Alternating Current Machines: Application to Practical Problems*, New York, NY: Halsted, 1975.
[18] P. L. Alger, *Induction Machines: Their Behavior and Uses*, 2nd ed., New York, NY: Gordon and Breach, 1995.
[19] M. Popescu et al., "Modern heat extraction systems for electrical machines: A review," in *Proc. 2015 IEEE Workshop on Elect. Mach. Design, Control and Diagnosis*, Torino, Italy, 2015, pp. 289–296.
[20] W. Tong, "Motor cooling," in *Mechanical Design of Electric Motors*, 1st ed., Boca Raton, FL: CRC Press, 2014, ch. 8.
[21] J. F. Gieras, "High power density machines," in *Advancements in Electric Machines*, New York, NY: Springer-Verlag, 2008, ch. 3, pp. 71–80.
[22] A. M. El-Refaie, "Fractional-slot concentrated-windings synchronous permanent magnet machines: Opportunities and challenges," *IEEE Trans. Ind. Electron.*, vol. 57, pp. 107–121, 2010.
[23] M. A. Rahman, A. Chiba, and T. Fukao, "Super high speed electrical machines: Summary," in *Proc. IEEE Power Eng. Soc. General Mtg.*, Denver, CO, 2004, pp. 1268–1271.
[24] D. Gerada et al., "High-speed electric machines: technologies, trends, and developments," *IEEE Trans. Ind. Electron.*, vol. 61, pp. 2946–2959, 2014.
[25] A. Tenconi, S. Vaschetto, and A. Vigliani, "Electric machines for high-speed applications: Design considerations and tradeoffs," *IEEE Trans. Ind. Electron.*, vol. 61, pp. 3022–3029, 2014.
[26] A. Binder and T. Schneider, "High-speed inverter-fed AC drives," in *Proc. IEEE Int. Aegean Conf. on Elect. Mach. and Power Elect.*, Bodrum, Turkey, 2007, pp. 9–16.
[27] A. Borisavljevic, H. Polinder, and J. A. Ferreira, "On the speed limits of permanent-magnet machines," *IEEE Trans. Ind. Electron.*, vol. 57, pp. 220–227, 2010.
[28] R. R. Moghaddam, "High speed operation of electrical machines, a review on technology, benefits and challenges," in *Proc. IEEE Energy Convers. Congr. and Expo.*, Pittsburgh, PA, 2014, pp. 5539–5546.
[29] R. D. van Millingen and J. D. van Millingen, "Phase shift torquemeters for gas turbine development and monitoring," in *Proc. ASME Int. Gas Turbine and Aeroengine Congr. and Expo.*, Orlando, FL, 1991, pp. 1–10.
[30] X. Yi, A. Yoon, and K. S. Haran, "Multi-physics optimization for high-frequency air-core permanent-magnet motor of aircraft application," in *Proc. IEEE Int. Elect. Mach. and Drives Conf.*, Miami, FL, 2017, pp. 1–8.

[31] T. M. Jahns, "The expanding role of PM machines in direct-drive applications," in *2011 Proc. IEEE Int. Conf. on Elect. Mach. and Syst.*, Beijing, China, 2011, pp. 1–6.

[32] ThinGap, Camarillo, CA, "Products: TG-Series Motors."Accessed: December 20, 2018. [Online]. thingap.com. www.thingap.com/standard-products/.

[33] Joby Motors, Santa Cruz, CA, "Joby JM1 and JM1S electric aircraft propulsion motors. Accessed: June 2017. [Online]. " jobymotors.com. www.jobymotors.com/public/views/pages/products.php.

[34] M. Mekhiche et al., "High speed motor drive development for industrial applications," in *Proc. IEEE Int. Conf. Electr. Mach. and Drives*, Seattle, WA, 1999, pp. 244–248.

[35] G. Munteanu et al., "No-load tests of a 40 kW high-speed bearingless permanent magnet synchronous motor," in *Proc. IEEE Int. Symp. Power Electr., Electr. Drives, Automation and Motion*, Pisa, Italy, 2010, pp. 1460–1465.

[36] L. Zheng et al., "Design of a superhigh-speed cryogenic permanent magnet synchronous motor," *IEEE Trans. Magn.*, vol. 41, pp. 3823–3825, 2005.

[37] D. Acharya et al., "Systems design, fabrication, and testing of a high-speed miniature motor for cryogenic cooler," *Int. J. Rotating Mach.*, vol. 2009, pp. 1–9, 2009.

[38] B. C. Mecrow et al., "Design and testing of a four-phase fault-tolerant permanent-magnet machine for an engine fuel pump," *IEEE Trans. Energy Convers.*, vol. 19, pp. 671–678, 2004.

[39] G. J. Atkinson et al., "The analysis of losses in high-power fault-tolerant machines for aerospace applications," *IEEE Trans. Ind. Appl.*, vol. 42, pp. 1162–1170, 2006.

[40] J. F. Gieras, R. -J. Wang, and M. J. Kamper, *Axial Flux Permanent Magnet Brushless Machines*, New York, NY: Springer-Verlag, 2008.

[41] EMRAX Innovative E-Motors, Kamnik, Slovenia, "EMRAX 208 and 228 electric motor technical specifications." emrax.com. emrax.com/products.html.

[42] YASA, Kidlington, UK, "YASA standard products technical specifications." Accessed: December 20, 2018. [Online]. yasa.com. www.yasamotors.com/products.

[43] AVID Technology, Northumberland, U.K. "EVO axial flux electric motors." Accessed: December 20, 2018. [Online]. avidtp.com. https://avidtp.com/product/evo-motors.

[44] G. Long, "High efficiency, high power density electric motors."Accessed: December 20, 2018. [Online]. http://cafefoundation.org/v2/pdf_tech/MPG.engines/HE_HP_electric_motors_Long_20090929.pdf.

[45] R. Sanchez et al., "Mechanical validation of a high power density external cantilevered rotor," *IEEE Trans. Ind. Appl.*, vol. 54, pp. 3208–3216, 2018.

[46] N. J. Renner et al., "Development of form-wound air-core armature windings for high-frequency electric machines," in *Proc. IEEE Int. Elect. Mach. and Drives Conf.*, Miami, FL, 2017, pp. 1–8.

[47] W. L. Soong et al., "Novel high-speed induction motor for a commercial centrifugal compressor," *IEEE Trans. Ind. Appl.*, vol. 36, pp. 706–713, 2000.

[48] J. F. Gieras and J. Saari, "Performance calculation for a high-speed solid-rotor induction motor," *IEEE Trans. Ind. Electron.*, vol. 59, pp. 2689–2700, 2012.

[49] J. Pyrhönen et al., "High-speed high-output solid-rotor induction-motor technology for gas compression," *IEEE Trans. Ind. Electron.*, vol. 57 pp. 272–280, 2010.

[50] J. Pyrhönen et al., "High-speed, 8 MW, solid-rotor induction motor for gas compression," in Proc. 18*th* IEEE Int. Conf. on Elect. Mach., Vilamoura, Portugal, 2008, pp. 1–6.

[51] T. C. O'Connell, J. R. Wells, and O. A. Watts, "A lumped parameter off-axis-capable dynamic switched reluctance machine model including unbalanced radial forces," *IEEE Trans. Energy Convers.*, vol. 30, pp. 161–174, 2015.

[52] J. F. Gieras, "Multimegawatt synchronous generators for airborne applications: A review," in *2013 Proc. IEEE Int. Elect. Mach. and Drives Conf.*, Chicago, IL, 2013, pp. 626–633.

[53] T. A. Burress et al., "Evaluation of the 2010 Toyota Prius hybrid synergy drive system," Oak Ridge National Lab., Oak Ridge, TN, Tech. Rep. ORNL/TM-2010/253, 2011.

[54] R. Bowman, private communication, March 2014.

[55] X. Zhang and K. S. Haran, "High-specific-power electric machines for electrified transportation applications-technology options," in *2016 Proc. IEEE Energy Convers. Congr. and Expo.*, Milwaukee, WI, 2016, pp. 1–8.

[56] GreatPlanes, Champaign, IL, "Rimfire Outrunner Brushless Motors (GPM4805)." Accessed: December 20, 2018. [Online]. greatplanes.com. www.greatplanes.com/motors/gpmg4505.php.

[57] Lange Aviation, Zweibrucken, Germany, "Antares 23E Technical Data." Accessed: December 20, 2018. [Online]. lange-aviation.com. www.lange-aviation.com/en/produkte/antares-23et/techn-daten/.

[58] Sineton Research and Development, Maribor, Slovenia, "Electric Motor A30K016. Accessed: December 20, 2018. [Online]." sineton.com. www.sineton.com/web/index-9.html.

[59] Remy Electric Motors, Pendleton, IN, USA. "REMY HVH250–115-DOM Electric Motor Brochure." (2011). Accessed: December 20, 2018. [Online]. Available: https://cdn.borgwarner.com/docs/default-source/default-document-library/remy-pds—hvh250–115-sheet-euro-pr-3-16.pdf?sfvrsn=ad42cd3c_11.

[60] J. Grotendorst, "Hannover Messe 2013 Preliminary Press Conference: Expert Talk. Accessed: December 20, 2018. [Online]." siemens.com. www.siemens.com/press/pool/de/events/2013/industry/2013-03-hannovermesse-pk/expert-talk-inside-e-car-e.pdf.

[61] Y. Wang et al., "Design and experimental verification of high power density interior permanent magnet motors for underwater propulsions." in *Proc. IEEE Int. Conf. Elect. Mach. and Syst.*, Beijing, China, 2011, pp. 1–6.

[62] T. Kosaka et al., "Experimental drive performance evaluation of high power density wound field flux switching motor for automotive applications," in *Proc. IET 7th Int. Conf. Power Electron., Mach., and Drives*, Manchester, UK, 2014, pp. 1–6.

[63] G. M. Raimondi et al., "Aircraft embedded generation systems," in *Proc. IET Int. Conf. Power Electron., Mach., and Drives*, Santa Fe, NM, 2002, pp. 217–222.

[64] S.-I. Kim et al., "A novel rotor configuration and experimental verification of interior PM synchronous motor for high-speed applications," *IEEE Trans. Magn.*, vol. 48, pp. 843–846, 2012.

[65] S. Kuznetsov, "Machine design and configuration of a 7000 HP hybrid electric drive for naval ship propulsion," in *Proc. IEEE Int. Elect. Mach. and Drives Conf.*, Niagara Falls, ON, Canada, 2011, pp. 1625–1628.

[66] Maxx Products International, LLC, Lake Zurich, IL. "HIMAX Brushless Motors, 500 W and Up." (2018). Accessed December 20, 2018. [Online]. Available: www.maxxprod.com/mpi/mpi-2601.html

[67] J. A. Haylock et al., "Operation of a fault tolerant PM drive for an aerospace fuel pump application," *IEE Proc. Elect. Power Appl.*, vol. 145, pp. 441–448, 1998.

[68] LaunchPoint Technologies, Goleta, CA, "Electric Machines for Propulsion." Accessed: December 20, 2018. [Online]. launchpnt.com. www.launchpnt.com/portfolio/aerospace/hybrid-electric-uav-motors.

[69] F.Crescimbini et al., "Compact permanent-magnet generator for hybrid vehicle applications," *IEEE Trans. Ind. Appl.*, vol. 41, pp. 1168–1177, 2005.

[70] F. Caricchi et al., "Axial-flux permanent-magnet generator for induction heating gensets," *IEEE Trans. Ind. Electron.*, vol. 57, pp. 128–137, 2010.

[71] T. Miura et al., "A ferrite permanent magnet axial gap motor with segmented rotor structure for the next generation hybrid vehicle," in *Proc. IEEE 19th Int. Conf. Elect. Mach.*, Rome, Italy, 2010, pp. 1–6.

[72] S. Chino et al., "Fundamental characteristics of a ferrite permanent magnet axial gap motor with segmented rotor structure for the hybrid electric vehicle," in *Proc. 3rd IEEE Energy Convers. Congr. and Expo*, Phoenix, AZ, 2011, pp. 2805–2811.

[73] R. F. Thelen et al., "A 2-MW motor and ARCP drive for high-speed flywheel," in *Proc. 22nd IEEE Appl. Power Electron. Conf. and Expo.*, Anaheim, CA, 2007, pp. 1690–1694.

[74] Y.-K. Kim et al., "High-speed induction motor development for small centrifugal compressor," in *Proc. Fifth Int. Conf. on Elect. Mach. and Syst.*, Shenyang, China, 2001, pp. 891–894.

[75] Y. Gessese and A. Binder, "Axially slitted, high-speed solid-rotor induction motor technology with copper end-rings," in *2009 Proc. IEEE Int. Conf. Elect. Mach. and Syst.*, Tokyo, Japan, 2009, pp. 1–6.

[76] Y. Gessese, A. Binder, and B. Funieru, "Analysis of the effect of radial rotor surface grooves on rotor losses of high speed solid rotor induction motor," in *2010 Proc. IEEE Int. Symp. Power Electr., Electr. Drives, Automation and Motion*, Pisa, Italy, 2010, pp. 1762–1767.

[77] H. Hofmann and S. R. Sanders, "High-speed synchronous reluctance machine with minimized rotor losses," *IEEE Trans. Ind. Appl.*, vol. 36, pp. 531–539, 2000.

[78] S. R. MacMinn and W. D. Jones, "A very high speed switched-reluctance starter-generator for aircraft engine applications," in *Proc. IEEE Nat. Aerosp. and Electron. Conf.*, Dayton, OH, 1989, pp. 1758–1764.

[79] A. V. Radun, "High power density switched reluctance motor drive for aerospace applications," in *Conf. Rec. IEEE Ind. Applicat. Soc. Annu. Mtg.*, San Diego, CA, 1989, pp. 568–573.

[80] C. A. Ferreira et al., "Design and implementation of a five-hp, switched reluctance, fuel-lube, pump motor drive for a gas turbine engine," *IEEE Trans. Power Electron.*, vol. 10, pp. 55–61, 1995.

[81] C. A. Ferreira et al., "Detailed design of a 30-kW switched reluctance starter/generator system for a gas turbine engine application," *IEEE Trans. Ind. Appl.*, vol. 31, pp. 553–561, 1995.

[82] C. A. Ferreira, S. R. Jones, and W. S. Heglund, "Performance evaluation of a switched reluctance starter/generator system under constant power and capacitive type loads," in *1995 Proc. IEEE Appl. Power Electron. Conf. and Expo.*, Dallas, TX, 1995, pp. 416–424.

[83] E. Richter and C. Ferreira, "Performance evaluation of a 250 kW switched reluctance starter generator" in *Conf. Rec. 1995 IEEE Ind. Applic. Soc. Annu. Mtg.*, Orlando, FL, 1995, pp. 434–440.

[84] A. V. Radun, C. A. Ferreira, and E. Richter, "Two-channel switched reluctance starter/generator results," *IEEE Trans. Ind. Appl.*, vol. 34, pp. 1026–1034, 1998.
[85] N. R. Garrigan et al., "Radial force characteristics of a switched reluctance machine," in *Conf. Rec. 1999 IEEE Ind. Applic. Soc. Annu. Mtg.*, Phoenix, AZ, 1999, pp. 2250–2258.
[86] E. F. Hammond, W. S. Neff, and W. J. Shilling, "A 2.5-MVA high-voltage lightweight generator.," *J. Aircr.*, vol. 16, pp. 55–61, 1979.
[87] M. van der Geest et al., "Power density limits and design trends of high-speed permanent magnet synchronous machines," *IEEE Trans. Transport. Electrif.*, vo. 1, pp. 266–276, 2015.
[88] N. Madavan et al., "A NASA perspective on electric propulsion technologies for commercial aviation," presented at the Grainger CEME/IEEE Workshop on Technology Roadmap for Large Elect. Mach., Urbana, IL, 2016.

4 Superconducting Machines and Cables

Thanatheepan Balachandran, Timothy Haugan, and Kiruba Haran

Introduction

As is clear from Chapter 3, significant advances are being made in conventional (i.e., non-superconducting [SC]) machines to achieve the high specific power and efficiency targets necessary to meet the requirements of electrified aircraft propulsion systems. However, Ohmic losses in conventional conductors, such as copper or aluminum wires, impose fundamental limits on the attainable performance. To obtain the full benefits of electrified aircraft propulsion (EAP), the large aircraft used in commercial aviation demand electric machines with power densities and efficiencies far beyond what today's state-of-the art technology can provide [1–2].

A generalization of Equation (3.14) describes the power produced by an electric machine P_{out} as

$$P_{out} = K \times \text{Volume} \times \text{Magnetic loading} \times \text{Electrical loading} \times \text{Speed}, \tag{4.1}$$

where K is a constant coefficient that depends on the units. For a given machine volume, power density can be increased by raising either the electromagnetic loading or the machine speed. In conventional machines, these types of increases introduce both thermal and mechanical challenges, respectively, due to the conductor Ohmic and gearbox friction losses and motor tip speed limits. Because SC wires eliminate Ohmic losses at cryogenic temperatures, employing them in the field and armature windings leads to significant increases in achievable electric and magnetic loading, resulting in high-power density machines without the need for a gearbox.

More broadly, SC machines are an attractive longer-term option for aircraft applications because they offer a number of advantages, among which are the following:

- High specific power

  In electrical machines with conventional coils or permanent magnet (PM) excitation sources, the magneto-motive force (MMF) that drives magnetic flux is limited by thermal and efficiency considerations. To stay within these constraints, the reluctance of the magnetic path is minimized with the use of ferromagnetic steel. Ohmic losses in the resistive coils, saturation of core material, and core losses pose challenges in increasing the flux density to higher value in conventional machines. SCs open up a new design space for electrical machines with significantly higher

MMF capability than conventional coils by being able to increase the current density without running into thermal or efficiency constraints. This capability can be used to eliminate ferromagnetic material in the machine and deal with a significantly higher magnetic reluctance (e.g., 10 times larger). With the iron and its saturation limit removed, designs with much higher airgap flux density (5 to 10 times current levels) can be generated. This can translate directly into a corresponding increase in torque and power density. In addition to higher operating magnetic flux density in the machine, the torque-producing component of the current could also be increased substantially without hitting efficiency and thermal limits with the appropriate use of SCs.

- High efficiency

  With the virtual elimination of Ohmic losses, which account for about 30–40 percent of the total losses in an electric machine, SC machines have improved efficiency both at nominal operating points and in off-design points. Air-core machine architectures also eliminate core losses in the ferromagnetic materials. Concept studies indicate efficiencies greater than 99 percent can be obtained.
- Reduced thermal management challenges

  This feature may be somewhat counterintuitive, because cryo-cooling is required to cool the SC components to lower than their critical temperatures. However, for large, megawatt-scale powertrains, system-level thermal management overhead can actually be reduced. The total heat load that must be managed in the electrical system can be cut by about a factor of five from the projected 10–20 percent of the total load for a non-cryogenic solution.
- Partial discharge and arcing

  A fully SC power system can enable the aircraft-wide distribution of tens of megawatts of power at significantly lower voltages than are possible with conventional conductors. Additionally, the SC components are contained in an isolated environment, typically within a vacuum, further reducing the risks of partial discharge and arcing at the ambient conditions present at altitude.

An introduction to the physics of SCs is given in Section 4.1. The important considerations relevant to SC machine design with a focus on electrified propulsion applications, and the various SC machine topologies being pursued by different research groups, including the current state of the art, are presented, respectively, in Sections 4.3–4.5. The physics and advantages of SC cables are detailed in Section 4.6. The chapter concludes with a look at SC technology trends in Section 4.7 and a summary in Section 4.8.

4.1 Superconductors

In this section, the basic physics of SCs and the importance of the three critical factors to achieving and maintaining the SC state – field, current, and temperature – are presented.

4.1.1 Basic Physics

SCs exhibit zero resistance to electrical current when cooled below a *critical temperature*, T_c, and can carry much larger currents than conventional conductors like copper or aluminum. Ever since superconductivity was observed in mercury by Kamerlingh Onnes in 1911, there has been a quest to apply it in electrical equipment, but engineering challenges associated with the need to cool the conductors down to very low temperatures has limited its adoption thus far. The earliest discovered SCs had T_c values below 10 K. In recent decades materials considered to be SC at temperatures as high as 130 K have been discovered [3]. Modern SCs can be grouped into low-temperature SCs (LTSs) and high-temperature SCs (HTSs) based on their value of T_c. LTSs have relatively low T_c, typically below 15 K, while HTSs have higher T_c values around 77 K or above. The critical temperatures of newly discovered SCs have been going up gradually since the discovery early in the last century. A significant jump in the critical temperatures occurred in the late 1980s when copper-oxide-based HTSs were discovered. Practical wires based on these new conductors are now readily available.

In addition to electrical resistance vanishing below a critical temperature, SCs exhibit another interesting property known as the *Meissner effect*. This is the ability of the material to spontaneously expel a magnetic field as it transitions to the SC state. External magnetic fields below a certain value will be perfectly opposed by induced currents, preventing the field from entering the material. This ability to generate screening currents and expel magnetic field even when there is no change in the externally applied field is a characteristic feature of SCs.

SCs are now routinely manufactured in long lengths in the form of round wires or tapes, facilitating the fabrication of coil windings for a number of practical applications. Mature SC conductors that are commercially available are listed in Table 4.1. In the table, H_{c2} is the *upper critical field*, which will be discussed in Section 4.1.2. Note that it is common to multiply the critical field values (upper and lower) by the *vacuum permeability* $\mu_0 = 4\pi \times 10^{-7}$ H/m so that the critical values are in units of Teslas (T), as shown in the rightmost column of Table 4.1.

Another form of SCs that are prepared by a melt process in a large single-grain form, are known as *bulk SCs*. These are usually fabricated as short cylinders with

Table 4.1 Common commercially available SCs.

SC	T_c Type	T_c [K]	$\mu_0 H_{c2}$ at 0 K [T]
NbTi	Low	9.2	15
Nb_3Sn	Low	18.3	30
Nb_3Al	Low	18.9	33
MgB2	Intermediate	39	15–20
Bi-2212	High	95	≈150
Bi-2223	High	108	≈180
YBCO	High	93	≈120

volumes of a few cubic cm that do not suffer from grain boundary effects that can limit the achievable current. Bulk SCs have the potential to "trap" high magnetic fields due to their outstanding ability to pin magnetic flux and support large current. For example, a stack of two silver-doped GdBCO bulk SC material samples with 25 mm diameter trapped a record field of 17.6 T at 26 K [4]. The magnetized bulk SCs are referred to as "trapped field magnets," "cryo-magnets," or "super magnets" and can be used as high-flux-density PMs in electrical machines. This is an emerging area of research that is making rapid progress.

4.1.2 The Critical Surface

Once a conductor is in the SC state, its electrical properties are affected by the following factors, each of which can disrupt and/or end the SC state:

- Applied magnetic field
  An applied magnetic field above a critical value will cause the material to exit the SC state. This *critical magnetic field* is a function of the operating temperature and correlates strongly with the SC critical temperature. SC materials are classified into two categories based on how they make the transition out of the SC state as the external magnetic field is gradually increased. In Type I SCs, the transition is perfect when the *critical field*, H_c, is reached. This means that the external field that was being expelled from the material can now penetrate. In contrast, in Type II SCs, when the external field exceeds the *lower critical field*, H_{c1}, a mixed SC and normal state is observed, allowing the field to gradually penetrate into the material. The material completely exits the SC state only after the *upper critical field*, H_{c2}, is reached.
- Current
  The current flowing through an SC material impacts its superconductivity. When the current reaches the *critical current*, I_c, while the material is in the SC state, the latter will exit the SC state and behave as a normal conductor. In SC applications, operating far below the critical current is advisable, especially in coils carrying a large amount of current, in order to avoid the damaging phenomenon known as "quench." This is detailed in Section 4.3.3.
- Temperature
  An SC must be maintained below T_c to remain in the SC state. The main design consideration in SC machines is to ensure that the coils remain at the required cryogenic temperatures.

These three factors – applied magnetic field, current, and temperature – are captured with the B–J–T characteristic known as the *critical surface*. Critical surfaces for NbTi, Nb_3Sn, MgB_2 and YBaCuO are shown in Figure 4.1. If any of the three factors exceed the limits of this surface, the material will exit the SC state and become normally conducting. These surfaces illustrate how I_c and H_c both decrease substantially with increasing temperature. For this reason, many SCs are operated at temperatures far below T_c to maximize their operating range within the critical surface.

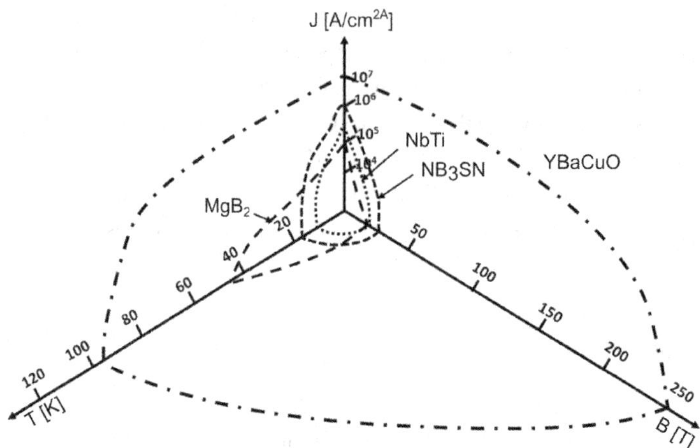

Figure 4.1 Critical surfaces of NbTi, Nb_3Sn YBaCuO, and MgB_2.

4.2 Sources of ac Loss in Superconducting Wires

Conductors used in SC machines commonly take the form of tapes or multifilament wires. In this section equations are provided to evaluate alternating current (ac)-based losses for these types of wires. Alternating current losses in SC wires consist of different loss components, including hysteresis, eddy current, and coupling. Two models are typically used to estimate these losses: the *critical state model* (CSM) and the *power law*. The estimation of ac loss components using these two models is the topic of the remainder of this section and its subsections.

Alternating current losses are a form of heat generated in SC wires when they are exposed to ac current or an alternating field. They are cyclic and increase with frequency. When SCs are utilized as ac machine windings (e.g., the armature windings in synchronous machines), they experience both the alternating field and current and generate losses in the form of heat. This heat, which serves as an extra load on the cryogenic system, must be removed so that the temperature remains below T_c to avoid quenching and potential machine failure. Therefore, ac losses adversely impact the overall system efficiency and weight and need to be predicted during the design phase for proper sizing of the cryogenic system.

Alternating current losses mainly depend on the properties (frequency, amplitude, and harmonics) of the alternating source and the geometry of the conductors. When considering geometry, the shape (cylinder or slab), radius or thickness, and number of filaments are some of the factors that influence the ac losses. Therefore, there are no general equations to evaluate ac losses across all types of conductors. The ac loss model must be carefully chosen before using it to estimate the conductor losses.

Once the conductor losses at a given field and operating current are evaluated, the total ac losses can be estimated by integrating over the machine stack length.

When estimating these losses, special attention must be given to the field variation across the armature slot. The conductors closer to the airgap experience higher field strength and generate higher ac losses than those that are farther away. Therefore, to increase the ac loss estimate accuracy, the field density across the armature slot must be evaluated ,and the corresponding ac losses evaluated according to the armature conductor location.

4.2.1 Hysteresis

Because of the Meissner effect, SCs produce screening currents that completely expel an applied weak magnetic flux. The CSM can be used to model these screening currents, which occupy a finite region in the material's cross section, as shown in Figure 4.2.

As the applied *magnetic field* H, in A/m, increases, additional screening currents are generated to prevent it from penetrating the SC. They create *magnetization* M, in A/m, in the SC, which expels the magnetic field. When the entire cross section is occupied by the screening current, further increases in H are unable to create compensating screening currents. Thus, no further magnetization within the SC occurs, and increments in H can penetrate the SC.

When the applied field is alternating, the screening currents will also alternate to compensate. As the field direction changes, areas that were previously magnetized in one direction will now require energy to reorient their magnetization based on new screening currents. This energy is dissipated as heat loss and referred to as *hysteresis loss*. The *penetration field limit* H_p is dependent on the properties of an SC. For a circular cross section of SC, H_p can be evaluated as

$$H_p = \frac{4J_c r_{\text{wire}}}{\pi}, \tag{4.2}$$

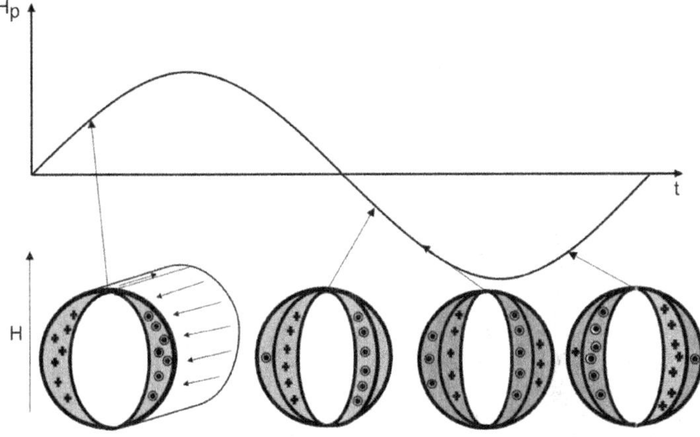

Figure 4.2 Screening current area with oscillating magnetic field, left to right.

where J_c, in A/m^2, and r_{wire}, in m, are the *critical current density* and *SC wire radius*, respectively. The hysteresis power loss in an SC section, P_{loss}, in W, can be expressed as

$$P_{\text{loss}} = \oint HdM = \oint MdH. \tag{4.3}$$

The field and magnetization magnitudes vary based on the geometry of the SC cross section.

4.2.2 Eddy Currents and Coupling

Practical SC wires are usually manufactured in the form of multifilamentary composites. SC filaments are embedded in a normal metal component such as copper or a copper alloy, as shown in Figure 4.3. This outer sheath of copper protects the wire from burnout from any disturbance that may lead parts of the wire to become unstable. In such cases, the metal shares the current and allows the SC wire to recover. The downside to this is that the metal sheath generates eddy currents – and their associated losses – in the presence of an ac field. These losses are analogous to the iron core eddy current losses found in motors and transformers.

Coupling losses also occur in the metal matrix that couples the filaments together. These are a type of eddy current loss and greatly increase the ac losses as the frequency increases. In the copper matrix, normal metal eddy currents pass through or around the filaments depending on the resistance between the filaments and the metal matrix. Coupling losses are calculated for a multifilament wire in Section 4.2.4.

4.2.3 Alternating Currents

The CSM assumes that the total *operating current* I_o, in A, in an SC can be expressed as a combination of regions with J_c and with zero current density across the SC cross section. For example, if the wire is carrying 60 percent of I_c, then 60 percent of the wire cross section would be occupied by J_c and the rest with zero current density.

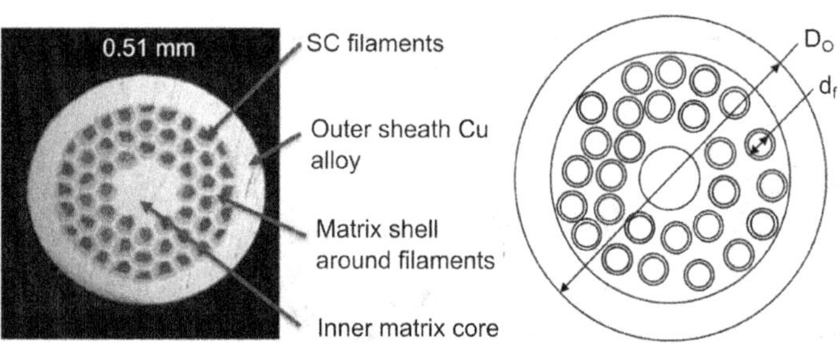

Figure 4.3 Cross-sectional area of a MgB2 wire. Courtesy HyperTech Research

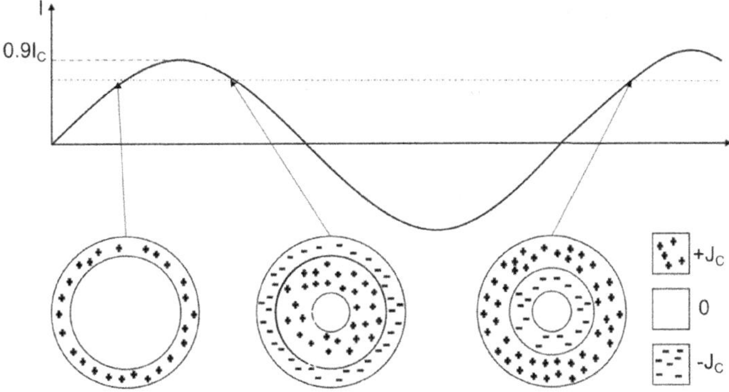

Figure 4.4 Current density distribution in the cross section of a circular SC wire, measured at different instants for the same instantaneous current value.

Due to the Meissner effect, the current-carrying regions initially crowd around the SC outer edges (the "skin") at low currents. But as the current increases, the size of the current-carrying regions increases to compensate. This construct is straightforward for direct current (dc); however, for ac currents, a negative term is introduced to account for the oscillating sign of the current. The total SC current density in this case is the aggregate of regions with $+J_c$, $-J_c$, and zero current density, with their relative proportions varying with the phase of the cycle. This is illustrated in Figure 4.4. As shown, more of the wire cross section is occupied by $+J_c$ while the current increases, but as it decreases, the region is replaced by $-J_c$.

Much like oscillating external fields, ac current flowing through the SC produces small magnetic moments. As currents change, they push against these magnetic moments and expend energy in realigning them, resulting in hysteresis loss. Thus, ac current–based ac SC losses are, in fact, hysteresis losses. However, because they are caused by the current in the SC itself, rather than by external ac fields, they are treated as separate losses.

4.2.4 Multifilament Wires

Simplified calculations for evaluating ac losses across multifilament wires based on established methods in the literature [5, 6] are provided here. For a cylindrical SC filament placed perpendicular to an applied rotating magnetic field with constant (peak) magnitude of B_m, in T, and frequency f, in Hz, assuming the applied field exceeds H_p, the *hysteresis loss density* p_h, in W/m^3, can be evaluated as

$$p_h = \frac{4}{3} B_m J_c d_f f, \tag{4.4}$$

where d_f is the *filament diameter* in m. The penetration field limit for this case is

$$H_p = 0.4 J_c d_f. \tag{4.5}$$

Eddy current loss density p_e, in W/m^3, can be evaluated from

$$p_e = \left(\frac{\pi^2}{k\rho_{\text{eff}}}\right)(B_m D_o f)^2, \tag{4.6}$$

where $k = 4$ is a constant for multifilament wires, ρ_{eff}, in Ω-m, is the copper *effective transverse resistivity matrix*, and D_o, in m, is the SC *outer diameter*.

Coupling loss density p_c, in W/m^3, can be evaluated from

$$p_c = \left(\frac{1}{n\rho_{\text{eff}}}\right)(fL_p B_m)^2, \tag{4.7}$$

where $n = 2$ is a constant and L_p is the *twist pitch*, in m, of the SC wire.

Transport current loss density, in W/m^3, can be evaluated from

$$p_t = \left(\mu_0 \frac{f}{\pi}\right) I_c^2 \frac{\left(1 - \frac{I_o}{I_c}\right)\log\left(1 - \frac{I_o}{I_c}\right) + \frac{I_o}{I_c} - \frac{1}{2}\left(\frac{I_o}{I_c}\right)^2}{\pi(D_o/2)^2}, \tag{4.8}$$

where I_o is the *operating ac current amplitude*, in A, and I_c is evaluated as

$$I_c = \lambda\pi\left(\frac{D_o}{2}\right)^2 J_c, \tag{4.9}$$

where λ is the *fill factor*. Typically, $I_o \leq I_c/2$, to maintain stability during disturbances.

If a conductor of the type shown in Figure 4.3 operates in the design space for EAP and wind turbines, the losses shown in Figure 4.5 result. In the two

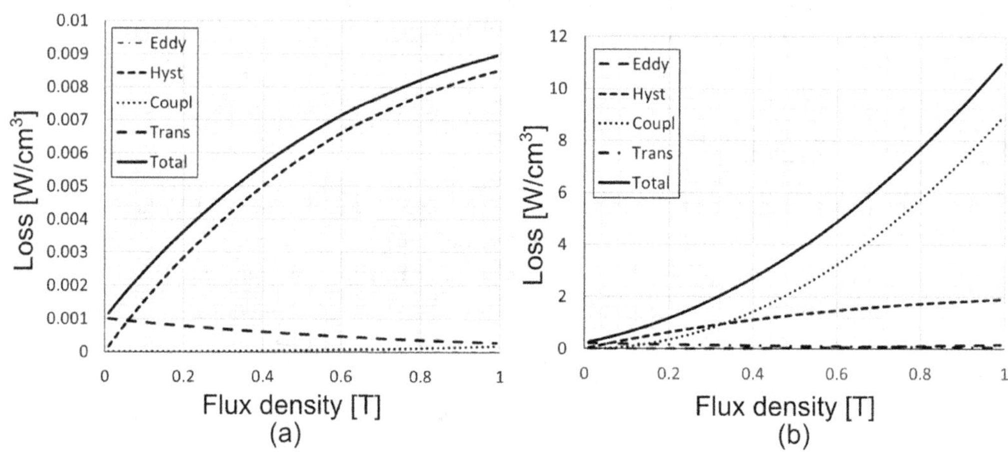

Figure 4.5 Variation of different ac loss components for varying flux density at (a) 1.33 Hz and (b) 300 Hz.

sub-figures, the flux density is varied, and the loss components are evaluated. Note the large difference in the y-axis scale between the two sub-figures. At low frequency (Figure 4.5a; 1.33 Hz, which is the case in a 16-pole, 10 rpm wind generator), hysteresis losses are dominant and total losses are insignificant. When the frequency increases, (Figure 4.5b; 300 Hz , which is the case in an 8-pole, 4,500 rpm EAP motor), the coupling losses dominate, and total losses become significant.

It is evident that for low-frequency applications, high airgap flux density (far exceeding 1 T) could be explored. However, since the total losses depend on the size of the motor, it is critical to determine the optimal airgap flux density. As the frequency increases, losses become significantly higher. Therefore, to have feasible designs, airgap flux density needs to be decreased substantially. Machines with airgap flux density up to 0.8 T could be explored at frequencies in the hundreds of Hertz with available low-ac-loss conductors. The following examples further illustrate this constraint.

Example 1 Evaluate the ac losses, in W/cm^3, for the conductor described in Table 4.2 when operated in a 0.4 T, 200 Hz sinusoidal field, applied perpendicular to the conductor. Assume the operating current is 50 percent of the critical current evaluated at the given field.

Assume that $I_0 = I_c/2$. Using Equations (4.4)–(4.9), the loss densities in each conductor can be evaluated. The results are given in Table 4.3.

Clearly, the conductor construction has a significant influence on the ac losses. Conductors with small diameter and tight twist pitch (e.g., Conductor I) result in relatively low losses.

Table 4.2 Low ac loss MgB_2 conductor data.

	Eqn. notation	Conductor I	Conductor II	Conductor III
Outer diameter [m]	D_o	3.2×10^{-4}	8.5×10^{-4}	8.5×10^{-4}
Filament diameter [m]	d_f	1×10^{-5}	2.8×10^{-5}	2.8×10^{-5}
Number of filaments	n	114	114	114
Twist pitch [m]	L_p	5×10^{-3}	1×10^{-2}	5×10^{-3}
Current density at 0.4 T and 20 K [A/m^2]	J_c	6.6×10^9	6.6×10^9	6.6×10^9
Resistivity matrix [Ω-m]	ρ	3.65×10^{-7}	3.65×10^{-7}	3.65×10^{-7}
Effective transverse resistivity [Ω-m]	ρ_{eff}	1.25×10^{-7}	1.25×10^{-7}	1.25×10^{-7}
Fill factor	λ	0.15	0.15	0.15
Effective fill factor	λ_{eff}	0.49	0.49	0.49

Table 4.3 Losses across a 1 m wire.

Conductor	Eddy current p_e [W/m^3]	Hysteresis p_h [W/m^3]	Coupling p_c [W/m^3]	Transport current p_t [W/m^3]	Total losses [W/cm^3]
I	1.29×10^4	7.04×10^6	6.4×10^5	1.79×10^5	7.87
II	9.13×10^4	1.97×10^7	2.56×10^6	1.26×10^6	23.6
III	9.13×10^4	1.97×10^7	6.4×10^5	1.26×10^6	21.7

Table 4.4 Machine specification.

Specification	Value
Power	2.5 MW
Pole number	8
Speed	4,500 rpm
Airgap	5×10^{-3} m
Outer radius	0.25 m
Armature inner radius	0.151 m
Armature slot height	2.51×10^{-3} m
Gap between two slots	1×10^{-3} m
Slot fill factor	0.5
Stack length	0.661 m
Operating temperature	20 K
Armature current density at 2 T	2.0×10^8 A/m^2
Field current density at 2 T	2.0×10^8 A/m^2
Shield current density at 2 T	2.0×10^8 A/m^2
Machine weight	≈60 kg

Example 2 Assume a fully SC air-core motor designed for EAP employing the MgB_2 Conductor I. The motor specifications are tabulated in Table 4.4. The motor cross section and flux density evaluated along three paths in the armature slot are given in Figure 4.6. The flux densities along paths 1, 2, and 3, respectively, are 0.46, 0.47, and 0.48 T. Evaluate the approximate average ac loss in the motor. Using the data for the cryo-cooler input power and weight provided in the next section, determine the required cryogenic power and overall system efficiency.

The *active armature conductor area per path* is found as

$$A_{ar} \simeq \frac{1}{3}\left(\pi\left((r_{ar}+h_{ar})^2 - r_{ar}^2\right) - h_{ar} n_{\text{slot}} n_{\text{pole}} g_{\text{slot}}\right), \tag{4.10}$$

where r_{ar}, h_{ar}, and g_{slot}, all in m, are the machine *armature inner radius*, the *armature slot radial height*, and the *gap between armature slots*, respectively, and n_{pole} and n_{slot}, respectively, are the numbers of *poles* and *slots per pole*. The *end winding armature conductor length*, in m, is found as

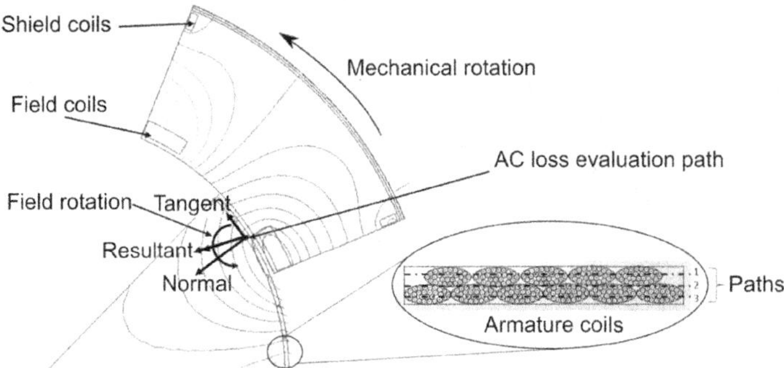

Figure 4.6 Motor cross section, flux path, and flux rotating direction for Example 2.

$$L_{\text{end}} \simeq \pi\left(r_{ar} + \frac{h_{ar}}{2}\right)\sin\left(\frac{\pi}{n_{\text{pole}}}\right), \tag{4.11}$$

and the *armature total length per path* is calculated as

$$L_{ar} \simeq \frac{A_{ar}ff_c(L_{\text{stack}} + L_{\text{end}})}{\pi D_o^2}, \tag{4.12}$$

where ff_c is the *slot fill factor* and L_{stack}, in m, is the motor *stack length.*

Path	Flux density [T]	Loss [W/m³]	Loss [W/m]	Armature conductor length [m]	Total loss [kW]
1	0.46	3.14×10^6	0.25	2.92×10^3	0.73
2	0.47	3.28×10^6	0.26	2.94×10^3	0.76
3	0.48	3.29×10^6	0.26	2.95×10^3	0.77
Total loss (all paths)					2.26

The final calculation in this example requires information from Section 4.3. It is recommended that the reader skip this and come back after completing the later material. However, for organization and simplicity in presentation, the material is included here "early." Using data in Figure 4.8, and assuming Brayton cycle cryo-cooler is used, the required input power to achieve 1.5 W of cooling at 20 K is 80 W. Therefore, the required input power to extract the above ac loss from the machine is

$$\frac{2.3 \times 10^3 \text{ W}}{1.5 \text{ W}} \times 80 \text{ W} = 123 \text{ kW},$$

the system efficiency is

$$\frac{(2500 - 123) \text{ kW}}{2500 \text{ kW}} \times 100\% = 95.1\%,$$

and the system weight is

$$60\ \text{kg} + \frac{2.3\ \text{kW}}{1.5\ \text{W}} \times (10\ \text{kg per } 1.5\ \text{W}) = 15.4 \times 10^3\ \text{kg} = 15.4\ \text{tons}.$$

This exercise shows that using these types of fully SC machines in electric airplanes is not practical with currently available conductors. However, taking advantage of liquid hydrogen technology, such as fuel cells, and cooling the motor with the available latent heat of vaporization, might alleviate these constraints.

Example 3 Repeat Example 2 for a wind turbine generator, with specifications given in Table 4.5. Assume the flux density values are 1.03, 1.04, and 1.05 T along the paths shown in Figure 4.6.

The required input power to extract the ac loss from machine is

$$\frac{809\ \text{W}}{1.5\ \text{W}} \times 80\ \text{W} = 43.1\ \text{kW},$$

Table 4.5 Example wind generator specification.

Specification	Value
Power	10 MW
Pole number	20
Speed	10 rpm
Airgap	80×10^{-3} m
Outer radius	2.196 m
Armature inner radius	1.885 m
Gap between two slots	1×10^{-3} m
Slot fill factor	0.5
Stack length	2.476 m
Operating temperature	20 K
Armature current density at 2 T	2×10^8 A/m^2
Field current density at 2 T	2×10^8 A/m^2
Shield current density at 2 T	2×10^8 A/m^2
Machine weight	$\approx 11.7 \times 10^3$ kg

Path	Flux density [T]	Loss [W/m^3]	Loss [W/m]	Armature conductor length [m]	Total loss [W]
1	1.03	2.05×10^4	1.6×10^{-3}	161.4×10^3	258
2	1.04	2.06×10^4	1.7×10^{-3}	161.8×10^3	275
3	1.05	2.06×10^4	1.7×10^{-3}	162.2×10^3	276
Total loss (all paths)					809

the system efficiency is

$$\frac{(10,000 - 43.1)\ \text{kW}}{10,000\ \text{kW}} \times 100\% = 99.5\%,$$

and the system weight is

$$11.7 \times 10^3\ \text{kg} + \frac{809\ \text{W}}{1.5\ \text{W}} \times (10\ \text{kg per } 1.5\ \text{W}) = 17.1 \times 10^3\ \text{kg} = 17.1\ \text{tons}.$$

Clearly, there are significant benefits to using fully SC machines in low-frequency applications like wind turbines, where the fraction of the system weight due to the cryo-cooler is relatively small and the achievable efficiency is very high. In high-frequency machines, the ac loss in a unit length of the conductor is very large. However, since the motor is fairly compact, the total armature SC length is low, and the total losses in the machine are on the order of kilowatts. In wind turbine applications, the ac loss in a unit length is very low, but the total armature SC length can be significant. Because of the much lower frequency, the total ac losses in the machine may still be quite modest.

The examples provided here neglect spatial and time harmonics present in the field and current experienced by the armature winding in a motor. In practical machines, these factors need to be included in the ac loss estimation.

4.3 Superconducting Machines: Design Considerations

SC machines are like traditional machines but with SC wires. Clearly, cooling the SCs to below T_c is the biggest challenge in SC machine design. However, practical SC machine designs must also address other challenges, such as mechanical integration, quench protection, short circuit performance, and cost. These will all be discussed in this section.

4.3.1 Cryogenic Cooling

The biggest challenge in the practical implementation of SC machines has been the necessity of cooling the SCs to their cryogenic operating temperatures. Many cooling approaches are possible to achieve these temperatures. This includes closed-loop cooling with cryo-coolers, as well as open loop cooling with cryogenic liquids (e.g., helium, hydrogen, helium, and nitrogen) and solids, such as frozen carbon dioxide dry ice and frozen nitrogen. Hybrid cooling systems utilizing both cryo-coolers and liquid cooling – e.g., using liquid or slush N_2 at 63–77 K for the upper stage of heat rejection – are also possible.

4.3.1.1 Cryo-coolers

Cryo-coolers are a well-established and mature technology with many systems used in commercial applications with a wide range of cooling powers from 10^{-2} W to 10^6 W.

Varying cryo-cooler types and thermodynamic cooling cycles are employed. While properties such as reliability and mean time between failure (MTBF) have improved significantly over the years, their size and weight have not changed much.

High power density is especially important for aircraft applications and can become a deciding criterion if other factors such as coefficient of performance are similar. Examples of state-of-art cryo-coolers that could provide reasonably high power density and efficiency for cooling SC coils of motors/generators are listed in Table 4.6. Many of these coolers are available as commercial off-the-shelf products. The variety of cryo-cooler options to choose from is large; typically, one manufacturer will provide 10–20 options for one cryo-cooler type, optimized for performance at varying temperatures and cooling loads. For aircraft applications, Brayton cycle cryo-coolers are attractive for 20–110 K operation with high cooling powers greater than 100 W. Brayton and Stirling cryo-coolers have met additional requirements including the ability to work in different orientations, ultra-high reliability (greater than 99.99 percent) for mission-sensitive applications, robustness to electromagnetic interference, and logistics considerations, such as ease of use.

Cryo-cooler power density, efficiency, and size can all become critical factor affecting system viability. The mass-specific inverse power density (or inverse of specific power) of the cryo-coolers in Table 4.6 are shown in Figure 4.7, plotted for varying input powers, indicating significant differences in machine designs and impacts on power densities. In general, as the power levels increase for a given cryo-cooler type, the inverse power densities decrease, reaching an ultimate low limit. For input powers less than 500 W, the Stirling cryo-coolers provide higher power density than other cryo-coolers and are optimal. However, for input powers greater than 1 kW, the Brayton-cycle cryo-coolers are the best option.

The impact of increasing operation temperature to improve cryo-cooler size, weight, and power (SWaP) properties is illustrated in Figure 4.8. Figure 4.8a shows the required cryo-cooler input power to achieve 1.5 W of cooling at varying operating temperatures. The cryo-cooler weight required to meet the same 1.5 W of cooling at varying temperatures is shown in Figure 4.8b. These curves indicate a dramatic reduction of SWaP, by a factor of approximately 100–500, obtained by increasing the operation temperature from 5 K to greater than or equal to 50 K. This demonstrates the strong importance and advantage of improving YBCO wire performance and increasing the operating temperature.

Cryo-cooling is a well-established solution, with dozens of commercial manufacturers and multiple technology choices that can be optimized for different operating temperatures and system requirements.

4.3.1.2 Closed-Loop Cooling

In addition to closed-loop cooling with cryo-coolers, potential weight- and energy-saving options utilizing the heat capacity of cryogenic fuels or other liquids or solids at cryogenic temperatures that can be renewed by resupply on the ground or regenerated with a ground-based cryo-cooler may be possible depending on the vehicle system design and mission profile. Examples include airplanes with liquefied natural gas (LNG)

Table 4.6 State-of-the-art cryo-coolers [Extracted from [7]].

Cryo-cooler type	Company	Input power [kW]	Mass [kg]	Output power [W] at 4.2 K	at 20 K	at 50 K	at 77 K	MTBF [h]
Stirling 1-Cycle	Cryo-tel	0.24	4.1			5.5	16	200,000
	"	0.16	3.1			3.5	11	200,000
	"	0.08	2.1			1.2	5.2	
	STI Conductus	0.135	2.8				9	>1,000,000
Stirling 2-Cycle	Stirling Cryogenics & Refrigeration (SCR)	12.6	551			450		
	"	54	1,150			1,800		
	"	11	551		75			
	"	45	1,160		300			
Thermoacoustic Stirling	Chart/Qdrive	0.25	15				8	125,000+
	Chart/Qdrive	0.6	23				25	125,000+
	Chart/Qdrive	2.75	80				140	125,000+
	Chart/Qdrive	4.5	115				175	125,000+
	Chart/Qdrive	20	388				1,000	125,000+
Brayton	Aerospace	0.16	11		3.5	9.8		
	Industry	0.2	15.5		4.4	12.3		
	Aerospace	0.25	10.5		5.5	15.4		
	"	0.4	20.8		8.8	24.6		
	"	3	21.0		65.7	184.4		
	"	20.8	249.6		455.5	1278.7		
	Industrial	1,000	8,000		21,898	61,475		
	Creare	10	30		219.0	614.8		
	"	32	96		700.7	1,967.2		
	"	160	480		3,503.6	9,836.1		
Gifford McMahon	ARS	3.6	80.1	0.1	2.3		4	
	"	3.6	80.1	0.2	8.0		14	
	"	7.7	128	0.8	16.0		26	
	"	6.8	121	1.5		40		
	SHI Cryogenics	1.5	49.2	0.1		5		15,000
	"	3.8	84	0.5		4		"
	"	3.8	111	0.4		33		"
	"	3.8	111	0.4		40		"
	"	7.5	178	1		65		"
	"	7.5	188.5	1.5		45		"
	Cryomech	3.3	88.2			70.0	116.4	15,000
	"	7	130.6			200.0	310	"
	"	10.4	212.5		63.0	227.0	279	"
	"	10.4	212.5		40.0	170.0	250	"
GM Pulse-Tube	Cryomech	2.9	76.7			28.0	57	25,000
	"	4.3	118.2			48	86	"
	"	9.2	215.5	1.5				"

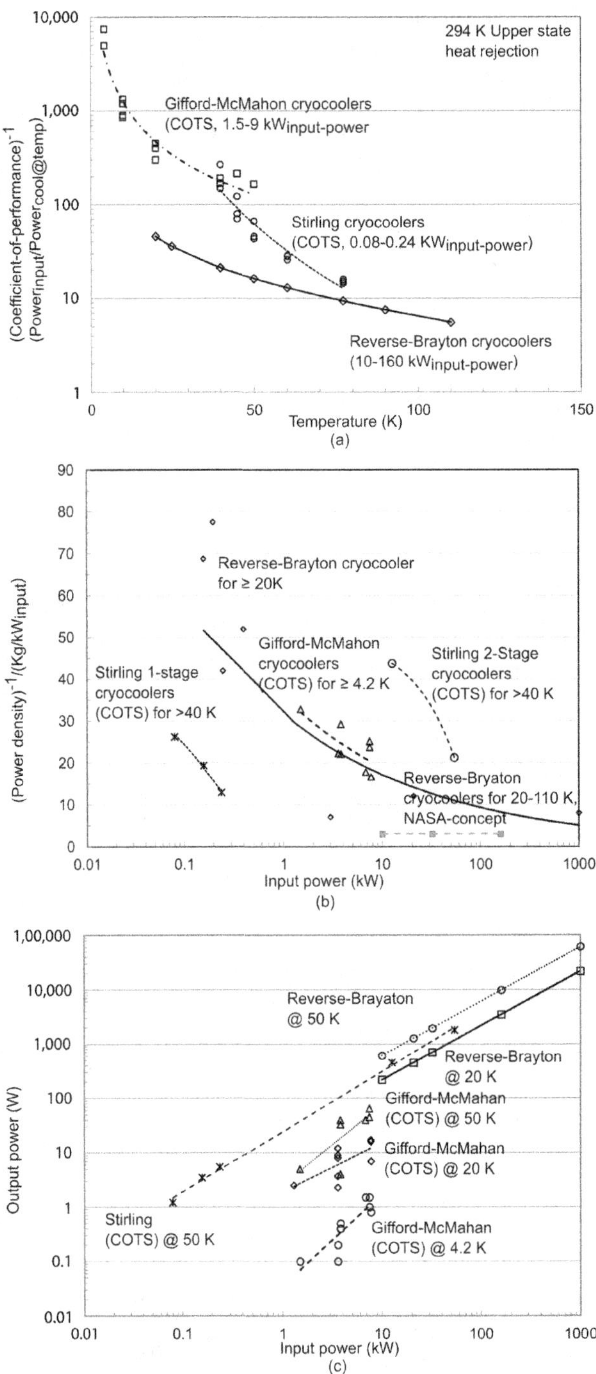

Figure 4.7 Cryo-cooler properties: (a) Cooling power versus cooling temperatures, (b) mass-specific inverse power density versus input power, and (c) output versus input power.

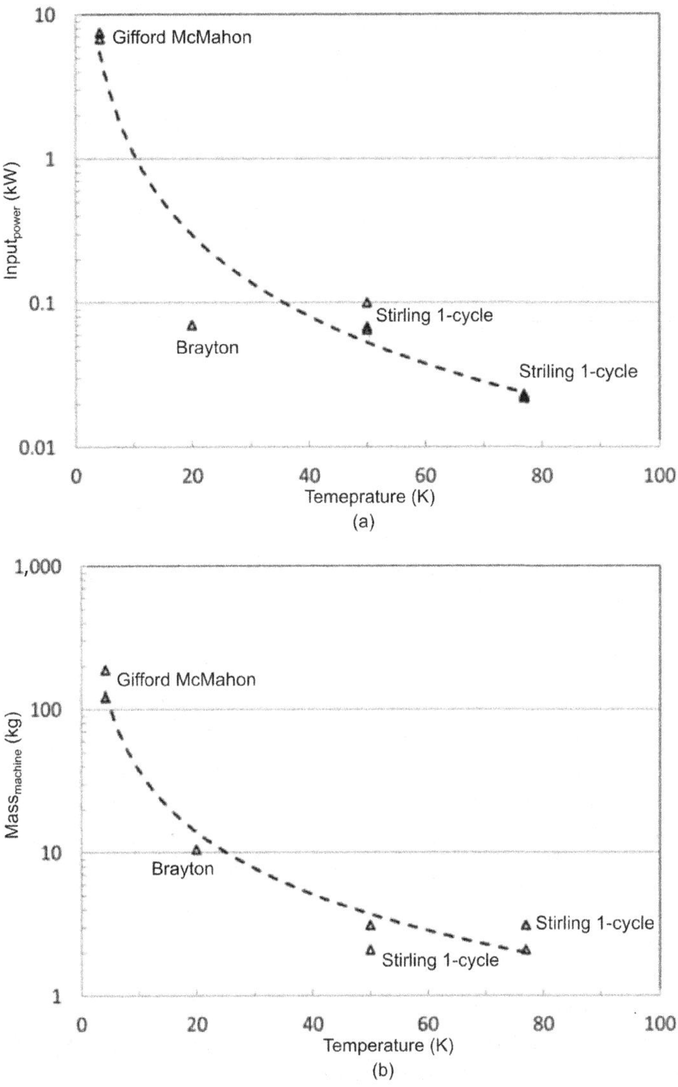

Figure 4.8 (a) Plot of cryo-cooler input power to achieve 1.5 W, and (b) plot of the cryo-cooler weight to meet the same to 1.5 W of cooling.

or liquid hydrogen fuel, for use with fuel cells to either generate electric power directly or to power a turbine. LNG has a boiling point at the atmospheric pressure of 112 K and freezing point of 91 K and can be used directly as a heat sink for cryogenic components such as power inverters and for thermal shields of lower-temperature components like SC machines and power transmission cables. It can also be employed as the upper temperature of cryo-coolers to improve Carnot efficiencies by up to approximately four times.

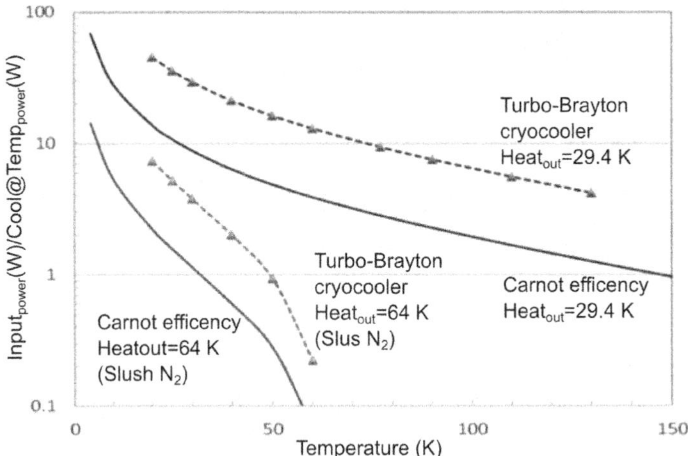

Figure 4.9 A hybrid-cooling system combining a cryo-cooler with liquid N_2 cooling. The ratio of input-power/output-cool-power is shown for varying temperatures of upper stage rejection. For reference, 294 K is room temperature, and 64 K is typical of slush N_2.

Hydrogen has a boiling point of 20 K and can be used to directly cool SC coils composed of MgB2 or HTS conductors. This method is analogous to the thermal management system in conventional airplanes where heat from the electrical components is dumped into the fuel, with some additional benefit coming from preheating the fuel.

An example of hybrid cooling, combining cryo-coolers with liquid nitrogen cooling, is shown in Figure 4.9, where it is possible to dramatically reduce cryo-cooler weights at 20 K by using liquid nitrogen in partial vacuum (slush N_2) at 63–65 K as an upper stage for heat rejection. Figure 4.9 shows the impact on Carnot efficiency and how the power input required can be reduced by a factor of five from 40 W to 8 W (to achieve 1 W of cooling at 20 K). And this further reduces the machine weight by a factor of five or more.

The continued improvement of SC wire performance at relatively high temperatures, approaching liquid nitrogen temperatures (≈65–77 K), will have a significant impact on reducing cryo-cooler weights and energy losses. SC wire technology advancements for YBCO and fine filament MgB2 are already significantly impacting motor/generator system performance, and this trend is expected to continue over the next few years.

4.3.2 Mechanical Design

The SC coil assembly has to be suspended mechanically in a manner that ensures structural integrity of the coil and maintains the strain within the conductor below critical levels while minimizing the heat leak from the ambient environment. Mechanical strain is invariably obtained in components that experience mechanical stresses, such as wires in a large magnetic coil. This constraint becomes more

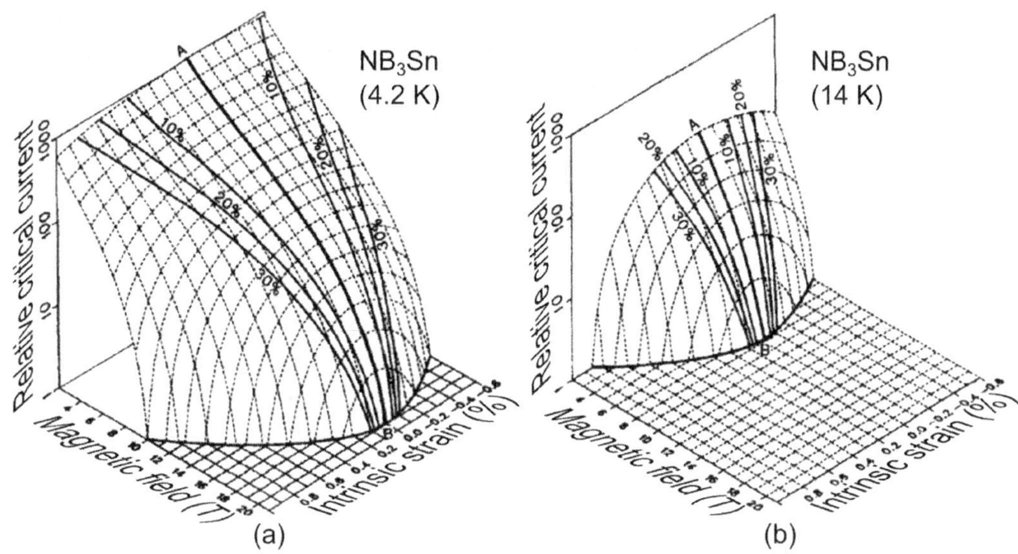

Figure 4.10 Critical surface comparison of Nb_3Sn at (a) 4 K and (b) 14 K under varying levels of mechanical strain. Reprinted from [9], with the permission of AIP Publishing

challenging as the industry migrates to machines operating at higher magnetic field levels. Careful attention has to be paid to the coil and coil support design to ensure the coil is not degraded as it is ramped up.

While T_c, H_c, and I_c are commonly measured and known for various SCs, only some general engineering guidelines exist for SC critical strain limits. Reasonably good data is available for popular conductors being considered at large scales, mostly for high-energy physics projects.

Figure 4.10 shows the impact of strain on Nb_3Sn conductor performance plotted at 4 K and 14 K. Based on these data, the ITER magnet team has specified the maximum permissible tensile strain of Nb_3Sn as 0.2 percent [8].

Additionally, the large torques generated by the machine must be transmitted without increasing the heat leak to the ambient components. Experience exists in practical suspension systems of large SC coils from the high-energy physics and magnetic resonance imaging (MRI) industries. The need to transmit torque is unique to electrical machines. Practical solutions have been developed with the use of relatively thin torque tubes made of high-strength/low-thermal-conductivity titanium alloys and carbon/glass fiber composites.

4.3.3 Quench Protection

Any disturbance that pushes a part of the SC coil into a resistive normal state results in Ohmic losses being generated in that zone and can lead to a thermal runaway situation with potentially severe consequences. This energetic failure mode is known as *quench*. Depending on the speed of quench propagation, most of the stored magnetic energy in the SC coil could be dissipated in a relatively small region. This can create excessive temperature rise and damage the coil. Practical means of detecting and dissipating the energy in a safe manner would be required to provide fault ride-through capability in electrical machines. One way to handle this to ensure that the normal zone propagates rapidly, through either active or passive means, so that the energy is dumped in a large thermal mass, thereby limiting the temperature rise. It is relatively hard to do this in HTS than in LTS coils. Since the trend in ongoing research is to drive coil flux densities to relatively high values, stability and quench protection would be important considerations in the design effort. Experience from high-energy physics and the MRI industry can be used to mitigate these risks.

4.3.4 Short Circuit Protection

Short circuit protection is a significant challenge in SC machines because they are invariably slotless and typically coreless, leading to extremely low synchronous reactance ($\approx$10 percent) composed mostly of the armature leakage. The risk of excessive short circuit currents has to be mitigated with rapid control of the excitation, likely utilizing an energy dump option rather than ramping the field down slowly with the exciter.

4.3.5 Cost

The high power density machines being sought after for aircraft applications are generally of the air-core topology and will likely have aggressive SC requirements. High-field designs that have recently been proposed have further increased the amount of SC conductor needed. This is especially challenging if HTS materials are utilized because the SC cost will then be a significant fraction of the total machine cost. How each of these aspects is addressed depends on the specific topology of the machine. Key features of the different types of machines being explored are provided in what follows.

4.4 Superconducting Machine Topologies

In general, all these machine types take advantage of the fact that Ohmic losses in the conductors are virtually eliminated, offering opportunities for increased efficiency and higher power density. This benefit has to be traded off against the need to maintain the SC coils at cryogenic temperatures, and system-level size, weight, efficiency, reliability, and cost have to be evaluated considering the machine and cryogenic thermal management infrastructure.

Various machine topologies have been proposed and demonstrated using SCs to partially or fully replace conventional conductors in the design. The main machine

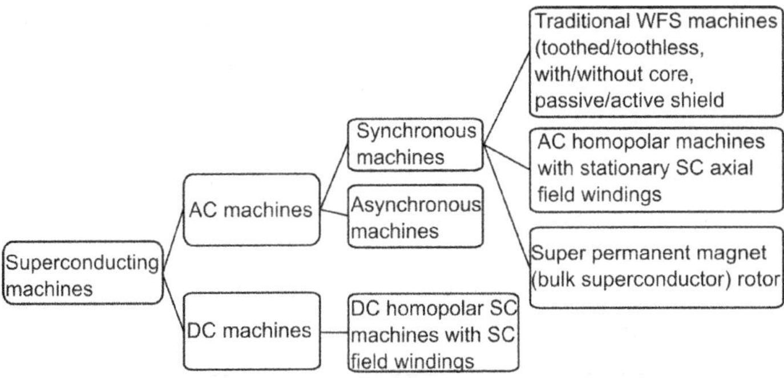

Figure 4.11 SC machine topologies.

topologies being investigated at the scale of interest for airplane application are summarized in the chart shown in Figure 4.11. Each of these can be either partially SC or fully SC.

4.4.1 Fully Superconducting Machines

Fully SC synchronous machines utilize SCs for both the dc field winding and ac armature windings. This type of machine is expected to be superior to a partially SC machine that uses SCs only in the dc field windings. Therefore, in addition to high fields, high efficiency can be obtained in fully SC machines by significantly reducing the armature Ohmic losses. It may also be possible to increase armature electrical loading substantially. While this topology has been worked on for a long time [10], there has been limited success because of a lack of good ac-capable conductors.

Specifically, the SC motors and generators required for turboelectric propulsion rotate at high speeds in the 2,000–8,000 rpm range and therefore operate at frequencies of hundreds of Hertz. In order to minimize weight, these machines are invariably of the slotless type, with the armature winding fully exposed to the airgap flux. Given the resulting high rate of change of flux density, the ac losses in the SC armature winding could be substantial with commercially available conductors. Based on the projected properties, the optimum magnetic flux density provided by the rotor is relatively low in these machines, usually between 0.5 and 1.2 T, to limit ac losses in the SC composite wire in the stator and external back-iron shield. Additional challenges include the need for nonconducting cryostats on the stator for many of the configurations considered. If sufficient cooling is available – for example, from onboard liquid hydrogen stored for use as fuel in a turbine – the SC stators may still make sense for some missions. It may also be an attractive option if the rest of the electrical system, like the power distribution cables, is at cryogenic temperatures, eliminating the need for large current leads that need to transition from the ambient environment to the cryogenic components.

There is extensive research ongoing to develop conductors that can help address the ac loss challenge. These losses can be reduced by using small-diameter wire with twisted fine SC filaments embedded in a resistive metallic matrix. Drawing wires down to obtain filament sizes less than 10 micrometers to reduce ac losses appears to be easier with metallic SCs like NbTi, but because of lower critical temperatures, the allowable ac losses to limit cryo-cooler requirements is also relatively low. On the other hand, HTS conductors could allow for a higher thermal budget, but at present there are no technologies available to get the ac losses in HTS conductors to levels comparable to the multifilamentary LTS conductors. BSCCO has been produced in multifilamentary form but without key features required for low ac loss (fineness of filaments, high resistivity matrix, and tight transposition). Attempts are being made to construct low-ac-loss YBCO conductors, but except for preliminary studies this superconductor has been produced mainly in the form of a thin, wide tape, which has prohibitively high ac losses. Cable concepts like Conductor on Round Core [11] hold some promise, but their feasibility and the effectiveness in an electrical machine application are yet to be proven.

At this time, MgB_2 appears to have the right balance of the aforementioned properties for the ac armature winding. Figure 4.3 shows a cross section of an example multifilamentary MgB_2 conductor under development. The MgB_2 filaments appear as black regions. The balance of the cross section is composed of a variety of normally conducting materials. In addition to managing ac losses, all other technical challenges associated with any SC machine needs to be dealt with.

In summary, implementing fully SC machines for electric aircraft requires developing practical conductors with significantly improved ac-loss properties and a high-operating temperature. While this is being actively pursued by research groups, the first application may be in relatively low-frequency machines like wind turbine generators. Fully SC machines for aircraft applications may take a few years to materialize. In the meantime, the partially SC machines described in what follows may provide most of the benefits in terms of specific power and efficiency with available conductors.

4.4.2 Partially Superconducting Wound-Field Synchronous Machines

In a partially SC wound-field synchronous machine (WFSM), the dc field winding is made of SC coils while the ac armature windings are generally composed of copper or aluminum coils. The machine is commonly constructed with an air-core (i.e., non-magnetic) rotor and no iron teeth in the stator, enabling the airgap field to be increased without core losses and avoiding the saturation problems inherent in iron-core stator and rotor. In conventional machines, resistive copper losses are typically about equal for the field and armature windings. Thus, a replacement of copper field winding with SC windings, along with the reduction/elimination of core losses, can potentially improve the machine efficiency by 1–2 percent, even over highly efficient baseline designs. The reduction in total losses also relieves system-level thermal management challenges. This can be traded off for higher loading of the conventional armature winding to obtain additional power density advantages.

A partially SC machine employing SC field winding on the rotor is shown in Figure 4.12. The normal conductor armature winding is placed in a slotless assembly

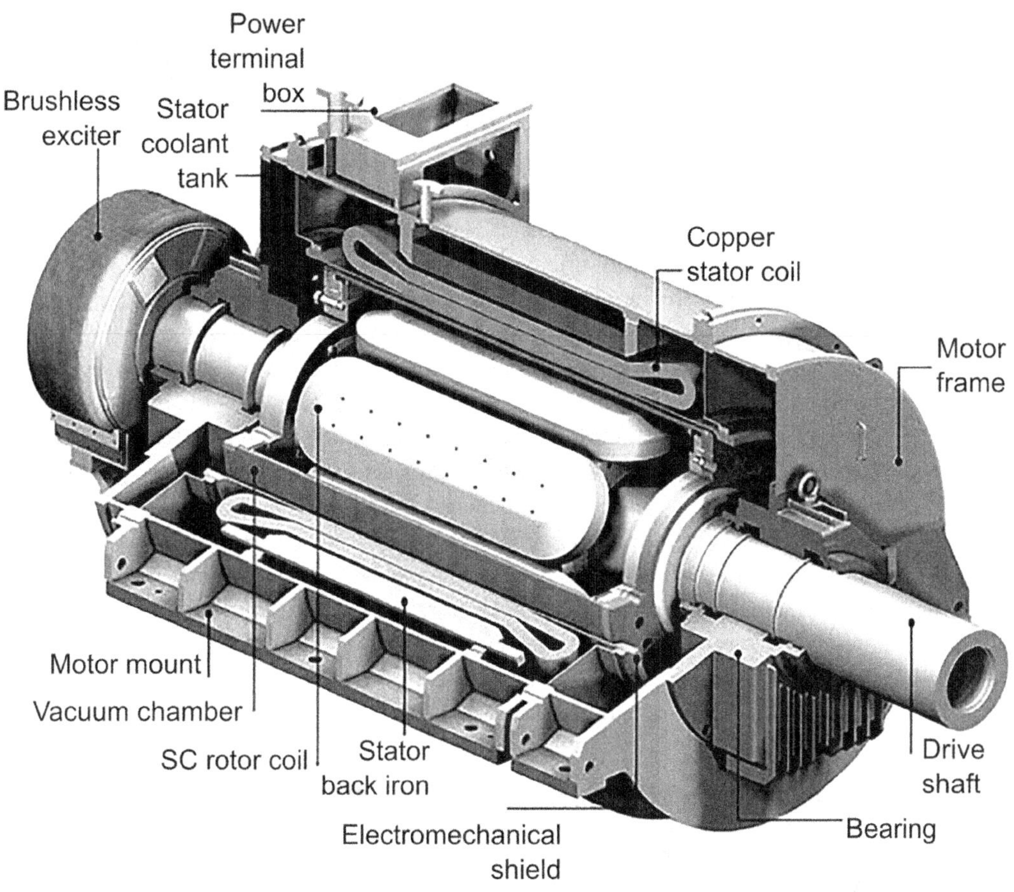

Figure 4.12 Partially SC synchronous machine employing SC field winding on the rotor. © [2006] IEEE. Reprinted, with permission, from [12]

within the magnetic airgap. In the steady state, the rotor speed is synchronous with the rotating field created by the armature currents, and the SC field winding experiences only dc magnetic fields. Under transient conditions that can result from load or source changes, the rotor moves with respect to the rotating fields from the armature and is exposed to ac field harmonics. Harmonic content in steady-state stator currents also lead to non-dc field on the rotor. SC motors may require inverters and filters that provide relatively clean current waveforms with low total harmonic distortion (THD). An electromagnetic (EM) shield is also usually employed to protect the SC field winding from these ac fields. A refrigeration system, which uses cold circulating cryogen (e.g., liquid or gaseous helium, neon, or nitrogen) in a closed loop, maintains the SC field winding at a cryogenic temperature. Conduction cooling with direct thermal links to a cryocooler is also being considered. In order to reduce eddy-current losses, the slotless armature winding is commonly fabricated with finely stranded and transposed (e.g., Litz) conductors. Because of a large effective airgap (between field and armature coils), the back electromotive force (EMF) in the airgap stator winding is nearly a pure sine wave – the harmonic field components are much smaller than those observed in the conventional machines. This can enable significant reduction in torque ripple and noise.

A number of partially SC wound field machines have been built and tested since the 1970s, when multifilamentary NbTi SCs became readily available. Large synchronous generators for utility applications were demonstrated by General Electric, Westinghouse, Super-GM, and others. The Super-GM project in Japan tested a series of 70 MVA generators with different NbTi field windings in the 1990s. A number of technical issues were reported: poor reliability of the cryogenic cooling system, stability issues relating to a small temperature operating range (4.2–5.5 K), protection of the NbTi field windings from harmonic fields generated by the stator during normal and abnormal operations, and complications and costs of building airgap stator windings. Most of the development work employing low-temperature NbTi SCs was discontinued after this effort. Recent projects have concentrated on HTSs with simplified cooling systems. HTS conductors operating at much higher temperatures (25–77 K) simplify refrigeration-cooling systems, and the HTS winding operating temperature range is much wider. Higher thermal heat capacity of materials employed in the windings also enables absorption of much larger transient thermal loads with little temperature rise. Cost considerations are reviving interest in LTS machines for applications like wind turbines, but for aircraft systems with emphasis on performance and reliability, HTSs remain the preferred choice.

4.4.3 Alternating Current Homopolar Superconducting Machines

In ac homopolar machines, armature voltage is induced using the alternating airgap flux generated by variation of the rotor reluctance. An example ac homopolar machine is shown in Figure 4.13a. The field coil drives the axial flux through the rotor and

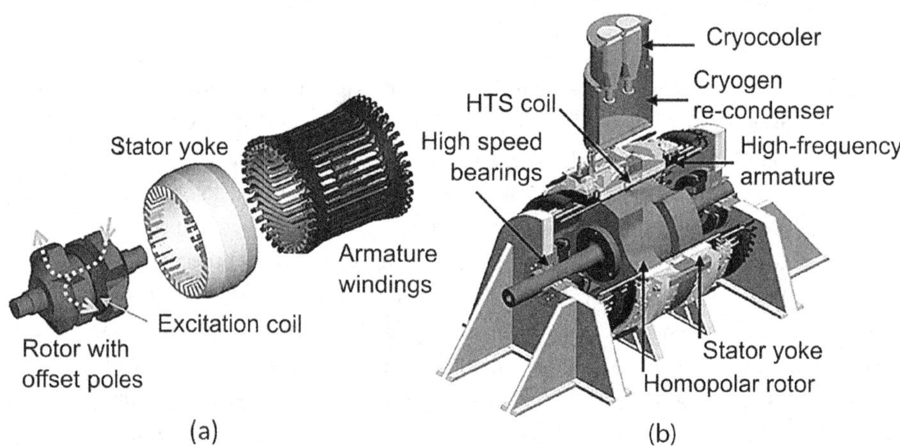

Figure 4.13 (a) Schematic of an ac homopolar machine; (b) detailed configuration of a megawatt-scale homopolar inductor alternator with HTS field winding [13].

creates north and south poles at the rotor ends. The yellow lines indicate the flux path through part of the rotor. Armature coils on one end of the machine use the flux from the rotor south pole, and the coils at the other end use flux from the rotor north pole. Because the salient poles are offset 180° electrically, the rotor rotation creates circumferential flux variation in the stator and induces the voltage in the armature coils. The pole arrangement sums the induced voltages in the armature winding in both stacks to generate the terminal voltage. This machine is called homopolar because the flux direction in either stack does not change with time. Since the rotor has no field winding, it can be manufactured from a single piece of high-strength steel. Such solid steel rotors make homopolar machines attractive for high-speed operation, including electric aircraft. An attractive feature of this machine is the possibility to employ a stationary field coil located between the two axially offset sets of salient poles. This greatly simplifies the suspension and cooling of the SC coil. The cryostat of the coil is stationary. There is no need for a transfer coupling to introduce a cooling medium into the rotating cooling circuit, and good thermal insulation can be provided in the stationary construction without the need to deal with centrifugal forces. The arrangement also eliminates brushes and slip rings and simplifies the design.

In SC ac homopolar machines, field coils are replaced with the SCs to provide high magneto-motive force capability and increased airgap flux density. Alternating current homopolar machines are attractive for employing SC field winding for two reasons. First, the stationary field windings do not experience any centrifugal forces that rotating coils experience in synchronous machines. Second, the stationary field coils simplify the associated cryogenic cooling system to maintain the SC field coils below the critical temperature. A practical implementation of an SC ac homopolar machine comprised of a stationary HTS field excitation coil, a solid rotor forging, and an advanced but conventional stator is illustrated in Figure 4.13b. The armature consists of a liquid-cooled airgap winding placed within an advanced iron yoke made of laminations oriented in three dimensions. The cryo-coolers and cryogen recondenser unit are mounted on top of the generator in a simple assembly.

A megawatt-scale version of this machine was developed by General Electric (GE) under contract with the Air Force Research Laboratory (AFRL) and successfully load tested in 2007. This type of machine has been observed to be capable of achieving a specific power greater than 13 kW/kg and efficiency greater than 98 percent in a robust architecture. This demonstration led to increased interest in SC machines for aircraft applications ranging from auxiliary systems to prime power.

4.4.4 Asynchronous Superconducting Machines

Asynchronous motors, or induction motors, are one of the most widely used types of motors in the industry. In this type of motor, power is supplied to the rotor by means of electromagnetic induction from the rotating magnetic field produced by ac currents in the stator. An induction motor's rotor can be either wound type or squirrel cage type. SC induction motors typically employ HTS conductors in a squirrel cage type rotor.

A group in Kyoto University in Japan has successfully demonstrated a 20 kW fully SC induction motor comprising a toroidal HTS stator and HTS rotor bars fabricated by stacking BSCCO tapes [14]. The HTS rotor is similar to the squirrel cage rotor of conventional induction motors, but the copper/aluminum bars and end rings are replaced with HTS conductors. It follows from induction machine theory that the rotor resistance has to be nonzero to develop electromagnetic torque. In an SC induction machine, torque is developed because the SC rotor experiences relatively high frequency and high current at starting, which drives the SC winding "normal" resulting in a nonzero rotor resistance. When the rotor approaches the synchronous speed, the current magnitude and the frequency drop, and the rotor resistance approaches zero. Under these conditions, the rotor can be pulled into synchronism, leading to relatively high-efficiency operation. The key components of the demonstrated motor are shown in Figure 4.14. The group is working on scaling it up to higher power levels.

4.4.5 Direct Current Homopolar Machines

The dc homopolar machines are a unique type of electrical machine in that they operate with a dc field interacting with a dc current to generate a steady torque. A sliding contact is required. The basic concept can be described with the Faraday's disk shown in Figure 4.15a. In this configuration, a disk with a vertical shaft is placed in a vertical magnetic field. A dc voltage is applied between the shaft and disk rim using brushes to create a radially directed current. Since the current and the magnetic

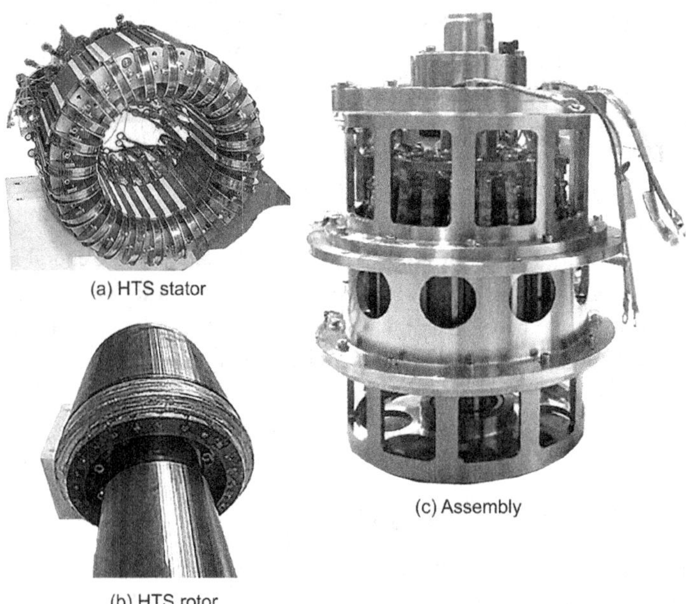

Figure 4.14 Fully SC induction motor: (a) HTS stator; (b) HTS rotor; (c) assembly. [14] Reprinted with permission of the author

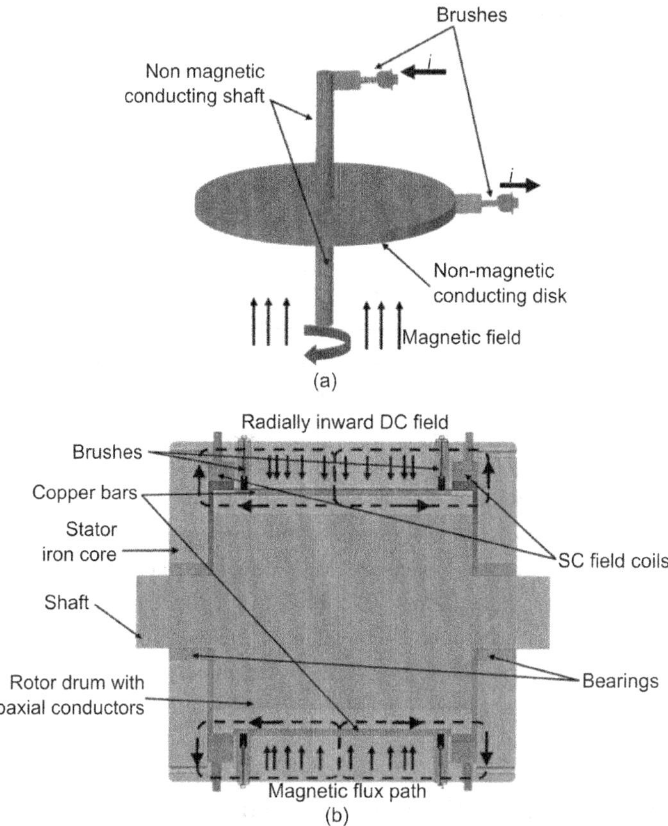

Figure 4.15 (a) Faraday's disk motor; (b) drum-type dc homopolar machine.

fields are perpendicular to each other, a force tangent to the circle of the disk will be induced, resulting in motor torque. These machines can also be operated as a dc generator by rotating the shaft. Because the magnetic field is constant, a constant dc voltage will be induced in the disk. SC field coils can be employed to produce a high magnetic field and increase torque and efficiency. As in ac homopolar machines, the field coils are simple solenoidal windings located on the stationary part and experience no motor torque. This simplifies the design of the SC field winding and its cryostat. In addition to the stationary field coils, the current in the disk is also dc. These characteristics make the dc homopolar machines an ideal candidate for employing SCs.

The disk and brush arrangement is basically a single-turn coil, which leads to low voltage and high current. This high current must be collected from the rotating members with brushes. Efficient current collection is challenging and prevents scaling up these machines. A more practical implementation is a drum rotor configuration, as shown in Figure 4.15b. This has several concentric drums connected to a shaft. The drums use a brush set on each end for handling the current. The advantage the drum rotor machine as compared to the disk rotor machine is its length; an increased drum length leads to increased induced voltage. The concentric rotor drums experience the same field applied radial to the rotor's outer surface using two cylindrical SC field coils located near the ends of the drum. The only major force experienced by the SC field coils is their mutual repelling force directed along the rotational axis of the machine. A magnetic shield is employed to reduce the impact on the surrounding equipment due to the high magnetic fields generated by the SC field coils. The motor housing can be made from a solid magnetic iron since the fields are only dc.

These drums need to be connected in series to build up to relatively high voltages. A brush on one end of a drum must be connected to the brush on the opposite end of the adjacent drum using copper bus bars. These bars are employed in the motor frame parallel to the rotor axis. The copper in the rotor drums and bus bars is designed to operate at the highest possible current density, within efficiency and thermal management constraints, to obtain the most compact machines. It is also necessary to provide active cooling of these copper components for heat management. This is a complicated process due to the introduction and recovery of coolant to the individual rotor drums while maintaining electrical isolation and uniform temperature distribution [15].

Direct current homopolar machines have also been attempted worldwide, beginning with significant development work by Appleton in England [16] in the late 1960s. Similar effort continued in the United States (US) until the first decade of the twenty-first century. These machines were mostly handicapped by challenges associated with brushes needed for current transfer to the rotor.

4.5 State-of-the-Art Superconducting Machines

Many SC motors and generators have been demonstrated. Some large-scale demonstrations are listed in Table 4.7. The majority of machines built and tested to date have been with partially SC WFSM, because it is considered to have a good balance of risks versus

Table 4.7 SC machine demonstrations.

Year	Speed [rpm]	Power [kW]	Conductor type	Cryogenic cooling	Specific power [kW/kg]	Efficiency [%]	Developer
1978	3,600	20,000	NbTi	Liquid He	0.17	99.3	GE
1978	3,600	300,000	NbTi	Liquid He	1.83	99.4	Westinghouse
1978	3,600	1,200,000	NbTi	Liquid He	2.97	99.6	Westinghouse
1980	7,000	20,000	NbTi & Nb_3Sn	Liquid He	0.06	99.5	GE
2001	1,800	3,725	BSCCO	Neon boil-off	0.745	97.7	AMSC
2001	1,500	400	1G-HTS	Ne-thermosiphon		97.0	Siemens
2003	3,600	1,500	BSCCO	Gaseous He			GE
2004	3,600	4,000	1G-HTS	Ne-thermosiphon	0.58	98.7	Siemens
2008	10,000	1,300	BSCCO	Neon boil-off	8.8	98.0	GE

benefits. It provides advantages in size, weight, and efficiency over existing machines with copper field winding and avoids ac loss challenges in the superconductor.

With these demonstrations, many of the key technology issues have been addressed at very high levels. However, the technology readiness level (TRL) needs to be advanced and risk reduced, especially on some of the more aggressive designs, such as fully SC machines and the high field air-core topologies, before they are ready for system integration. Currently, two major hurdles for wide adaption of superconductor-based technologies are the cost of SCs and their cooling system and workable superconductor mechanical properties. Government subsidies are needed to continue development of this technology in order to keep pace with this emerging technology.

4.6 Superconducting Cables

Power transmission cables are essential elements of every electric drivetrain and transportation power system, which utilize hybrid or full electric power for propulsion. When incorporated with other cryogenic/SC components, it is a synergistic technology that makes a fully SC power system an attractive option for electric aircraft. Somewhat surprisingly, power cables are by far the largest component weight of any aircraft electric power system and can reach greater than 3 percent of aerospace vehicle weight, especially in lighter vehicles such as spacecraft and satellites. Thus, the use of SC/cryogenic power cables can provide some of the largest benefits and reduction of mass for aerospace platforms.

Aircraft requirements for power cables include both dc and ac, and voltage levels for aerospace vehicles are limited to less than or equal to ±270 Vdc or 230 Vac because of corona discharge at high altitudes and electric arcing. Primary ac frequencies are up to 400 Hz, and 800 Hz for more-electric aircraft to achieve higher power densities – however, at a penalty of lower efficiencies for components including inverters. A detailed listing of the types and voltages of power transmission cables required by a Boeing 787 aircraft is provided in Table 4.8.

Cable requirements are dictated by the current needed for the application. However, in addition, the cabling process provides many benefits, including (1) high currents (2) high compactness, which also enables high *engineering current density* J_e; (3) full transposition of filaments or wires, which is required to reduce ac loss; (4) dimensional accuracy, which is required for some types of ultra-precise magnets and to lesser degree for motors/generators; (5) controlled interstrand resistance needed for fault-current limiter applications and quench management; (6) enhanced mechanical properties for high-stress applications such SC-magnetic-energy-storage (SMES), high-speed generators, fusion reactors and pulsed sources for particle accelerators; (7) increased strength to enable high-stress winding operations; and (8) robustness to protect against individual filament or wire failure. For high current or power levels, coaxial cables designs are of interest, to limit external magnetic fields almost to zero and also to limit very strong Lorentz attract/repel forces that can tear apart wires running jointly in one direction. For cryogenic operation, a coaxial or plus/minus

Table 4.8 Electrical distribution requirements for a Boeing 787 aircraft [17]. The total aircraft load is greater than 4 kA.

Power sources	**Size [kVA]**	**Number**
230 Vac generator	250	4
230 Vac auxiliary power unit (APU)	225	2
Distributions and loads	**Size [A]**	**Number of loads**
230 Vac	large	≈20
230 Vac → all Vdc inverters		
± 270 Vdc		≈10
115 Vdc	large (>10)	≈25
115 Vac, 28 Vdc	small (<10)	≈850
28 Vdc	large (>10)	≈150
Power inverters	**Size [A]**	**Conversion function**
Auto transformer rectifier Units (ATRUs)		230 Vac to ±270 Vdc
Auto transformer units (ATUs)		230 Vac to 115 Vdc
Transformer rectifier units (TRUs)		230 Vac to 28 Vdc
Remote power distribution units (RPDUs)	≈17	115 Vdc to 115 Vac 115 Vdc to 28 Vdc 28 Vdc to 28 Vdc
Motor controllers – adjustable Speeds	≈10	

codirection bundled-wire cable is of interest, to enable cooling of both the to and from power lines in one cryo-flex tubing.

Technology options for SC/cryogenic conductors are shown in Figure 4.16, which can include any wire that is commercially manufactured in long lengths (typically longer than 500 m). Also plotted are the allowable current densities for metallic conductors – including Cu – from the National Electric Code, which electricians use for wiring. This demonstrates how the current density for Cu wire decreases sharply as the current level increases above 10 A and reduces J_e below 4 A/mm^2, compared to YaBaCuO cables at 50 K achieving J_e greater than 1,000 A/mm^2. In addition to higher current density, SC cables have the additional benefit of zero heat loss, and only thermal management of a cooling system is required.

The reason for the sharp decline of J_e for metallic conductors in Figure 4.16a is that the current capacity of metallic cables is fundamentally limited by the Ohmic heat losses (which, remember, are not present in SC cables) that must always be accounted for to prevent thermal runaway and system failure. Ohmic heat losses of metallic conductors are written from Ohm's law as

$$P_{\text{loss}} = RI^2, \tag{4.13}$$

where *resistance* R, in Ω, is

$$R = (\rho L_{\text{wire}}/A_{\text{wire}}) \tag{4.14}$$

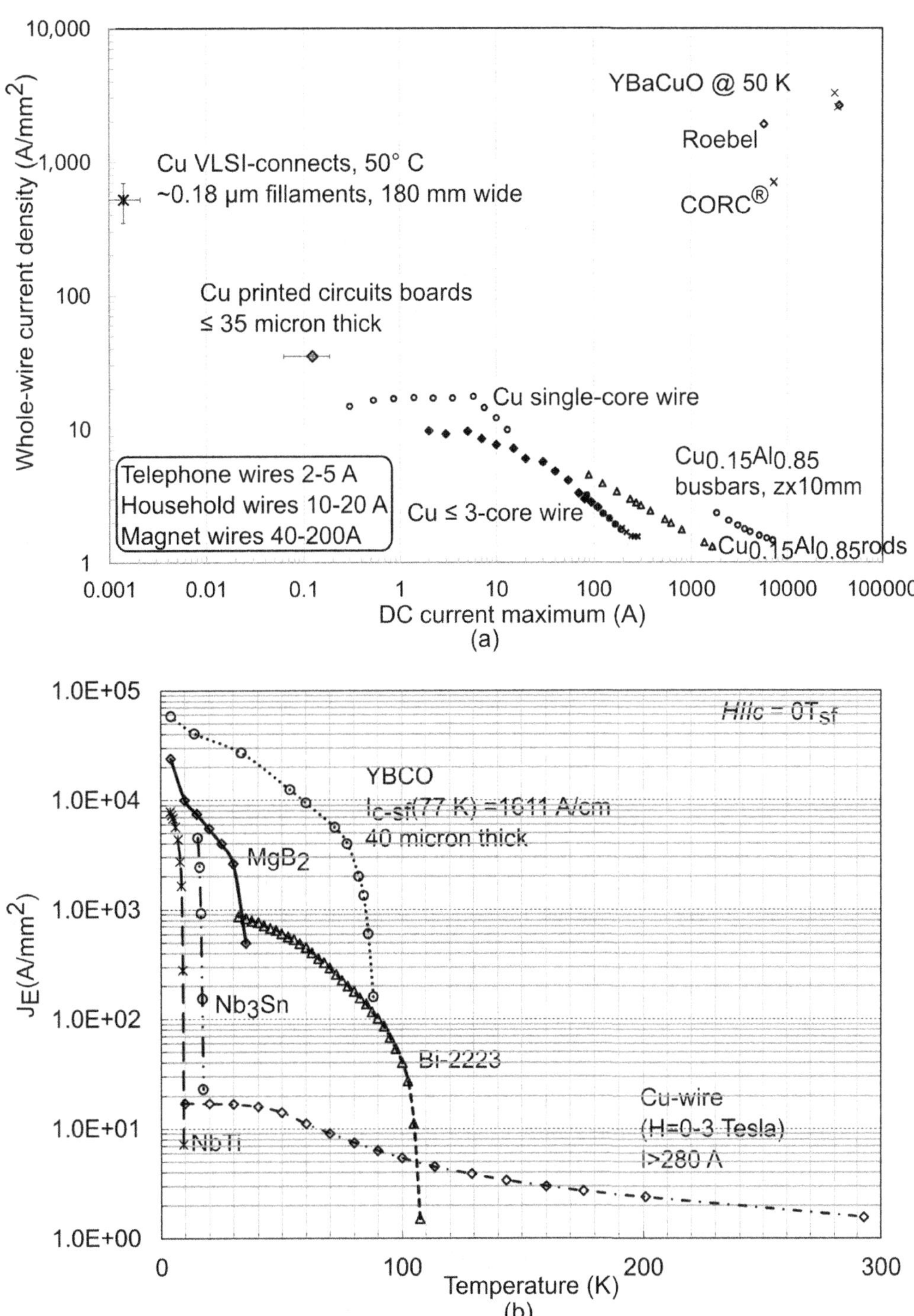

Figure 4.16 (a) Whole-wire current density of copper and $Cu_{0.15}Al_{0.85}$ conductors (Cuponal™), which are state of the art for aircraft such as the Airbus A380, compared to compact YBaCuO cables at 50 K. (b) Whole-wire current density versus operation temperature of commercially available SC wires, compared to standard copper wire capable of transmitting high currents above 280 A.

ρ is the *resistivity*, in Ω-m, L_{wire} is the *wire length*, in m, and A_{wire} is the *wire cross-sectional area*, in m^2. Thus,

$$P_{\text{loss}} = I^2(\rho L_{\text{wire}}/A_{\text{wire}}). \tag{4.15}$$

At the *maximum current*, $I = I_{\max}$, the current density is

$$J_e = \frac{I_{\max}}{A_{\text{wire}}} \tag{4.16}$$

so that (4.15) becomes

$$P_{\text{loss}} = J_e I_{\max} \rho L_{\text{wire}}. \tag{4.17}$$

The dependence of J_e on P_{loss} is thus

$$J_e = \frac{P_{\text{loss}}}{\rho L_{\text{wire}} I_{\max}}, \tag{4.18}$$

which indicates that J_e is directly proportional to the Ohmic losses. Power cables have finite ability to dissipate heat by air cooling, and liquid cooling only allows three to four times more heat than air cooling, so $P_{\text{loss}}/L_{\text{wire}}$ becomes constant in (4.18); ρ is also constant at a fixed operating temperature. Therefore,

$$J_e = \frac{\text{const}}{I_{\max}} \tag{4.19}$$

and to maintain constant heat loss and prevent runaway temperature, J_e must be decreased inversely to $I_{\max}$. It is demonstrated in Figure 4.16a that the J_e of Cu wires varies almost inversely to $I_{\max}$ (the x-axis, labeled "DC Current Maximum"), as $I_{\max}$ increases from approximately 10 A to 1,000 A, confirming (4.19).

For aircraft electrical power systems, the upper limit for dc voltage is ± 270 V due to corona discharge. Since voltage is limited, the only possible way to increase power is to increase the current, which decreases the current density and increases the wire mass. To further highlight the impact of current on wire mass, it is noted that

$$\rho = \frac{m_{\text{wire}}}{V_{\text{wire}}}, \tag{4.20}$$

where m_{wire} is the *wire mass*, in kg, and V_{wire} is the *wire volume*, in m^3. Equation (4.18) can then be written as

$$m_{\text{wire}} J_e = \frac{P_{\text{loss}} V_{\text{wire}}}{L_{\text{wire}} I_{\max}} \tag{4.21}$$

and

$$V_{\text{wire}} = A_{\text{wire}} L_{\text{wire}}. \tag{4.22}$$

Now, substituting (4.19) for J_e on the left and (4.22) for V_{wire} on the right in (4.21) and canceling like terms gives

$$m_{\text{wire}} \times \text{const} = P_{\text{loss}} A_{\text{wire}}. \tag{4.23}$$

Assuming P_{loss} is maximized, (4.15) (with $I = I_{\max}$) can be substituted into (4.23) for P_{loss}. Dividing both sides of the resulting equation by the constant yields an equation for the wire mass:

$$m_{\text{wire}} = \frac{I_{\max}^2 \rho L_{\text{wire}}}{\text{const}}. \tag{4.24}$$

Equation (4.24) reveals that, for a given wire of fixed length, m_{wire} increases quadratically with $I_{\max}$. The fundamental Ohmic loss-based limitations to using metallic conductors for higher power level EAP as current levels increase are encompassed in (4.19) and (4.24).

Some compact SC cable designs that are in different stages of development are illustrated in Figure 4.17. At the bottom of the figure, two types of conductors carrying $\approx$ 4,000 A dc are compared: 10 MCM 750 gauge copper cables and a single YBaCuO stacked cable operating at about 20 K and packaged into Nexan's cryo-flex tubing of about 25 mm diameter.

Other cable designs that are less compact are also being developed, especially for ac power transmission. Other ac cable designs are for high voltage (1 kV to 300 kV), contain large dielectric insulation, and can be up to 20 cm in diameter. The current capacities and physical properties of compact YBaCuO cables being developed are shown in Table 4.9. The ac loss properties of electric conductors are critically important for many applications, including ac power cables, magnets, transformers, and motor/generators. The ac loss properties of the three YBaCuO SC cables are plotted in Figure 4.18, based on studies available in the literature. The losses shown have been used to accurately estimate losses of Roebel cables in single loops of 1 MVA transformers and to predict losses during the charge or discharge of a magnet made out of those cables.

4.7 Technology Trends

Based on the successful demonstration of SC machines over the last few decades, SC machines can now be considered to be technically feasible. However, their long-term reliability and cost effectiveness for various commercial applications have yet to be confirmed. Every application has its own specific requirements and trade-offs. For aerospace applications, important attributes include high specific power, high reliability, and fault tolerance.

Since SC coils that can operate at higher temperatures are less prone to sudden quenching, SC machine reliability can be improved by adopting HTSs, which are becoming more readily available. These conductors will also enable simplified cooling schemes, further improving the reliability. Most future development work on SC machines is expected to be with HTS and MgB_2 conductors.

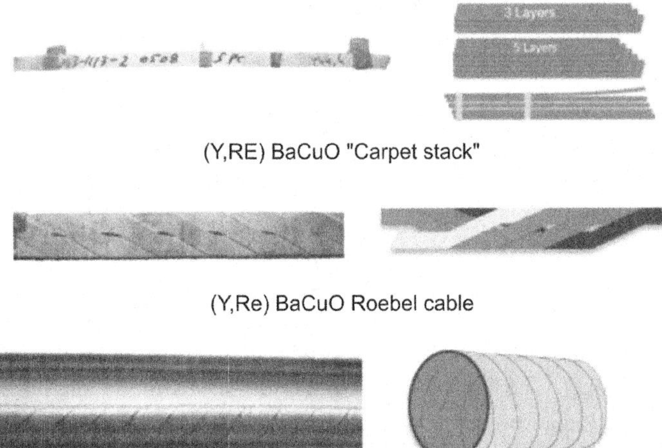

(Y,RE) BaCuO conductor-on-round-core (CORC®) conductors

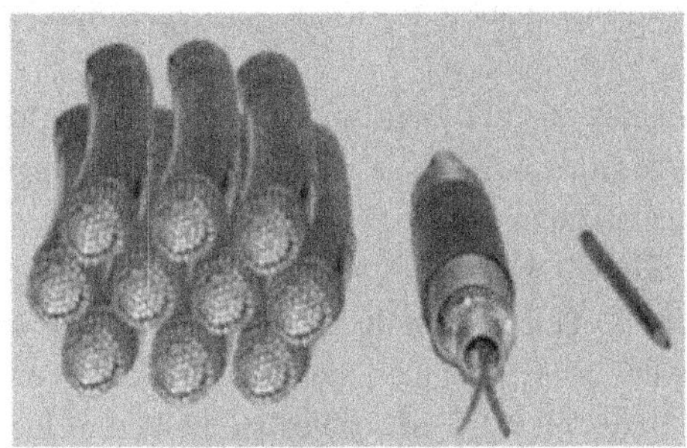

Cu-cable, made by handling together ten MCM 750 gauge wires, compared to YBCO tapes in a Nexans cryoflex vacuum tubing

Figure 4.17 Types of SC transmission cables.

Table 4.9 Critical current and physical properties of electric conductors shown in Figure 4.17.

Conductor type	I_c (77 K, self-field) [A]	I_c (50 K, self-field) [A]	**Diam. or width [mm]**	**Number of tapes**
"Carpet Stack" – 5 layers	≈400–600	≈1,800	4	N/A
Roebel	2,000	9,000	13	15
CORC® Cable [18]	10,950	49,200	≈8	50, 3 mm width
CORC® Wire	2,400	10,800	≈3.7	11, 2 mm width
Cu-Wire, MCM 750 Gauge	400	N/A	29	N/A

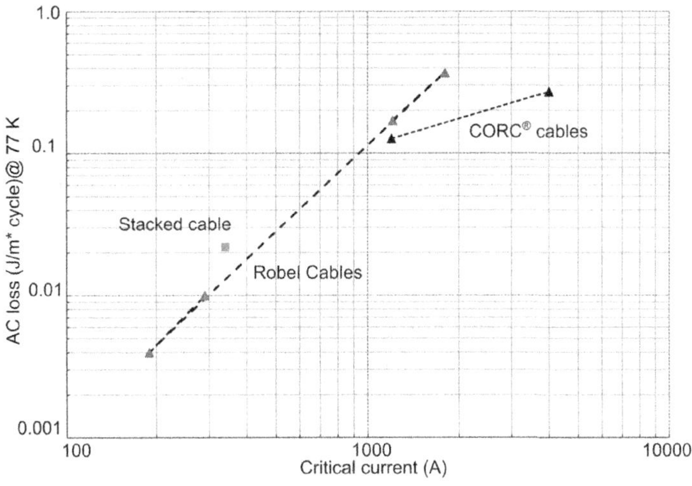

Figure 4.18 Ac loss of the YBaCuO SC cables listed in Table 4.9, at 77 K.

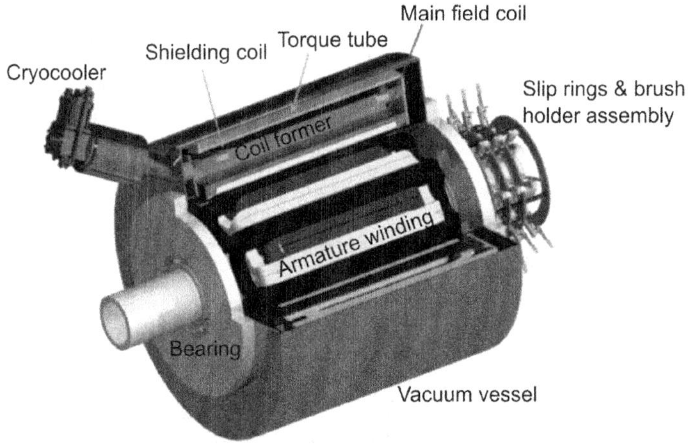

Figure 4.19 Configuration for actively shielded air-core SC machine concept [19].

In addition to increasing the maturity of proven machines concepts, new approaches can lead to even higher specific power and efficiency than is possible today. Partially SC machines have the potential for increased specific power by employing even higher operating flux density. A completely air-core machine with active shielding coils currently being investigated is illustrated in Figure 4.19. This type of machine is projected to give motor specific power exceeding 50 kW/kg [19]. For the full potential of these topologies to be realized, HTS conductors capable of high fields at high temperatures need to be matured and composite structures developed to replace relatively heavy vacuum vessels and torque tubes.

Fully SC machines have the potential to attain the highest efficiency by minimizing losses in both the field and armature windings. To minimize these losses, however, significant advances in ac-capable SCs must be made. Since partially SC machines already are projected to achieve 99 percent efficiency, the ac losses in an SC armature must be significantly less than 0.1 percent of the machine rating to be attractive. Further, this must be achieved while maintaining ac flux on the order of 10^3 T/s to be competitive in power density with the partially SC machines.

Another area of active research is focused on addressing reliability and system integration challenges associated with their need for special cooling. One potential advance is motors with integrated cryo-coolers. The US National Aeronautics and Space Administration (NASA) Glenn Research Center is developing a partially SC, WFSM that combines a self-cooled SC rotor with a slotless stator, allowing the motor to achieve exceptional specific power and efficiency without the need for an external cryo-cooler and the weight and efficiency penalty that come along with it. The goal is to have a motor that can be integrated and operated like any traditional (non-SC) machine, but with all the advantages of having SC coils on the rotor. The integrated motor under development is targeting specific power of 16 kW/kg with an efficiency of 99 percent at the megawatt scale. The key features of this high-efficiency megawatt motor (HEMM) concept are illustrated in Figure 4.20.

Machines built with bulk SCs are another exciting area of active research, though they are currently at a lower TRL. Bulk HTS SCs show great magnetic

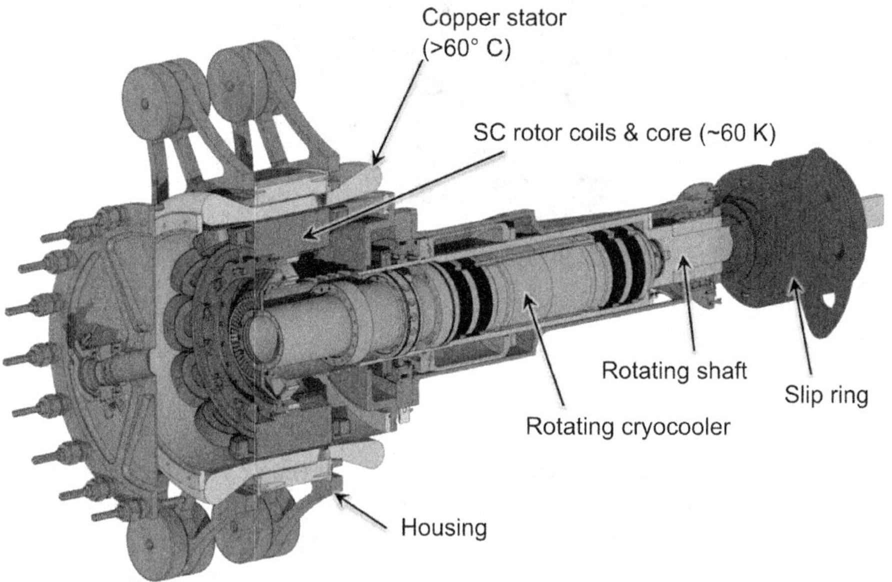

Figure 4.20 NASA's High Efficiency Megawatt Motor (HEMM) concept design [20, 21]. Reprinted with permission of NASA

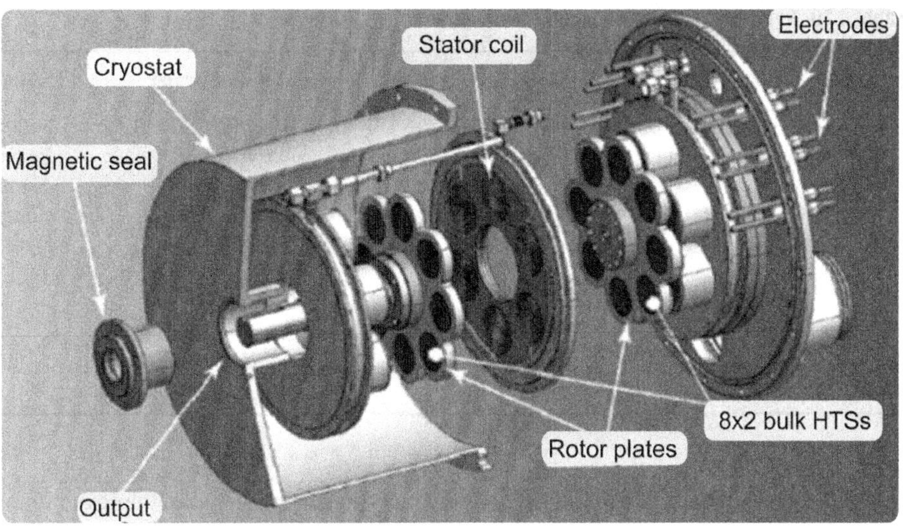

Figure 4.21 Axial-gap HTS synchronous motor with twin rotor [7]. © IOP Publishing. Reproduced with permission. All rights reserved

flux-trapping performance as cryo-PMs. The trapped flux is more than an order of magnitude higher than conventional Fe-Nd-B magnets. An axial-flux synchronous motor was developed as a prototype by using Gd-Ba-Cu-O family bulks as field poles. The demonstrator motor is shown in Figure 4.21. A key feature of this motor is the superior torque/field pole flux compared to conventional PM machines. Cooling was achieved using a thermosiphon of neon. An effective magnetization technique for HTS bulks cooled below T_c is the key to realizing a high magnetic flux density suitable for the field-pole application with an electric motor or generator. Plans are under way to demonstrate these machines at the tens-of-kilowatt power ratings. Improved bulk SCs capable of operation at liquid nitrogen temperatures and ways to magnetize the bulks at high fields would further increase the potential for these machine types.

A recent review paper considered these efforts and other technologies being developed around the world and generated a technology road map with projections for the specific torque of electric machines over the next several years [7]. How the specific torque of SC machines has increased over the last couple of decades and how it can be advanced further with some of the SC technologies described before are summarized in Table 4.10. While specific torque is a proxy for specific power and a metric closely related to the electromagnetic capability of the machine, trade-offs will have to be made in high-speed machines to obtain high specific power.

Though SC machine technology has to be matured before it is employed in large commercial aircraft systems, SC machines offer the opportunity to overcome major system-level challenges. For the promised benefits to be achieved, further progress

Table 4.10 Torque density advances in SC machines.

	Pre-1990s	Currently available	Medium term	Long term
Key features	LTS conductors Liquid Helium cooling	HTS conductors Gaseous He or liquid H_2/Ne cooling Airgap winding	High field SC coils Actively shielded – air core Bulk SCs Advanced cryo-coolers Composite structure	Fully SC: Low ac loss conductors Nonmetallic cryostats
Torque density	<1 Nm/kg	Up to 40 Nm/kg	Up to 60 Nm/kg	>100 Nm/kg

must be made in conductor technology (higher temperature, improved pinning, lower ac losses), cooling (cryo-coolers adapted to electric machine cooling needs), and protection and control. Additional operational prototypes must be built and tested in relevant environment to prove performance and reliability.

4.8 Summary

SC machines have the potential to achieve very high power density and efficiency and are an attractive option for electric aircraft, especially for use in large commercial transport applications, where conventional machines cannot meet the stringent performance demands. However, the TRL of SC machines must be advanced before they are considered for commercial applications. More work is also required in system integration because these machines have very different operating characteristics and protection needs than conventional ones. Reliability and safety questions will remain until long-term endurance tests are performed on such systems.

In some ways, the megawatt-scale SC machine demonstrated in 2007 by AFRL helped reinvigorate efforts to electrify large aircraft propulsion. Further advances and demonstrations of SC machines and propulsion systems are sure to propel the field forward in future years.

Abbreviations

ac	alternating current
AFRL	Air Force Research Laboratory
APU	auxiliary power unit
ATRU	auto transformer rectifier unit
ATU	auto transformer unit
CSM	critical state model
dc	direct current

EAP	electrified aircraft propulsion
EM	electromagnetic
EMF	electromotive force
GE	General Electric
HEMM	high-efficiency megawatt motor
HTS	high-temperature superconductor/superconducting
LNG	liquefied natural gas
LTS	low-temperature superconductor/superconducting
MMF	magnetomotive force
MRI	magnetic resonance imaging
MTBF	mean time between failure
NASA	National Aeronautics and Space Administration
PM	permanent magnet
RPDU	remote power distribution unit
SC	superconducting/superconductor
SMES	SC-magnetic-energy-storage
SWaP	size, weight, and power
THD	total harmonic distortion
TRL	technology readiness level
TRU	transformer rectifier unit
US	United States
WFSM	wound-field synchronous machine

Variables

A_{ar}	armature conductor area per path, [m]
A_{wire}	wire cross-sectional area, [m^2]
B, B_{ar}	flux density, at armature radius, [T]
B_m	peak magnetic flux density, [T]
D_o	conductor outer diameter, [m]
d_f	filament diameter [m]
f	frequency, [Hz]
ff_c	slot fill factor
g_{slot}	gap between armature slots, [m]
H, H_c, H_p	magnetic field, critical field, penetration field limit, [A/m]
H_{c1}, H_{c2}	lower and upper critical field, respectively, [A/m]
h_{ar}	armature slot radial height, [m]
I_o	operating ac current amplitude, [A]
I_c, $I_{\max}$	critical current, maximum current, [A]
J_c	critical current density, [A/m^2]
J_e	engineering current density, [A/m^2]
K	machine power coefficient [units vary]
k	eddy current constant (equal to four for multifilament wires)

L_{ar}	armature axial length per path, [m]
L_{stack}, L_{end}	stack length, end winding armature conductor length [m]
L_p, L_{wire}	twist pitch, wire length, [m]
λ	fill factor
M	magnetization, [A/m]
m_{wire}	wire mass, [kg]
μ_0	vacuum permeability, [$4\pi \times 10^{-7}$ H/m]
n	coupling constant (equal to two for multifilament wires),
n_{pole}	number of poles
n_{slot}	number of slots per pole
p_c, p_e, p_h	power loss density (coupling, eddy current, and hysteresis), [W/m^3]
P_{loss}	power loss (general), [W]
p_t	transport current loss density, [W/m^3]
R	resistance, [Ω]
r_a	armature inner radius, [m]
r_{wire}	wire (conductor) radius, [m]
ρ, ρ_{eff}	resistivity, effective transverse resistivity matrix, [Ω-m]
T_c	critical temperature, [K]
V_{wire}	wire volume, [m^3]

References

[1] National Academies of Sciences, Engineering, and Medicine, *Commercial Aircraft Propulsion and Energy Systems Research: Reducing Global Carbon Emissions*, Washington, DC: National Academy Press, 2016.

[2] R. Jansen et al., "Turboelectric aircraft drive key performance parameters and functional requirements," presented at the 51st AIAA/SAE/ASEE Joint Propulsion Conf., Reston, VA, 2015, Paper AIAA 2015-3890.

[3] G. S. Withers and J. R. Toggweiler, "Superconductivity above 130 K in the Hg-Ba-Ca-Cu-O system," *Nature*, vol. 363, pp. 56–58, 1993.

[4] P. Vanderbemden, "Processing and applications of (RE)BCO and MgB2 bulk superconductors: And introduction to the special issue," *Supercond. Sci. Technol.*, vol. 29, 060302, 2016.

[5] W. J. Carr, Jr., *AC Loss and Macroscopic Theory of Superconductors*, 2nd ed., New York, NY: Taylor and Francis, Inc, 2001.

[6] M. D. Sumption, "AC loss of superconducting materials in motors and generators for very high density motors and generators for hybrid-electric aircraft," presented at the 1st AIAA/IEEE Electric Aircraft Technol. Symp., Cincinnati, OH, 2018, Paper AIAA 2018-5001.

[7] K. S. Haran et al., "High power density superconducting rotating machines: Development status and technology roadmap," *Supercond. Sci. Technol.*, vol. 30, 123002, 2017.

[8] C. T. J. Jong, N. Mitchell and J. Knaster, "ITER magnet design criteria and their impact on manufacturing and assembly," presented at the 22nd IEEE/NPSS Symp. on Fusion Eng., Albuquerque, NM, 2007, DOI: 10.1109/FUSION.2007.4337879.

[9] J. W. Ekin, "Four-dimensional J-B-T-ε critical surface for superconductors," *J. Appl. Phys.*, vol. 54, pp. 303–306, 1983.

[10] P. Tixador et al., "Electrical tests on a fully superconducting synchronous machine," *IEEE Trans. Magn.*, vol. 27, pp. 2256–2259, 1991.

[11] D. C. Van der Laan et al., "Characterization of a high-temperature superconducting conductor on round core cables in magnetic fields up to 20 T, *Supercond. Sci. Technol.*, vol. 26, 045005, 2013.

[12] S. S. Kalsi et al., "The status of HTS ship propulsion motor developments," in *2006 Proc. IEEE Power Eng. Soc. General Meeting.*, Montreal, Que., Canada, 2006, DOI:10.1109/PES.2006.1709643.

[13] K. Sivasubramaniam et al., "Development of a high speed HTS generator for airborne applications," *IEEE Trans. Appl. Supercond.*, vol. 19, pp. 1656–1661, 2009.

[14] T. Nakamura et al., "Tremendous enhancement of torque density in HTS induction/synchronous machine for transportation equipments," *IEEE Trans. Appl. Supercond.*, vol. 25, 5202304, 2015.

[15] S. Kalsi, *Applications of High Temperature Superconductors to Electric Power Equipment*, Hoboken, NJ: Wiley-IEEE Press, 2011.

[16] A. Appleton, "Superconducting dc machines," in *Superconducting Machines and Devices*, Boston, MA: Springer, 1974.

[17] [Online]. Available: www.boeing.com.

[18] J. D. Weiss et al., "Introduction of CORC® wires: highly flexible, round high-temperature superconducting wires for magnet and power transmission applications," *Supercond. Sci. Technol.*, vol. 30, 014002, 2017.

[19] D. Loder, "Actively shielded air-core superconducting machines: optimization and design considerations," M.Sc. thesis, University of Illinois at Urbana-Champaign, Urbana, IL, 2016.

[20] R. Jansen et al., "High efficiency megawatt motor conceptual design," presented at the 54th AIAA/ASME/SAE/ASEE Joint Propulsion Conf., Cincinnati, OH, 2018, Paper AIAA 2018-4699.

[21] J. Scheidler, T. Tallerico, W. Miller, and W. Torres, "Progress toward the critical design of the superconducting rotor for NASA's 1.4 MW high-efficiency electric machine," presented at the 2nd AIAA/IEEE Electric Aircraft Technologies Symposium, Indianapolis, IN, August 2019, Paper AIAA 2019-4496.

5 Conventional Power Electronics for Electrified Aircraft Propulsion

Patrick Wheeler

Introduction

Previous chapters have introduced the concept of electrified aircraft propulsion (EAP) and its numerous proposed benefits. Indeed, this book is organized around that central theme. Chapters 2–4 have described the power architecture and electric machines necessary to enable EAP. This chapter continues the development, introducing the power electronic circuits necessary to enable these technologies. The importance of power electronics cannot be understated. Here, we introduce basic concepts, focusing on those circuits and devices that are crucial for EAP. This chapter is not intended to be used as a textbook or tutorial; rather, its purpose is to provide general concepts and information so that the reader will have a good top-level understanding of power electronics and the crucial role they play. From the more-electric aircraft (MEA) – an aircraft with some or all electrified subsystems, but traditional propulsion – to the all-electric aircraft (AEA) – an aircraft with fully electrified subsystems and propulsion – power electronics is the enabling technology for electrification.

5.1 Fundamental Concepts

Power electronic converters transform electrical energy between circuits at different voltage magnitudes and/or frequencies in an efficient and effective manner, which enables numerous advanced technologies. The ability to manipulate and control electrical energy is critical for its transmission, generation, and storage and for the control of electrical machines. Recent advancements in power electronics have made possible both the MEA and future EAP system concepts envisioned for AEA.

Power electronic converters are constructed using three basic elements: *inductors*, *capacitors*, and *switches*. Inductors and capacitors are used to store (and supply) energy; an inductor stores energy in the magnetic field created by its current, whereas a capacitor stores energy in the electric field that is the source of its voltage. Power electronic switches are semiconductor devices, which can be either controlled or uncontrolled. Controlled switches enable the controller to determine when the device conducts current (i.e., turns on) and when it blocks voltage (i.e., turns off). The two most common controlled semiconductor devices are the *insulated gate bipolar transistor* (IGBT) and *metal oxide semiconductor field effect transistor* (MOSFET).

In contrast, the turn-on and turn-off of an uncontrolled switch is determined by circuit conditions. The most common example is the *diode*, which turns on when forward biased and turns off when negatively biased.

When deciding on the appropriate power converter topology for a given application, the amount and form of the energy to be transferred is a crucial factor. For example, because the energy stored in an inductor is associated with its current, it cannot be transferred by connecting one inductor to another inductor unless their current magnitudes are equal. Therefore, it is common in power circuits to transfer energy from inductors to capacitors, which are voltage-based energy storage elements, and vice versa. The size, number, and layout of these components are chosen to meet the requirements of the application.

In order for a power electronic converter to fulfill its design function of changing the nature of the electrical energy to match system or load requirements, strict circuit waveform control is necessary. The control scheme sends commands to a modulation process for the converter, which in turn ensures that the output requirements of the converter are achieved. In modern power converters, this process is most often accomplished by a digital control platform; however, analog circuits are also a viable option. In fact, analog control platforms were common until relatively recently; they have been displaced by the flexibility and ease of programming offered by modern, cost-effective digital circuits and processors. Typical digital control platforms can consist of devices such as *field-programmable gate arrays* (FPGAs) – digital processors – or micro-controllers/-processors. The choice depends on the number and complexity of calculations required by the modulation and control schemes, as well as the power converter performance requirements, particularly the switching frequency and power quality (PQ) requirements.

Closed control loops are used to regulate the output waveforms such that the measured output variable converges to a reference value quickly and smoothly. To do this, most power converters require some sort of modulation to control the fully controlled switching devices; this is the power converter's modulation process. Sometimes these two operations are combined in one process, such as in *model predictive control*, but usually they are two separate functions, often implemented within the same control platform.

A general schematic of the control and modulation functions of power electronic converters is shown in Figure 5.1.

5.2 Target Metrics and Integration Techniques

The main components in any power converter topology are as follows: heat sinks and/or heat transfer elements, control platforms, power semiconductor devices (controlled and uncontrolled) and conductors for interconnections, housing and connectors, passive components (both within the converter and for input/output waveform filtering), and gate drives and sensors. These all impact the converter's size and weight, and these impacts can be qualitatively assessed. This section focuses on the biggest factors influencing converter volume and how they can be optimized for EAP applications.

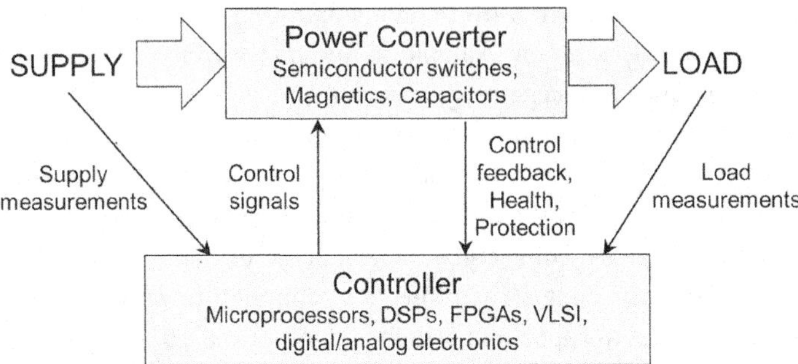

Figure 5.1 Control and modulation of power electronic converters.

5.2.1 Power Density

Power converter absolute power and volumetric density are difficult to quantify, as these are influenced by numerous factors. Here, we present some guidelines and a potential road map for predicting and improving the volumetric power density (i.e., converter output power divided by converter volume) of designs over time. These are based on a set of assumptions for a given application and should be considered as a general guide to the improvements that can be made rather than as a definitive timeline.

5.2.1.1 Heat Sinks and Heat Transfer Elements

Often, the heaviest and largest components in a power converter are the heat sink and heat transfer elements. The cooling medium (e.g., air, liquid coolant, or two-phase refrigerant) has a significant impact, with liquid cooling giving a more compact design at the cost of additional hardware elsewhere on the aircraft. It is therefore important that design choices are made to minimize the thermal design at the aircraft level through system optimization. Much more detail on this process is found in Chapter 8.

5.2.1.2 Control Platform

For relatively large power converters, the weight and volume of the control platform are usually insignificant relative to other components. Thus, the control platform will not be considered in detail here. However, for the interested reader, the technologies required to minimize the weight and volume of digital and communication circuits are well known from other applications, e.g., mobile phones. These concepts can be applied to power converter control platforms, if/when required by the application.

5.2.1.3 Power Semiconductor Devices and Conductors for Interconnections

Power semiconductor devices themselves are very thin and small with insignificant weight and volume. However, these devices require packaging and reliable connections, so they are typically housed in modules that do have significant volume and can contribute to weight. For aerospace applications, existing techniques that ensure lightweight and low-profile power modules are often employed. This is an area of active technology development for emerging wide-bandgap devices, as discussed in Section 5.4.4.

The design of power semiconductor modules is also essential for maximizing the reliability of power converters, as these are complicated mechanical systems with bonding and connections between many materials with different thermal expansions, something that leads to a host of fatigue conditions. This understanding of the factors and failure mechanisms that influence the lifetime of power converters is essential for predictive maintenance based on usage and environment as well as in the improvement or control of lifetime as a design criterion.

5.2.1.4 Housing and Connectors

Power converters often require some form of housing or box. This adds to the weight and often the volume of the converter and additionally leads to a requirement for connectors. In some applications, it is essential to protect the power converter from external threats, such as water and dust. In other applications, the power converter may be sited in areas where it is necessary to protect other equipment, or even people, in event of a failure. Therefore, housing often has to provide to exclude external threats and contain the converter if there is a catastrophic failure. However, it may be possible to minimize this housing in integrated designs or in applications where the location of the power converter on the aircraft removes this requirement.

5.2.1.5 Passive Components

The size of passive components, which reside both within the power converter topology itself and in input/output waveform filters, can be minimized in a number of ways. Two common methods are creative geometrical design and increasing the switching frequency. While increasing the switching frequency initially seems attractive, this comes with the penalty of increased switching losses and their associated larger cooling and heat transfer components. This trade-off must be assessed in design optimization to ensure that the overall design achieves a minimum weight and volume.

The minimum allowable size of filtering components is usually determined by the required PQ standards. MIL-STD-704F [1] or RTCA DO-160 [2] are the most common standards for low-voltage aerospace applications. PQ standards usually specify the allowable frequency components (i.e., harmonics) and their magnitudes present in the input and output waveforms. This information is used for filter design. If the standards are overly stringent and require lower levels of harmonics/frequency components than are strictly required to maintain system stability and operation, then the filters will be larger than necessary. Reviewing and creating PQ standards for aerospace applications is a process which takes time, but the impact of these standards on the weight and volume of power converters is significant.

5.2.1.6 Gate Drives and Sensors

Power electronic converters require a range of small circuits and sensors for their operation and control. These components all contribute to the weight and volume of the power converter, but, like the control platform, their size is usually insignificant at the power levels required for EAP. For circuits such as gate drives, designs using surface-mounted components and integrated circuits minimize the weight and volume. For sensors, the choice of technology has a significant impact, and minimizing insertion losses and sensor housing sizes is essential to creating compact designs.

5.2.2 Influence of Power System Voltages

The power semiconductor devices used in power converters usually come in a range of standard voltages. Those found in many existing applications use 1,200 V or 1,700 V devices, as these ratings well exceed the maximum transient voltages found in typical aircraft electrical power system (EPS) requirements. With the existing voltage ranges found on aircraft today – e.g., 115 V or 230 V for ac systems and 270 V or 540 V for dc systems – these devices are very suitable and can block the full electrical system voltage when needed. However, as described in Chapter 2, the cable size and to a certain extent the system losses rise as the power level – and, hence, the current – increase.

To briefly reiterate the points: the power P, in W, delivered by the electrical system increases in proportion with the current I, in A, if the system voltage V, in V, is constant:

$$P = IV. \tag{5.1}$$

In addition, the power distribution loss P_{loss}, in W, for a given cable resistance R_w, in Ohms, increases with the square of the current:

$$P_{\mathrm{loss}} = I^2 R_w. \tag{5.2}$$

Therefore, if the current in a system is to be increased, the resistance of the cables needs to be reduced, leading to far larger, heavier cables. To resolve this problem, the system voltage can be increased, thereby reducing the current, according to (5.1). However, if the system voltage on an aircraft is higher than the minimum voltage for air to breakdown and conduct, as specified by Paschen's curve, then design measures will need to be implemented to ensure that this breakdown is avoided. In the low-pressure environment typical of aircraft, this will result in additional measures being required in the design of the insulation and connection systems.

5.2.3 Application-Oriented Integration

In EAP systems, there are significant advantages in combining the housing and thermal design of the power converter with the electrical machine and its associated environment. By taking advantage of this integration, the complete system will become smaller and lighter compared to using stand-alone pieces of equipment, as much of the mechanical structure can be shared. The thermal design must be done in such a way so as to respect the requirements of all the components used, and the design has to consider effects such as vibration and temperature excursions/variations, which have an impact on reliability.

When a power electronic converter is housed within the space around an electrical machine, it is also possible to begin the integration of the design and operation of the motor and power converter, making machines more efficient in operation. Integration of the power electronic converter in close proximity to the motor load also has significant advantages in terms of filtering requirements, particularly at higher frequencies (electromagnetic interference [EMI] frequencies). The shorter cables associated with integrated designs significantly reduce radiated EMI emissions and the need for cable shielding as well as having advantages for conducted emissions due to simplified layouts and minimized parasitic components.

5.2.4 Power Density Targets

Power converter density trends and targets have to compare converters designed and tested to a common set of requirements and operating conditions. As EAP is a relatively new application area, data does not yet exist for historical aircraft platforms, so a range of assumptions and extensions must be used. This section is based on power densities of similar power converter designs and assumes that losses will be dissipated using an air cooling system. It is the trend and relative targets that are important, and near-term power density improvements are expected from integrated designs and new semiconductor device technologies.

The leading aerospace designs available today achieve power densities in the range of 5 kW/kg, when all filtering and thermal components are included. A review of relevant, application-orientated research literature suggests that demonstrators with high technology readiness level (TRL) are available at power densities in the region of 20 kW/kg, using technologies that will certainly be available for MEA applications within the next five years. Looking further out on a 10-year horizon and using integrated designs, a target of 50 kW/kg is certainly a good technology target that might be achievable using emerging power semiconductor device and packaging technologies as well as advances in thermal design.

The rate of improvement in the technology of passive components is considerably slower, as the designs are more mature and the disruptive technologies recently arising for power semiconductor devices (e.g., silicon carbide [SiC]) have not been forthcoming for passive components. Reduction in the weight and volume of passive components is therefore more likely to come from innovation and understanding, leading to better electrical system PQ requirements, rather than in the design of the passive components themselves.

5.3 Power Converter Topologies

Many options exist in the design and control of power electronic converters. This section considers some of the commonly considered power converter topologies for aerospace applications, including those that may be used at higher speeds and/or at high power levels. Most power converters fundamentally consist of switching power semiconductor devices (either fully controlled devices, usually IGBTs or MOSFETS, and uncontrolled devices, diodes) along with passive components (capacitors and inductors).

In aircraft EPSs in both traditionally designed aircraft and MEA, the most common system voltages and frequencies are

- 28 Vdc
- 270 Vdc (±135 Vdc)
- 540 Vdc (±270 Vdc)
- 115 Vac, 400 Hz
- 230 Vac, 400 Hz

These voltages and this frequency have become industry standards,[1] and a range of power converter topologies have been designed based on them. However, as we have seen, these voltages are generally too low to support EAP, and future aircraft will require higher voltages. The power converter topologies chosen for these future applications must therefore focus on the challenges of operating at high system voltages as well as the higher currents required for EAP.

Depending on the design of the electrical generation system and distribution network and the nature of the load, different types of power converters will be required. In this section, we consider some of the possible topologies for dc–ac, ac–ac and ac–dc power converters. In most proposed and in-development EAP systems, the electric machines (which are the main loads in EAP) require an ac waveform; therefore, dc–dc converters are not considered in this chapter.

For aircraft applications using ac systems at power levels greater than, say, 1 kW, a three-phase supply is always used. A three-phase system has the advantage that its power waveform is constant, unlike the pulsating power variation found in a single-phase system. Thus, as this chapter is focused on high-power converters for EAP, only three-phase power converter topologies are considered.

5.3.1 dc–ac Converters

5.3.1.1 Two-Level Power Converters

The most basic form of the dc-to-three-phase-ac power converter is the traditional inverter circuit, as shown in Figure 5.2a. This circuit consists of a relatively large dc capacitor (C_{dc} in the figure) and six fully controlled semiconductor devices (usually IGBTs or MOSFETs), which allow the formation of a switched output voltage waveform at the required fundamental frequency. The higher-frequency switching components in the waveform can be filtered out to give a sinusoidal output current. This filtering effect is often provided by the inductance of the load, which is the case in most electrical machine applications. The basic inverter circuit has many applications in ac motor drive applications as well as feeding ac power systems and loads

[1] A discussion of the history of why these particular values were chosen is given in Chapter 2.

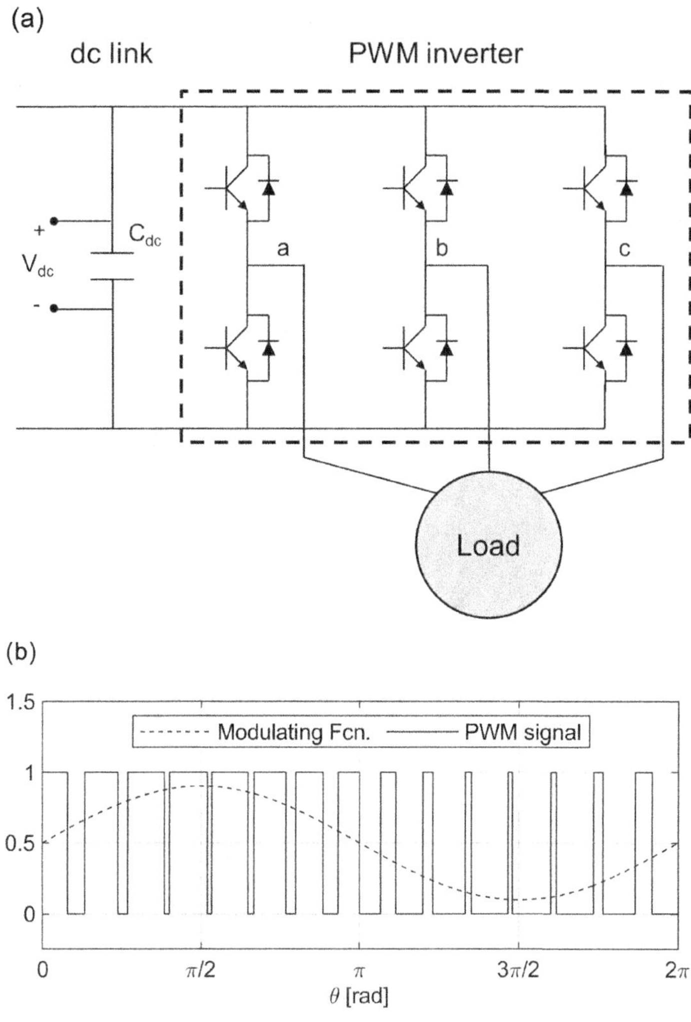

Figure 5.2 (a) The two-level inverter dc–ac topology, and (b) a normalized PWM voltage output waveform from this converter.

from dc sources. A pulse width modulation (PWM) switching scheme is common, and a normalized PWM voltage waveform for a single phase, as well as its modulation signal, are shown in Figure 5.2b. This particular PWM switching scheme is referred to as a *two-level* scheme because the output voltage switches rapidly between two levels.

5.3.1.2 Multi-level Power Converters

The basic two-level inverter circuit is suitable for many applications, but when the system has a voltage that is higher than the blocking voltage of readily available semiconductor switching devices, other solutions must be considered. One solution is to connect a number of semiconductor devices in series in order to support the required blocking voltage; however, this solution has a number of control problems, and it limits the maximum switching frequency of the power converter.

A cleverer solution divides the dc voltage into sections (i.e., levels) and uses these lower-voltage sections to build up an output waveform with a number of steps. The power converters in this family are usually referred to as *multi-level* power converters. A typical three-level power converter topology, the neutral point clamped (NPC) inverter, is shown in Figure 5.3a. Here, V_{dc} is the input dc voltage, which is divided

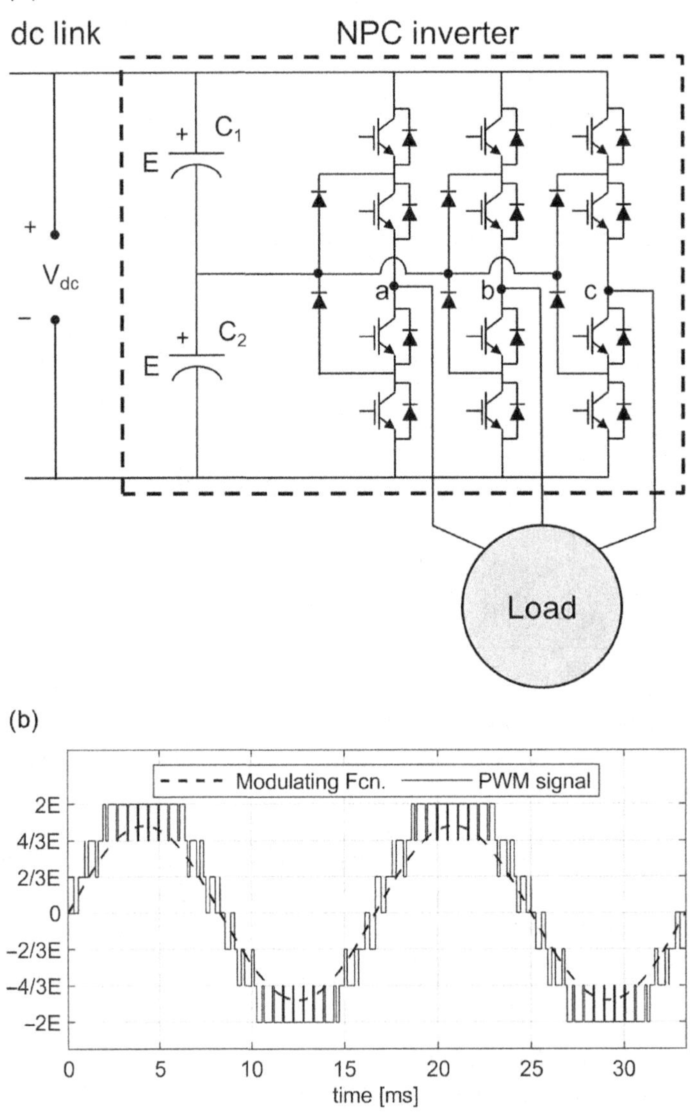

Figure 5.3 (a) The NPC three-level inverter dc–ac topology, and (b) a three-level NPC PWM voltage output waveform.

evenly between the two series input capacitors, C_1 and C_2. The voltage on each capacitor, E, is nominally half of V_{dc}, and the neutral point is "clamped" to all of the switches via six additional diodes. This topology can use a PWM scheme with a three-level waveform (three positive and three negative nonzero voltage levels), resulting in a significantly lower switching frequency for a given waveform quality. A notional waveform for one phase of the output voltage is given in Figure 5.3b.

The NPC topology is typically used for medium voltage/power levels (up to 2 or 3 kV) or in applications where higher-than-usual output frequencies are required in the ac waveform. The latter is a common requirement in aerospace applications, as system optimization based on weight and volume often dictates electrical machines with high rotational speeds. These, in turn, require higher electrical ac excitation frequencies and/or higher pole counts to reduce the diameter of the stator, which again multiplies the required ac excitation frequency. Simply increasing the switching frequency of a traditional inverter topology is always an option, but this leads to an increase in the semiconductor device losses, resulting in larger heat dissipation systems.

The NPC power converter can be extended to more levels, leading to a more complex circuit with significantly more semiconductor devices. This can be a good option if higher power levels and\or voltages are needed, but it will reduce the converter reliability and impose additional internal control requirements. Multi-level power converter topologies other than NPC exist, including the *capacitor clamped* and the *H-bridge* topologies. However, the NPC family is usually the most suitable for aircraft applications, as its design features often lead to an optimal system solution.

5.3.2 ac–ac Converters

For many EAP EPS designs, dc–ac multi-level power converter topologies are considered because the systems use dc supply voltages. However, if the electrical energy is derived from an ac source, it is also possible to directly convert the ac supply waveform to an ac load using an ac–ac converter. ac–ac topologies have the advantage of requiring minimal energy storage elements, usually just a small, high-frequency filter. This gives this solution a significant advantage over dc–ac converters in terms of the size and weight of the required passive components.

A good choice of topology for ac–ac power conversion is the *matrix converter*. This family of converters offers high power density if the switching frequency is large enough to minimize the size of the required input filter. There are two commonly considered types of matrix converters: *direct matrix converters* and *indirect two-stage matrix converters*. These operate in a similar fashion; however, they differ in their loss profiles, and choosing between them often comes down to the aircraft mission profile and its operating conditions.

5.3.2.1 Direct Matrix Converter

The direct matrix converter topology for a three-phase-to-three-phase application is shown in Figure 5.4. It consists of nine bidirectional switches connected so that any input line can be connected to any output line for any period of time. The modulation

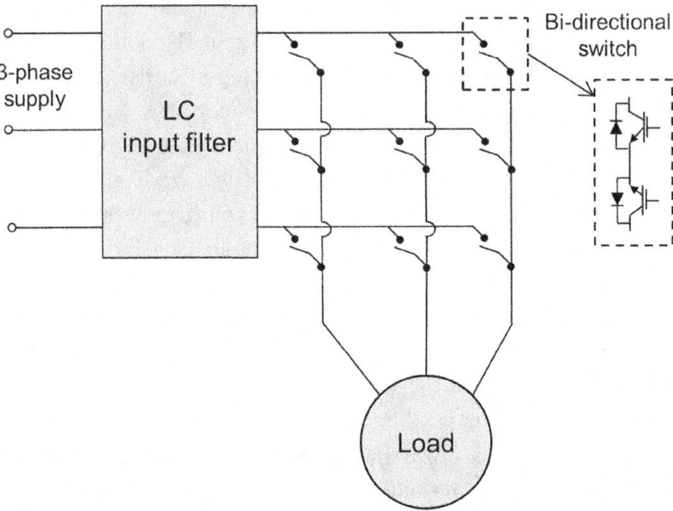

Figure 5.4 Direct matrix converter topology.

techniques used for this topology are similar to those used for dc–ac power conversion except that the converter has to modulate three ac input voltages rather than a single dc voltage. This complicates the modulation techniques slightly, but they are still well within the capabilities of modern control platforms.

The direct matrix converter is modulated in such a way that one and only one input line is connected to each output line at any point in time. In a given configuration, when switches are not transitioning between states, this condition is relatively easy to guarantee. However, during the transitions between configurations, the nonideal nature of semiconductor switching devices means that special consideration has to be given to maintain the requirement, leading to the implementation of a current commutation strategy in the practical power converter. This is not needed in converters in which the current path transfers from a fully controlled device to an uncontrolled device (diode) during a switching event because this leads to a natural current commutation process. However, in a matrix converter, the current path transfers between active, fully controlled semiconductor devices, which drives the requirement to fully control the current commutation process.

5.3.2.2 Indirect Two-Stage Matrix Converter

A typical indirect matrix converter topology for a three-phase-to-three-phase application is shown in Figure 5.5. It consists of a three-phase-to-two-phase direct matrix converter circuit followed by a traditional inverter (a cascade of Figure 5.4 – with a two-phase load and Figure 5.2a). It lacks large energy storage components, which means it has the same potential for high power density as the direct matrix converter. It can be modulated, controlled, and designed in a similar way to the direct matrix converter, and the input side converter also requires a current commutation strategy.

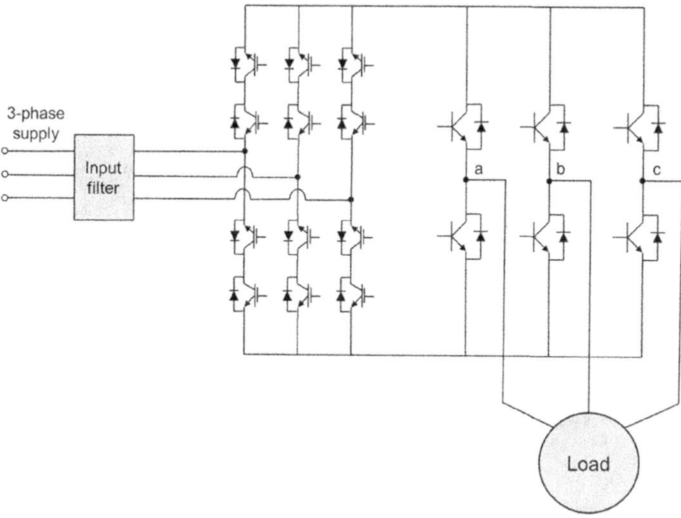

Figure 5.5 Indirect matrix converter topology.

Indirect matrix converters have a different power-loss characteristic compared to direct matrix converters. The full load power losses in a direct matrix converter are lower because only two semiconductors are in the circuit between the input and output of the converter, compared to three in the indirect matrix converter. However, at part loads, particularly under high-torque/low-speed conditions, the losses in the indirect matrix converter are lower. This is because the reactive current is contained in the inverter portion of the circuit, and it does not circulate in the matrix, as it does in the direct matrix converter.

For higher-voltage applications, a range of multi-level matrix converter topologies exists, offering a combination of the advantages of ac–ac power converters and multi-level power converters. Many of these topologies do require the use of some additional passive components and have an increased component count, but in higher-power applications these topologies can offer many advantages. These advanced circuits are outside the scope of this book.

5.3.3 ac–dc Converters

The simplest and most common ac–dc power converter for a three-phase system is the six-pulse diode rectifier bridge and capacitor circuit shown in Figure 5.6a. This circuit has numerous applications, but its input currents are far from sinusoidal and will not meet aerospace PQ standards for anything other than very low power levels. Its input current harmonic magnitudes, shown in Figure 5.6b as a fraction of the dc output current, demonstrate significant harmonic distortion on the ac side. Instead, if only unidirectional power flow is required, then it is common to find 12- or 18-pulse rectifiers (a 12-pulse is shown in Figure 5.6c) for many applications on the aircraft.

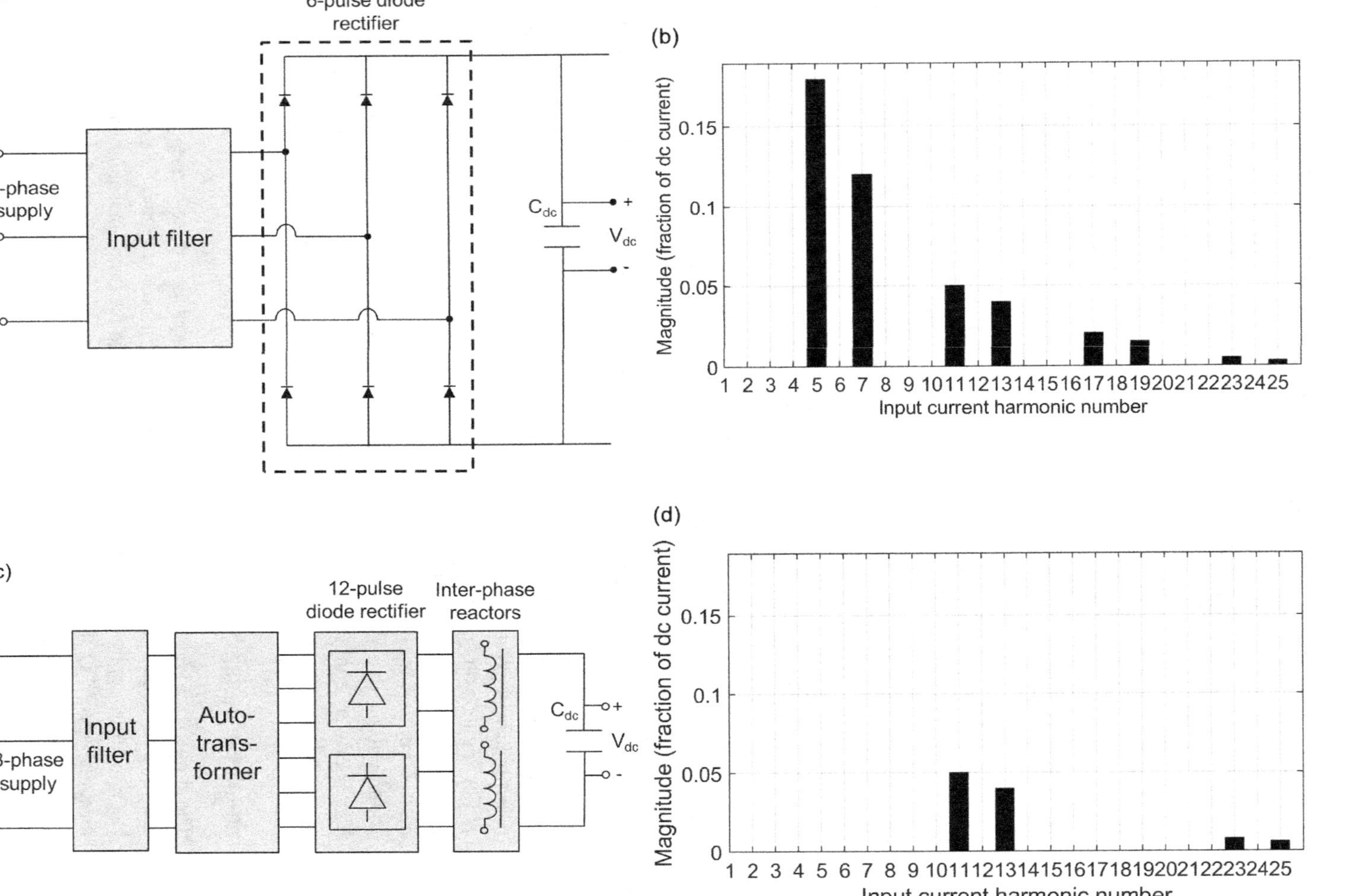

Figure 5.6 ac–dc diode bridge circuits. (a) Three-phase-to-six-pulse rectifier circuit, (b) harmonic content in (a), (c) three-phase-to12-pulse rectifier, (d) harmonic content in (c).

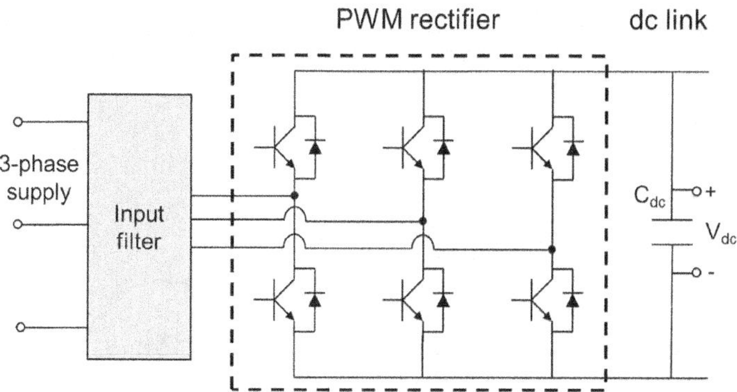

Figure 5.7 ac–dc controlled rectifier.

These shift the phase of the three-phase ac waveforms with an autotransformer and utilize additional diode bridges, which results in significant improvements in the input current waveforms. This is demonstrated by the improved harmonic content – compared to the six-pulse diode rectifier – shown in Figure 5.6d.

A basic inverter can also be used "in reverse" to rectify ac waveforms, as shown in Figure 5.7. This is often referred to as a *controlled rectifier*. The input to the converter must be inductive; therefore, input inductors are usually included and sized to ensure that the input currents meet relevant PQ standards. The passive components in this topology are often significantly larger than those required for the filtering in an ac–ac power converter topology, since the passive components in a matrix converter input filter work together, whereas those in this topology are arranged on either side of the switching components.

5.3.4 Power Converters for Multiphase Electrical Machine Designs

In many industrial applications, electrical machines have three-phase supplies and windings arranged in either a star or delta (wye) configuration. This gives a constant power flow if the waveforms are sinusoidal, resulting in low torque ripple from the machine and minimal low frequency filtering required in the power converter. However, in some high-performance applications or applications requiring relatively high availability, there are advantages at looking at alternative arrangements of machine windings and their associated power converter topologies. In this section, we briefly introduce *open winding machines* and *multiphase machines*.

5.3.4.1 Open Winding Machines

Instead of connecting electrical machine windings in a star or delta configuration, it is also possible to connect power converters to both ends of the windings and leave the windings in an open configuration, as shown in Figure 5.8. This arrangement provides a number of advantages to the converter operation because it is possible to achieve improved waveform quality without increasing individual device losses, albeit at the

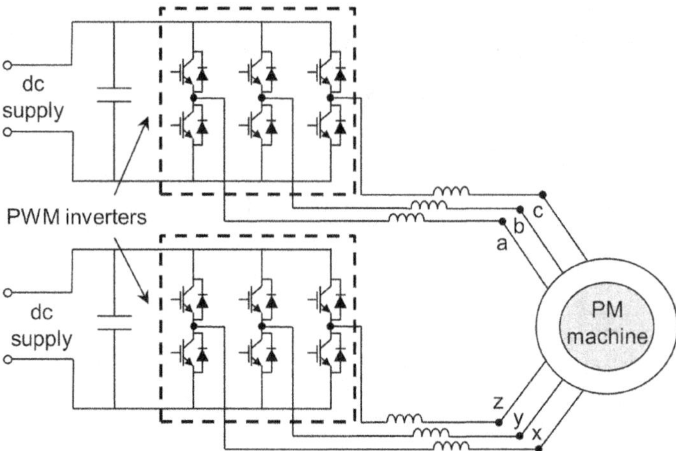

Figure 5.8 A power converter arrangement for an open winding three-phase permanent magnet (PM) machine.

cost of additional semiconductor devices. Feeding both ends of a winding also makes it possible to use unequal dc link voltages. This facilitates modulation patterns that impose multi-level voltage waveforms across the winding, reducing current ripple and losses in the electrical machine.

5.3.4.2 Multiphase Machines

As an alternative to three-phase machines, it is also possible to increase the number of phases in any electrical machine. The main motivation for this is to spread the high power required among more phases. This can provide many advantages, but in terms of the power converter, this allows a number of lower-rated devices to be used to drive one larger electrical machine, giving modularity in the converter manufacturing as well as a degree of fault tolerance. The number of phases chosen should be part of the integrated optimization of the power converter and electrical machine, and multiples of three are common. For the higher-power systems required for EAP, this modular approach also facilitates good design flexibility and modularity in the machine winding configurations (see Chapter 3). Combining these techniques with an open winding configuration has many advantages, and the resulting manufacturability is essential in aerospace applications.

5.3.5 Power Converter Topologies with Fault Tolerance

In many applications, it is beneficial to have a power converter topology that contains a degree of fault tolerance. In some designs, this can be arranged in conjunction with the electrical machine design, as is the case with multiphase machines. However it is also possible to include additional circuitry within the converter topology, to be used when a device fails elsewhere.

One simple solution for a fault-tolerant power converter topology is to provide an additional output leg of the power converter, a replica of the active output phase legs of the converter. This extra output leg can then be switched in to replace any one of the active legs in the event of a failure – using back-to-back thyristors, for example. While this arrangement does improve the availability of the power converter, it does have disadvantages in terms of increasing the complexity of the circuits required as well as having potentially dormant faults, and both characteristics may be undesirable in EAP applications.

5.4 Semiconductor Devices and Materials

In this section we provide an introduction to the most common fully-controlled semiconductor devices and the materials used in their construction. The circuit symbols of the two most common devices, the IGBT and the MOSFET, are shown in Figure 5.9a and b, respectively. The IGBT shown is a PNP transistor combined with an n-channel MOSFET, and the MOSFET is an n-channel device.

5.4.1 Metal Oxide Semiconductor Field Effect Transistor (MOSFET)

The MOSFET is a transistor that is commonly used as a switch in power electronics circuits. It has an insulated gate, which means that the voltage applied between the gate and source connections of the device determine the switching state of the device, a zero voltage turning the device off and a positive voltage turning the device on so that it conducts. One of the advantages of the MOSFET is that in the steady state, this voltage requires very minimal current to sustain the state of the device, although some current is required during the transients to charge/discharge the gate capacitance.

The losses associated with conduction of current in a MOSFET when it is on (commonly referred to as *conduction losses*) are largely associated with the device

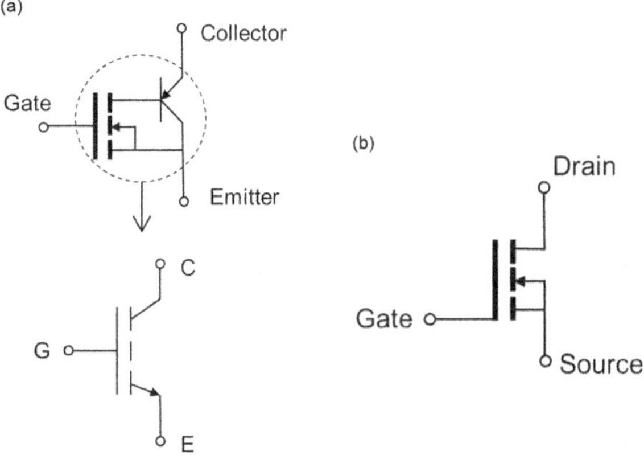

Figure 5.9 (a) IGBT and (b) MOSFET power semiconductor device circuit symbols.

resistance. This resistance is a characteristic of the semiconductor material used in the construction of the device. The average conduction loss can be estimated using Ohm's law, which is also (5.3) customized for the MOSFET variables):

$$P_{\text{loss}} = R_{DS(\text{on})} I_{RMS}^2, \tag{5.3}$$

where $R_{DS(\text{on})}$, in Ohms, is the effective on-state resistance of the MOSFET and I_{RMS}, in A, is its root mean square (RMS) conduction current.

5.4.2 Insulated Gate Bipolar Transistor (IGBT)

The IGBT is a semiconductor device that is almost exclusively used as a switch in power electronic circuits. Structurally, it is a combination of a bipolar junction transistor (BJT) and a MOSFET, having the current-carrying capabilities of the BJT but with the voltage-driven gate feature of a MOSFET. The IGBT was patented in the late 1960s but did not gain popularity until the 1990s. Today, it is the silicon (Si) device of choice in many larger power converter applications. It is available in devices with voltage ratings up to 6.5 kV and comes packaged in modules with devices arranged in parallel in order to conduct thousands of amps.

IGBT conduction losses are predominately associated with the forward voltage drop across the device, with a secondary effect of the device resistance. The conduction losses in an IGBT can therefore be estimated as

$$P_{\text{loss}} = \left(V_{CE(sat)} + R_{DS(\text{on})}\right) I_{RMS}, \tag{5.4}$$

where $V_{CE(sat)}$ is the on-state forward voltage drop across the IGBT (collector-to-emitter voltage).

5.4.3 Switching Losses

When semiconductor devices are switched, there is a transition time for both the current and voltage. This finite transition time results in voltage across and current through the device simultaneously, which leads to losses. In most power converter designs, these *switching losses* are proportional to the device switching frequency and can be estimated based on the turn-on and turn-off energies of the device:

$$P_{\text{loss}} = W_{\text{total}} \cdot f_{sw} = \left[\frac{V_{op}(I_{\text{on}} t_{\text{on}} + I_{\text{off}} t_{\text{off}})}{2}\right] f_{sw}, \tag{5.5}$$

where W_{total}, in J, is the total switching energy lost during one switch period; f_{sw}, in Hz, is the average device switching frequency; V_{op} is the operating voltage (assumed constant here); I_{on} and I_{off} are, respectively, the device currents at turn-on and turn-off; and t_{on} and t_{off}, in s are, respectively, the device turn-on and turn-off times.

The device switching and conduction losses are combined and summed for a given power converter topology under a defined operating condition that gives the total losses that must be processed by the thermal management system. An example is shown for a 5 kW converter in Figure 5.10.

5.4.4 Semiconductor Device Materials

For the last few decades, nearly all semiconductor devices for power electronic applications have been based on Si-based technologies. These include the IGBTs, MOSFETs, and diodes used in modern, switching power converters. However, Si-based devices have severe limitations when it comes to EAP, particularly in their temperature limits and corresponding large device areas required for high currents. Thus, in recent years, there has been increased interest and investment in alternative semiconductor device materials for power electronic applications. SiC and gallium nitride (GaN) are the two main emerging technologies. SiC is somewhat farther along in its development, and we briefly discuss it here. Although GaN is not covered here, it is quickly gaining ground on SiC and will likely play a large role in EAP.

For aerospace applications, SiC is an emerging wide-bandgap device material that has proven advantages. A comparison of Si and SiC material properties is given in Table 5.1. Compared to Si, the device area required for a given current level is

Table 5.1 Comparison of Si and SiC material properties.

Material Property	Si	SiC
Bandgap [eV]	1.1	3.25
Thermal conductivity [W/mK]	150	390
Maximum operating temperature [°C]	600	1,580
Breakdown electric field [MV/cm]	0.3	3.3
Electron saturation velocity [cm/s]	1.0×10^7	2.0×10^7

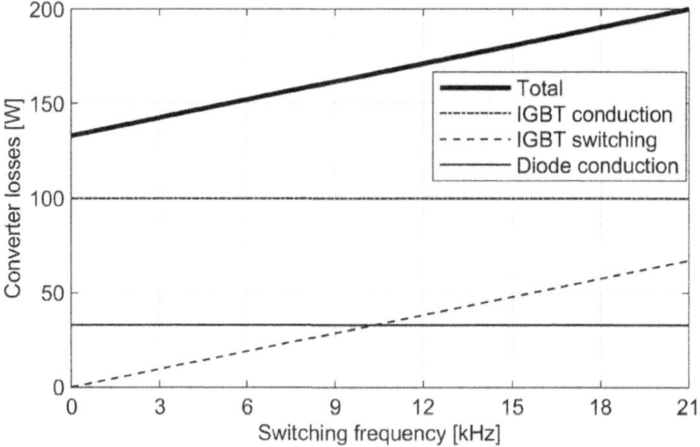

Figure 5.10 Device switching and conduction losses vs. switching frequency for a 5 kW converter.

significantly smaller, the device on-resistance is typically an order of magnitude lower (reducing conduction losses in MOSFETs), and the devices switch more quickly (reducing switching losses). The device material can also operate at significantly higher temperatures, which means that SiC devices are limited by their packaging and connection properties rather than the SiC material properties.

5.5 Summary

This chapter has introduced the power converter topologies and designs relevant to EAP. Critical design aspects for power electronic converters have been considered, and commonly used semiconductors have been described. Advances in both power electronic converter designs and the devices available for these designs have enabled large technology advancements across many electrified transportation platforms – including automotive, marine, and rail. Similar advances, enabled by power electronics, are expected in EAP. Power electronics is the enabling technology for all future electrical options in aircraft design. Indeed, the goals and forecasted benefits described in Chapter 9 rely fully on the power density and reliability targets of future power electronic converter solutions.

Abbreviations

ac	alternating current
AEA	all-electric aircraft
BJT	bipolar junction transistor
dc	direct current
EAP	electrified aircraft propulsion
EMI	electromagnetic interference
EPS	electrical power system
FPGA	field-programmable gate array
IGBT	insulated gate bipolar transistor
MEA	more-electric aircraft
MOSFET	metal oxide semiconductor field effect transistor
NPC	neutral point clamped
PM	permanent magnet
PQ	power quality
PWM	pulse width modulation
RMS	root mean square
Si	silicon
SiC	silicon carbide
TRL	technology readiness level

Variables

f_{sw}	switching frequency, [Hz]
I	current, [A]
I_{off}	current during device turn-off, [A]
I_{on}	current during device turn-on, [A]
I_{RMS}	RMS current, [A]
P	power, [W]
P_{loss}	power loss (ohmic), [W]
$R_{DS(\text{on})}$	on-state resistance, [Ohms]
R_w	cable/wire resistance, [Ohms]
t_{off}	device turn-off time, [s]
t_{on}	device turn-on time, [s]
V	voltage, [V]
$V_{CE(sat)}$	on-state forward voltage drop, [V]
V_{op}	operating voltage, [V]
W_{total}	total switching energy lost per switch period, [J]

References

[1] Department of Defense Interface Standard: Aircraft Electric Power Characteristics, MIL-STD-704F, 2004.

[2] RTCA: Environmental Conditions and Test Procedures for Airborne Electronic/Electrical Equipment and Instruments, DO-160G1, December 2014.

6 Cryogenic Power Electronics

Zheyu Zhang, Fei (Fred) Wang, Ruirui Chen, and Handong Gui

Introduction

The previous chapter emphasized the essential role power electronics play in aircraft electrification, providing a survey of widely used circuits and components. The push for electrification – in particular, electrification of the propulsion system – requires power circuits capable of power levels and efficiencies well beyond their current capabilities. It is this reality that motivates this chapter, an introduction to and survey of cryogenic power electronics. As we will see, this technology has to the potential to completely alter aircraft powertrain design.

The increasing electrification of aircraft functions will only intensify in the future. Hybrid power generation and distributed propulsion have been identified as transformative aircraft configurations for future commercial transport vehicles with reduced fuel burn and emissions. Unfortunately, a major hurdle to realizing the promised benefits is that the necessary components are not available in the required power, efficiency, and power density ranges for transport-class aircraft. Accordingly, the development of high-efficiency, high-specific-power (HSP) power electronics intended for eventual use in electrified aircraft propulsion (EAP) drives is critical [1].

Much like the cryogenic electric machines detailed in Chapter 4, cryogenic power electronics offer numerous game-changing benefits. This chapter focuses on several key perspectives for the development of cryogenic power electronics from the component up to converter level. First, the characterization of critical components, including power devices and magnetics, at cryogenic temperature is introduced to establish the basic knowledge necessary for cryogenic design and optimization. Second, special considerations, trade, and design studies of cryogenic power stage and filter electronics are discussed. Third, an example of a high-power cryogenically cooled inverter system for aircraft application is illustrated, with safety considerations and the protection scheme highlighted.

6.1 Benefits of Cryogenic Power Electronics

The main contributors to power loss and weight in aircraft power converters are the power stage and the electromagnetic interference (EMI) filter. As illustrated

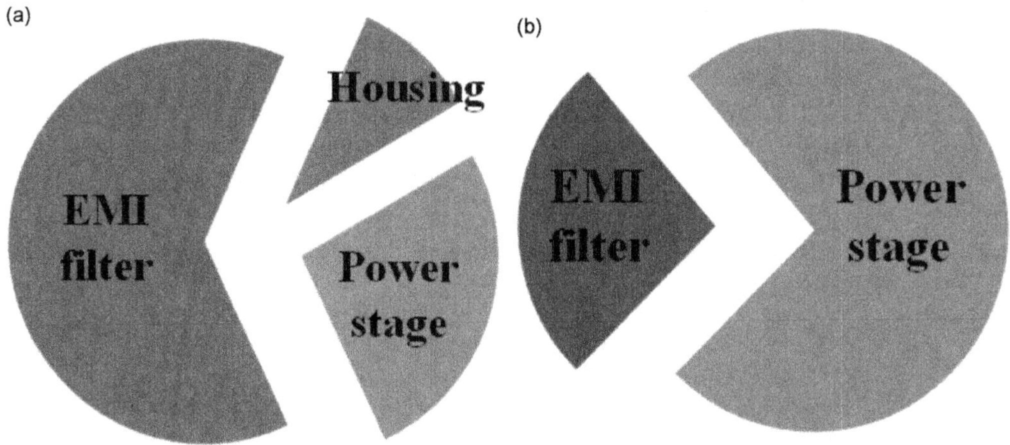

Figure 6.1 A design example of 1 MW inverter for aircraft application.

in Figure 6.1, an example 1 MW inverter system clearly shows these two components as the primary barriers for the system to achieve a desired efficiency and power density. Cryogenic power electronics, the focus of this chapter, provide benefits in both loss minimization and weight reduction thanks to the following features:

1. Numerous power semiconductor devices, both silicon (Si) and germanium (Ge) based, have improved performance at low temperature, offering decreased specific on-state resistance and increased switching speed [2–7]. Therefore, both conduction loss and switching loss of power devices using cryogenically cooled power electronics are reduced, leading to the improved power stage efficiency.
2. Power devices at cryogenic temperatures can be operated at faster switching frequencies, greatly reducing the need for passive – e.g., electromagnetic interference (EMI) – filtering; thereby reducing filter weight. Additionally, conductor (e.g., copper and aluminum, which are widely used as inductor windings) resistivity significantly reduces as temperature decreases. Hence, without redesigning the thermal management system, inductor windings can carry much higher current density, leading to decreased usage of the conductor materials. Furthermore, a superconducting inductor technique can be utilized at the dc side of the power electronics to reduce the weight and loss of EMI filter.
3. The cryogenic cooling system can further improve the efficiency and power density of inverter systems. This is because less cooling is required at extremely low ambient temperatures, and light and/or efficient busbar designs can be utilized due to the low resistivity of conductors at cryogenic temperature.

Beyond the component-level advantages highlighted before, system-level benefits are enabled by superconducting technologies. Cryogenic motors/generators, along with their supporting power electronics, will grow in importance. Integrating the power electronics into the superconductive motor/generator systems can avoid extra thermal insulation and temperature regulation systems, which in turn reduces the system complexity and improves the power density.

Generally speaking, the temperature can be called *cryogenic* when it is lower than 123 K. This low-temperature environment can be created with the help of liquefied natural gas (111 K), liquid nitrogen (77 K), liquid hydrogen (20 K), or liquid helium (4 K). Although cryogenic power electronics can be used in a high-performance power conversion system, development of high-density power electronics operating efficiently and reliably with a cryogenic cooling system is challenging due, in part, to the following:

1. There is relatively limited knowledge of component performance at cryogenic temperature.
2. Considering the previous point, special design considerations for both the power stage and filter components are needed.
3. The safety considerations and protection schemes are unique.

6.2 Component Characterization at Cryogenic Temperature

Key components in a power converter mainly include active power devices and passive filters. Thus, the following section summarizes the performance of power devices and magnetics, including the characterization method at cryogenic temperature and the performance of several typical candidates suitable for high-power aircraft applications.

6.2.1 Power Devices

For high-power aircraft propulsion, the candidate active power devices at standard ambient temperature include Si metal-oxide-semiconductor field-effect transistors (MOSFETs), Si insulated gate bipolar transistors (IGBTs), Si carbide (SiC) MOSFETs, and gallium nitride (GaN) high electron mobility transistors (HEMTs). Extensive literature exists detailing the characterization of these devices at cryogenic temperature. SiC MOSFETs have proven to have increased on-resistance at low temperatures, which makes them unsuitable for cryogenic applications [8–12]. GaN HEMTs have shown improved static on-resistance and constant breakdown voltage at cryogenic temperatures compared to that at room temperature [13–16]. However, additional switching loss has been observed when the temperature drops. Although not fully understood, this may be attributable to dynamic on-resistance caused by the current collapse phenomenon, which significantly deteriorates the efficiency improvement from the reduction of static on-resistance, especially at high switching frequencies [17–19]. Si IGBTs show improved switching performance at cryogenic temperatures and also lower forward voltage drop [4, 20, 21]. However, their switching speed and conduction loss still cannot rival that of MOSFETs, and it is hard to use them for the applications with switching frequency higher than 100 kHz.

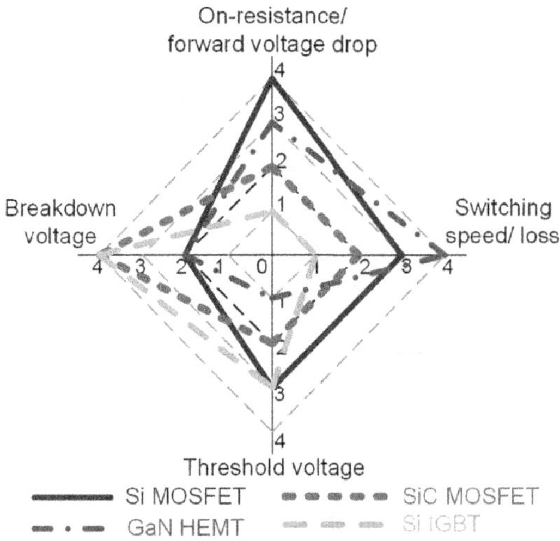

Figure 6.2 Comparison of typical power devices' performance at cryogenic temperature versus room temperature. © [2020] IEEE. Reprinted, with permission, from [26]

Yet Si MOSFETs show decreased on-resistance and higher switching speed at cryogenic temperatures [5, 22–25].

A comparison of the performance of different devices at cryogenic temperatures versus room temperature is illustrated in Figure 6.2. The higher the point is, the better the device performance at cryogenic temperatures. Accordingly, it can be observed that Si MOSFETs are good candidates to achieve high efficiency in high-power, high-frequency applications. Considering the limited data available for these devices at cryogenic temperature, additional comprehensive device characterization is crucial for full understanding of the available design options. In the next section, this will be addressed.

6.2.1.1 Power Device Characterization Testing

To characterize power devices under cryogenic temperature, a temperature-controlled environment and monitoring system are required. The following discussion focuses on the liquid nitrogen-based cryogenic temperature of 77 K.

A cryogenic chamber with an attached liquid nitrogen Dewar can be employed to establish the cryogenic testing platform, as shown in Figure 6.3a. The temperature inside the chamber is regulated by the temperature control panel on the front of the chamber. During the test, the liquid nitrogen from the Dewar is injected through an insulated aluminum duct, entering the chamber as vapor. Due to the mechanics of the temperature chamber, the controllable temperature range of the chamber for this setup is limited to a threshold above 93 K.

To evaluate device performance at 77 K, liquid nitrogen bath testing is conducted as illustrated in Figure 6.3b. During the test, a small amount of liquid nitrogen is first poured into a Styrofoam bucket so that the vaporized nitrogen gas fills the container and creates an inert environment to avoid condensation. Then the device under test (DUT) is

(a)

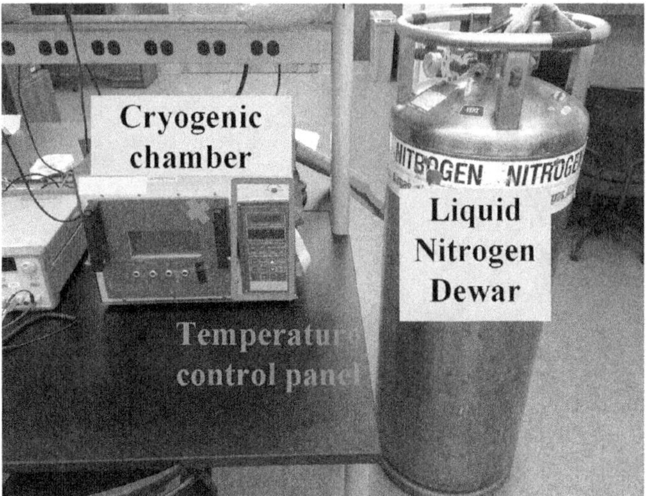

(b)

(c)

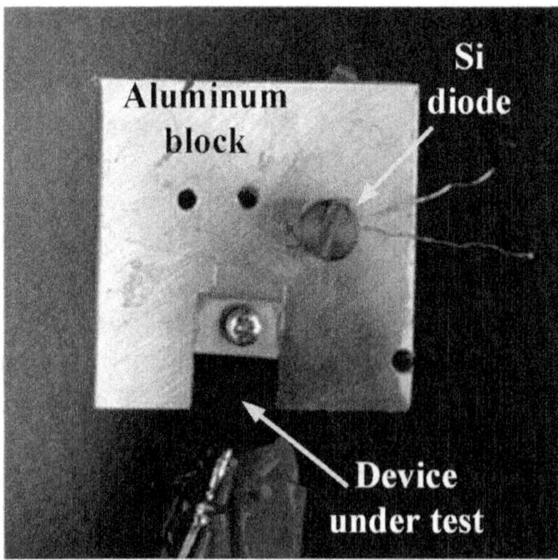

Figure 6.3 Temperature control for component characterization at cryogenic temperature. (a),(c) © [2017] IEEE. Reprinted, with permission, from [22]

placed in the container. Afterward, more liquid nitrogen is slowly added until the DUT is submerged to ensure the characterization is conducted at 77 K. A Si diode is used as the cryogenic temperature sensor to verify the testing temperature of the device.

To maintain the temperature of the DUT during the characterization, an aluminum block is used, as shown in Figure 6.3c. The device is attached to the top of the aluminum block with the temperature sensor in close proximity so that stable temperature with accurate measurement can be confirmed.

The characteristics of power devices at cryogenic temperature significantly affect both the static and dynamic performance of a cryogenically cooled power converter. Static characteristics include on-state resistance and body diode forward voltage drop (if applicable) for conduction loss calculation, leakage current, and breakdown voltage for voltage margin assessment. Dynamic characteristics include switching time for switching frequency and dead-time setting and energy loss for switching loss evaluation.

For the static characterization of the device in the current example test, a Keysight B1505A curve tracer is employed, and a Kelvin connection is used to guarantee the accuracy of the measurement as shown in Figure 6.4a. The testing platform is shown in Figure 6.4b.

(a)

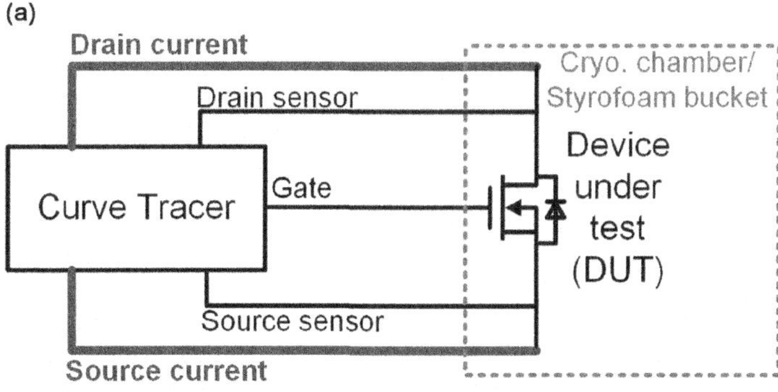

(b)

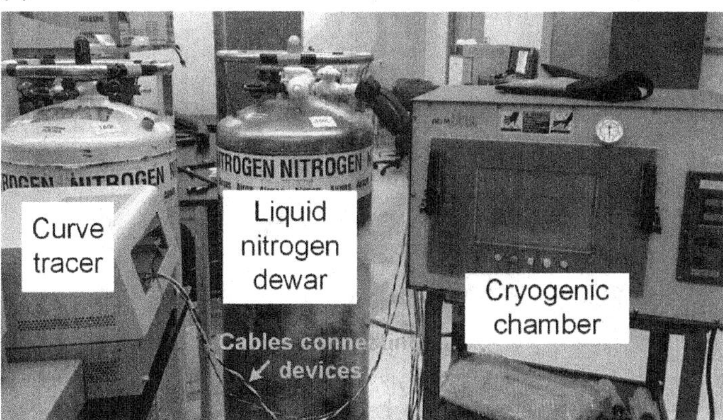

Figure 6.4 Static characterization at cryogenic temperature. © [2020] IEEE. Reprinted, with permission, from [26]

For the dynamic characterization, a double pulse test (DPT) is adopted to get the switching performance. The configuration and testing platform of the DPT is illustrated in Figure 6.5. It should be noted that in addition to the DUT, it is suggested to locate the gate drive, dc bus capacitors, signal isolator, and other auxiliary circuits inside the chamber at cryogenic temperature; otherwise, the parasitics introduced by interconnecting the DUT with the neighboring components can cause severe ringing and noise during the switching transient. Also, this setup mimics the actual cryogenically cooled power electronics converter.

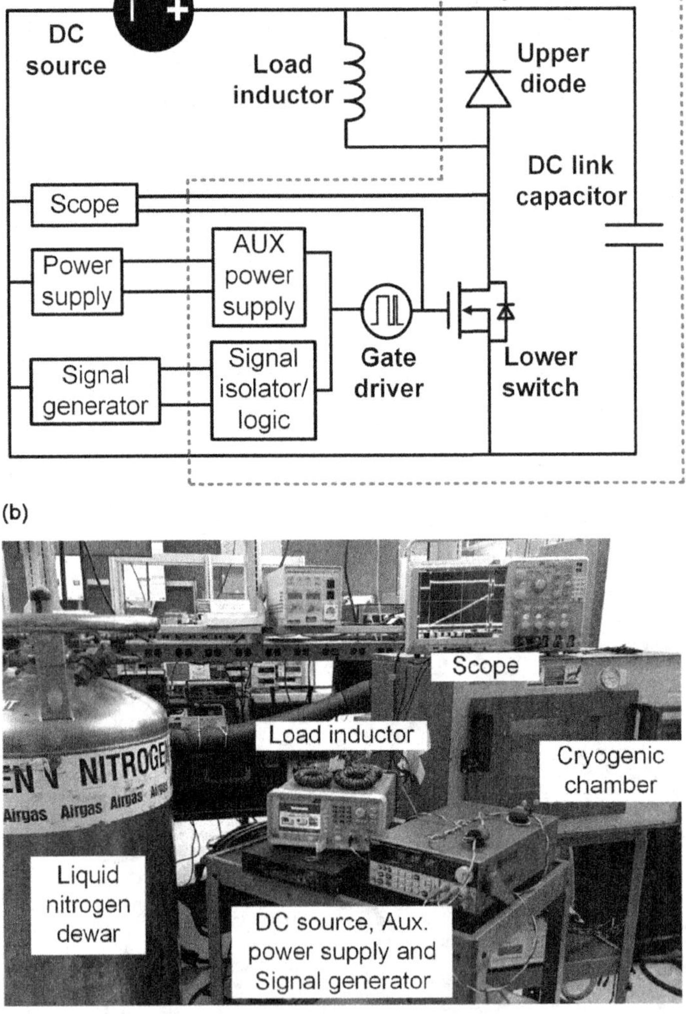

Figure 6.5 Dynamic characterization at cryogenic temperature. © [2020] IEEE. Reprinted, with permission, from [26]

6.2.1.2 Power Device Performance at Cryogenic Temperature

As previously discussed, among candidate semiconductor-based switching devices, Si MOSFETs exhibit preferred characteristics at cryogenic temperature. Therefore, performance of two Si MOSFETs from Infineon with different performance attributes are summarized and compared based on the aforementioned characterization method in Figure 6.6, while more comprehensive characteristics of the state-of-the-art SiC MOSFETs and GaN HEMTs can be found in [27]. Specifically, the two Si MOSFET presented are the 600 V, 50 A IPZ60R040 C7 CoolMOS and the 900 V, 35 A IPW90R120 C3 CoolMOS. In Figure 6.6, their performance is summarized with respect to the on-resistance, breakdown voltage, body diode forward voltage, transconductance, threshold voltage, and turn-on and turn-off losses.

The on-resistance dependence on junction temperature is illustrated in Figure 6.6a. It is observed that the on-resistance decreases from room temperature to around 100 K thanks to the increased carrier mobility at low temperature. However, the on-resistance starts to increase when temperature further reduces. This can be explained by the reduced number of carriers being available, which is referred to as carrier freeze-out [28]. Note that although the C3 CoolMOS exhibits much higher on-resistance than that of C7 CoolMOS at room temperature, it decreases faster as temperature decreases and is even lower than that of the C7 CoolMOS close to 77 K.

The body diode forward voltage under the drain current of 50 A is plotted in Figure 6.6b. It shows that both Si MOSFETs have an increased forward voltage trend as temperature decreases. This is because of an increased potential barrier height at the *P*+ –*N*- junction induced by a reduction in the intrinsic carrier concentration at the low temperature. Also note that the C3 CoolMOS has lower body diode forward voltage drop than that of the C7 CoolMOS.

The measured breakdown voltage is given in Figure 6.6c. The breakdown voltage decreases because the mean free path of the carrier increases at the low temperature, which contributes to higher-impact ionization efficiency. As a result, more electron–hole pairs with high energy are created to launch impact ionization and the avalanche is enhanced. According to the test results, the breakdown voltage of the tested Si MOSFETs at 77 K is only 80 percent of that at room temperature.

The threshold voltage of two MOSFETs, which increases at cryogenic temperature because of the reduction of intrinsic carrier concentration, is shown in Figure 6.6d. The transconductance, which first increases in the range from room temperature to about 100 K and then decreases when the temperature continues to decrease, is illustrated in Figure 6.6e. The reason is similar to that of on-resistance, which is caused by carrier freeze-out.

To illustrate switching performance, the switching loss of two DUTs with the same gate resistance is shown in Figure 6.6f. Note that due to their different voltage ratings, the dc bus voltage is different for the two MOSFETs. It is 250 V for the 600 V C7 MOSFET and 500 V for the 900 V C3 MOSFET. Results show that switching loss decreases as temperature drops, but the decreasing trend becomes slower after 125 K. This is influenced by both the inversion layer mobility and the transconductance.

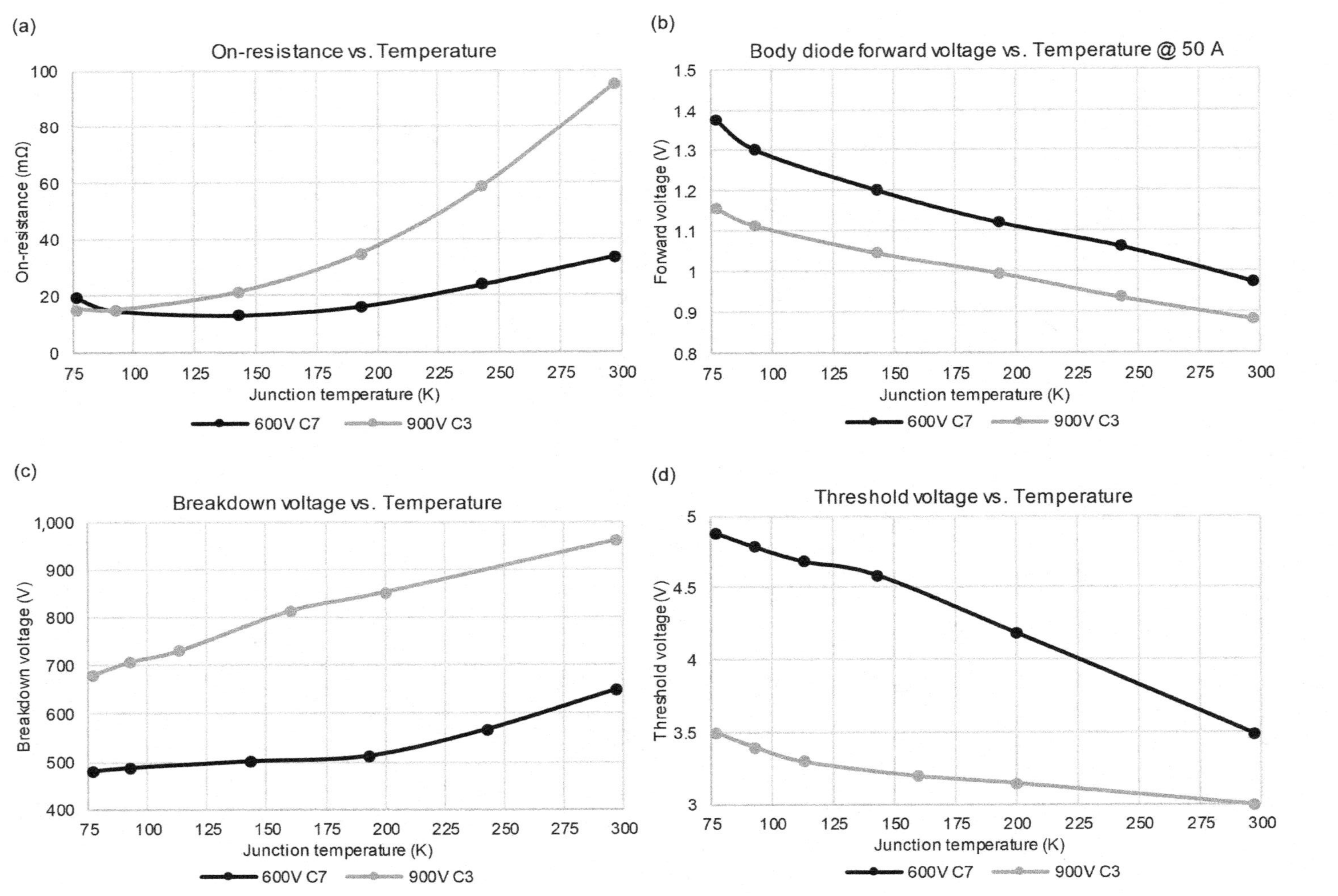
(a)
On-resistance vs. Temperature
On-resistance (mΩ)
0
20
40
60
80
100
75
100
125
150
175
200
225
250
275
300
Junction temperature (K)
600V C7
900V C3
(b)
Body diode forward voltage vs. Temperature @ 50 A
Forward voltage (V)
0.8
0.9
1
1.1
1.2
1.3
1.4
1.5
75
100
125
150
175
200
225
250
275
300
Junction temperature (K)
600V C7
900V C3
(c)
Breakdown voltage vs. Temperature
Breakdown voltage (V)
400
500
600
700
800
900
1,000
75
100
125
150
175
200
225
250
275
300
Junction temperature (K)
600V C7
900V C3
(d)
Threshold voltage vs. Temperature
Threshold voltage (V)
3
3.5
4
4.5
5
75
100
125
150
175
200
225
250
275
300
Junction temperature (K)
600V C7
900V C3

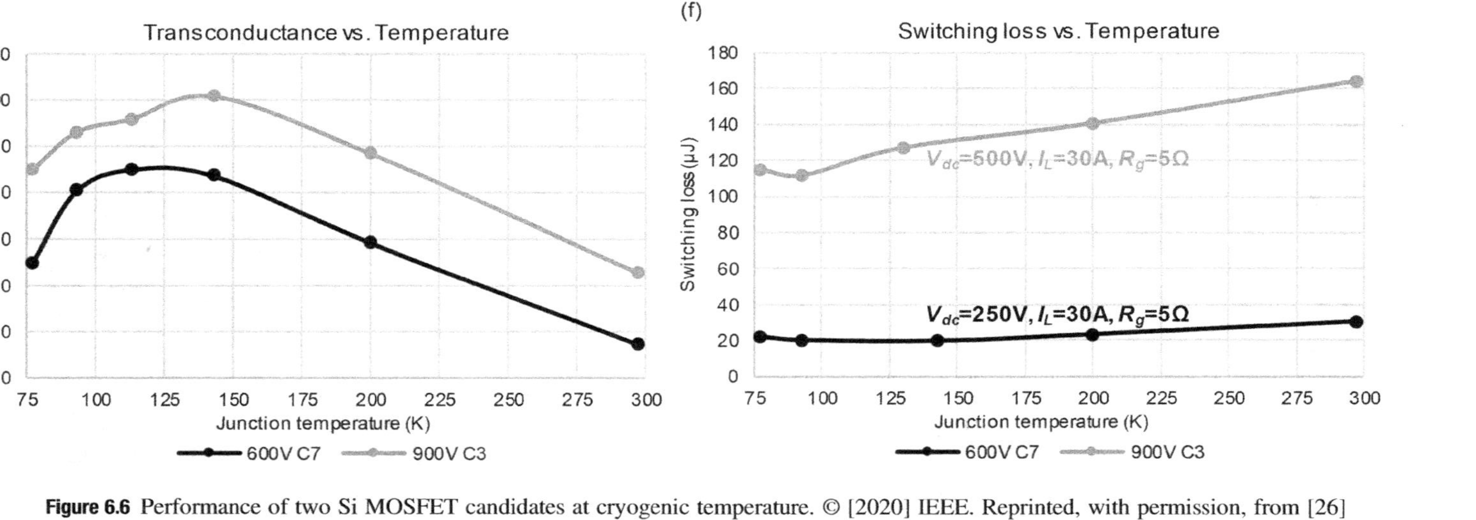

Figure 6.6 Performance of two Si MOSFET candidates at cryogenic temperature. © [2020] IEEE. Reprinted, with permission, from [26]

6.2.2 Magnetic Devices

Similar to power devices, the available data for magnetic components at cryogenic temperature are limited [29–32]. A characterization method for magnetic devices at cryogenic temperature is introduced in what follows. The performance of two typical magnetic core materials, ferrite and nanocrystalline, at cryogenic temperature are summarized in [33].

6.2.2.1 Magnetic Device Characterization Testing

The most important magnetic core characteristics for power electronic converter inductor design include *core loss* P_c, in W; *saturation flux density* B_s, in T; and *relative permeability* μ_r – all of which can be obtained from the B-H curve of the magnetic material. The test circuitry for magnetic core characterization is shown in Figure 6.7a. A sinusoidal excitation generated by the signal generator is applied to the primary winding of the core through a power amplifier, inducing a voltage on the secondary winding. The two windings need to be well coupled to reduce leakage inductance. The *magnetic field strength H*, in A-turns/m, can be calculated using the primary side current, using (6.1), where n_1 is the *number of primary turns*; I_1 is the *primary side current*, in A; and l_e is the *core mean magnetic path length.*, in m. *Flux density B*, in T, is obtained from the secondary side voltage using (6.2), where n_2 is the *number of secondary turns*; v_2 is the *secondary side voltage*, in V; and A_e is the *core*

(a)

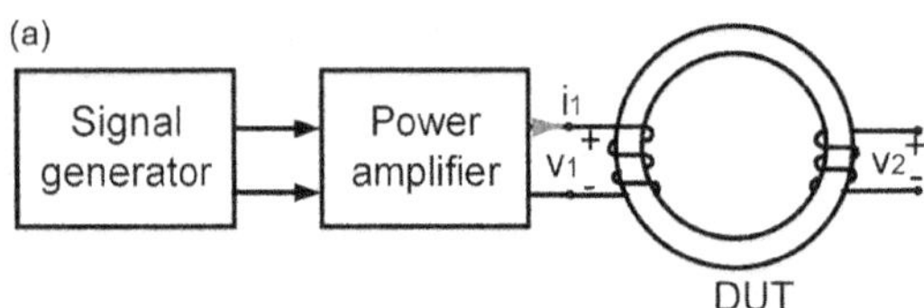

(b)

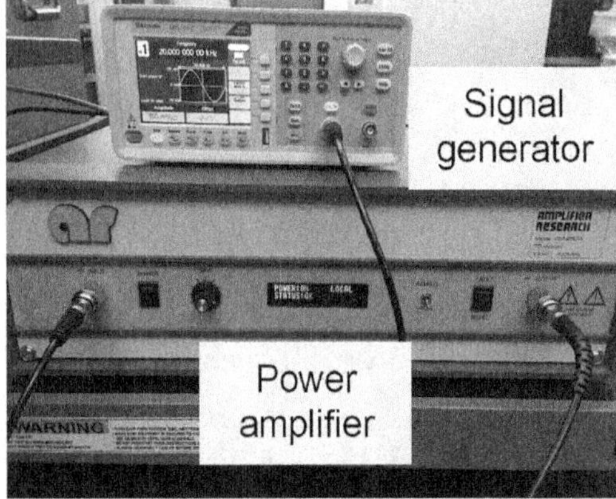

Figure 6.7 B-H curve characterization at cryogenic temperature. © [2018] IEEE. Reprinted, with permission, from [33]

effective cross-sectional area, in m^2. Hence, *core energy loss E*, in J, is calculated from (6.3), which is the loop area of the B-H curve:

$$H = \frac{\eta_1 I_1}{l_e} \tag{6.1}$$

$$B = \frac{1}{n_2 A_e} \int v_2 dt \tag{6.2}$$

$$E = \oint H dB. \tag{6.3}$$

To characterize the B-H curve at cryogenic temperature, similar to the setup illustrated in Figure 6.3, an environmental chamber and the liquid nitrogen Dewar are utilized. The testing platform for magnetic core characterization is shown in Figure 6.7b.

6.2.2.2 Magnetic Device Performance at Cryogenic Temperature

Two commonly used magnetic materials, ferrite N87 and nanocrystalline core Vitroperm 500 F, are selected for the following case study to demonstrate the performance of magnetic cores at cryogenic temperature. Since the B-H curve is a function of temperature, frequency, and flux density, two materials are tested in the range from room temperature to cryogenic temperature. Under each temperature point, the B-H curve test is conducted with varying sinusoidal exciting frequencies from 3 kHz to 500 kHz and different levels of flux density until the core is saturated. The influence of low temperature on magnetic core characteristics can thus be observed.

Figure 6.8a shows the Ferrite N87 B-H curves' dependence on temperatures under the operating point of 0.1 T and 100 kHz. The corresponding core loss as a function of temperature is presented in Figure 6.8b. It can be found that the core loss increases by a factor of 13 as temperature decreases from 298 K to 93 K.

Similarly, for the nanocrystalline Vitroperm 500F material, B-H curves at different temperatures at 0.5 T and 20 kHz are given in Figure 6.8c. The corresponding core loss as a function of temperature is summarized in Figure 6.8d. The core loss doubles as temperature decreases from 298 K to 93 K.

The core loss dependence on temperature of Ferrite N87 under more operating frequencies and flux density levels is summarized in Figure 6.9. A similar trend can be found that the core loss increases more than 10 times when the temperature decreases from room temperature to cryogenic temperature.

The core loss of Vitroperm 500F as a function of temperature and flux density at frequencies of 20 kHz, 70 kHz, 100 kHz, and 200 kHz is presented in Figure 6.10. Similar trends can be observed that core loss increases approximately between 1.5 and 2.5 times at the cryogenic temperature as compared to that at room temperature.

The saturation flux density B_s is identified from the B-H curve; it is the value of flux density that is approximately constant with respect to the exciting field. The B-H curve of Ferrite N87 at 20 kHz with a large exciting source applied at different low

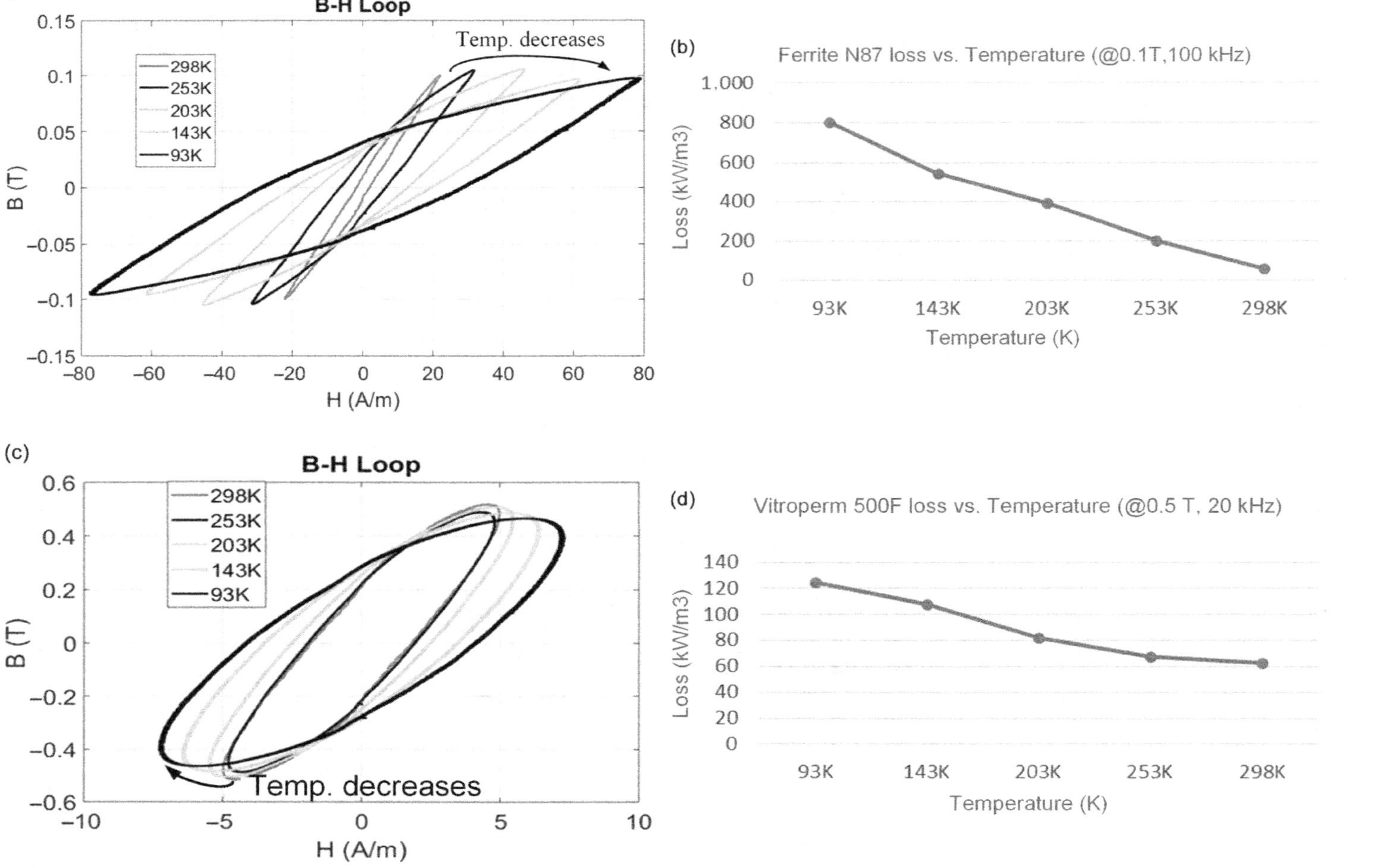

Figure 6.8 Magnetic characteristics dependence on temperature. © [2018] IEEE. Reprinted, with permission, from [33]

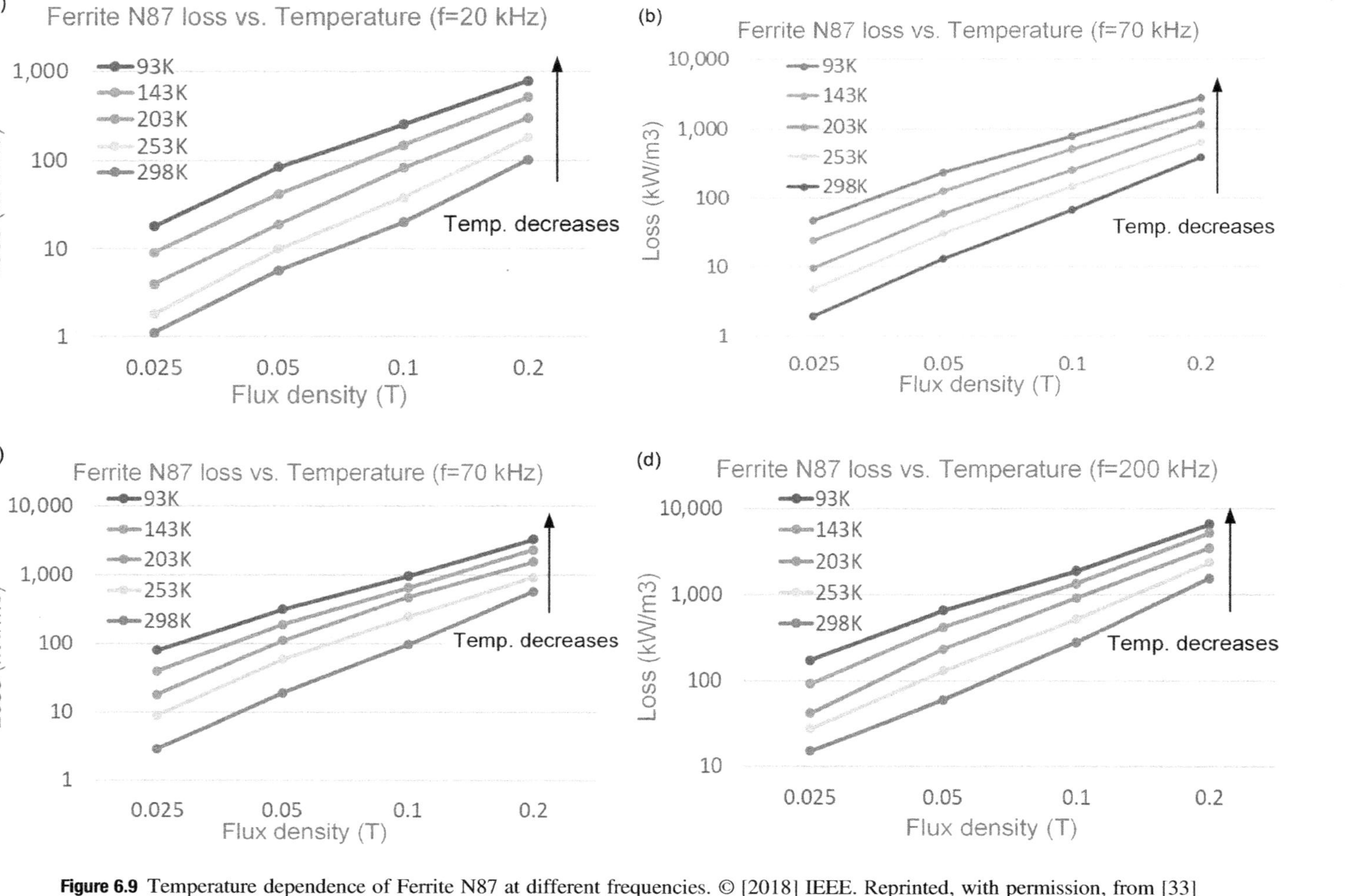

Figure 6.9 Temperature dependence of Ferrite N87 at different frequencies. © [2018] IEEE. Reprinted, with permission, from [33]

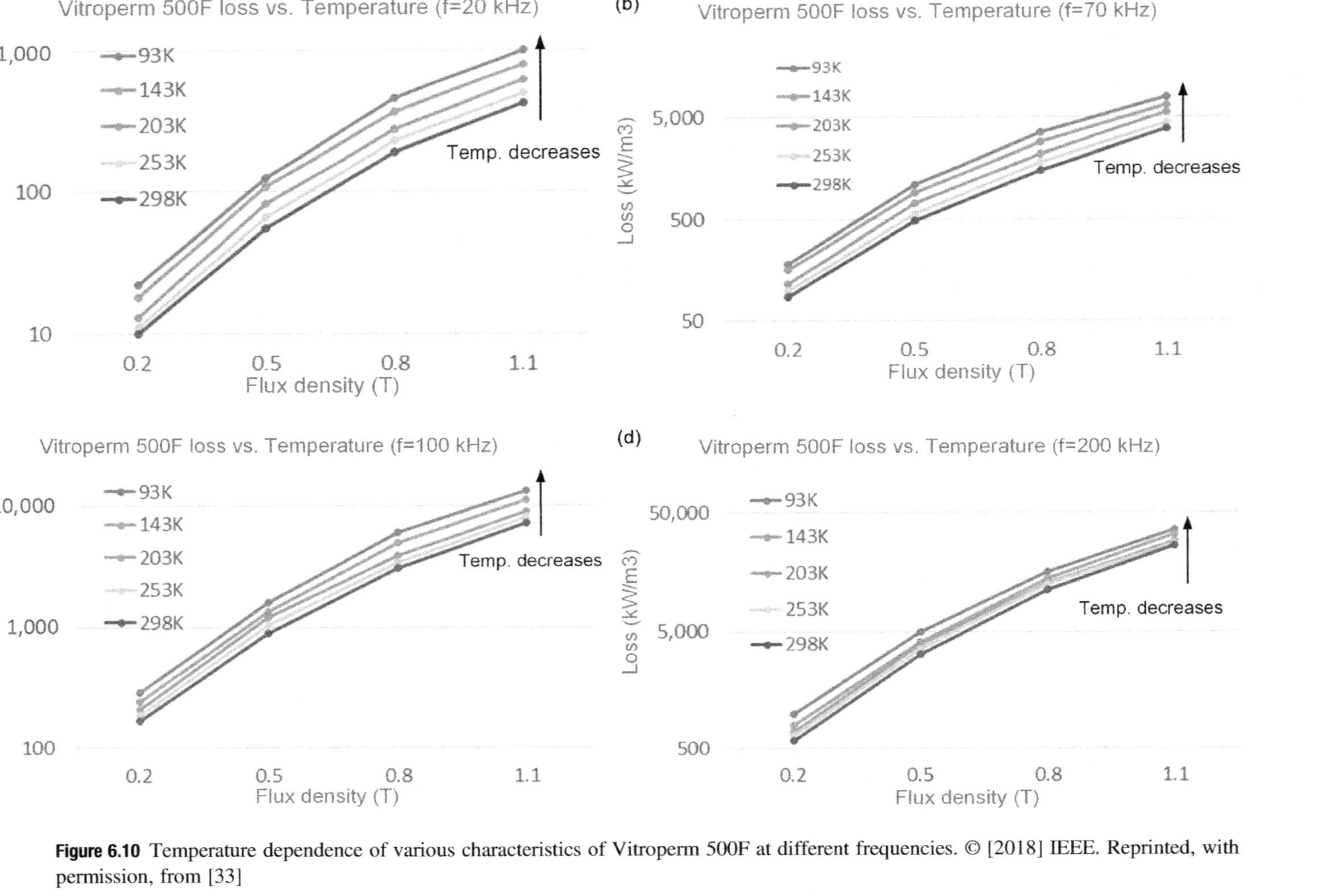

Figure 6.10 Temperature dependence of various characteristics of Vitroperm 500F at different frequencies. © [2018] IEEE. Reprinted, with permission, from [33]

temperatures is shown in Figure 6.11a. Although the core is not fully saturated in the tested curve due to the limitation of the testing equipment, it can be predicted that the saturation flux density will slightly increase when the temperature decreases from 298 K to 93 K, with a value around 0.4 T. The B-H curves of the nanocrystalline Vitroperm 500F at 20 kHz under different temperatures are presented in Figure 6.11b. It is observed that the saturation flux density remains in the range of 1.2 to 1.3 T when the temperature decreases from 298 K to 93 K.

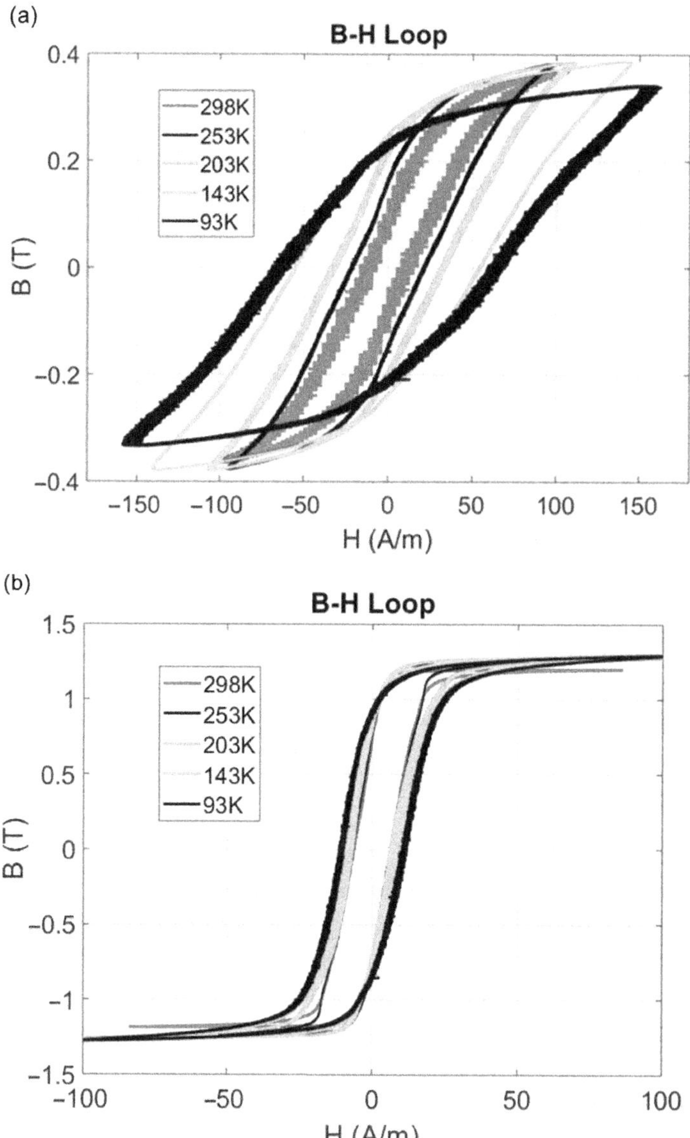

Figure 6.11 Temperature dependence of the saturation flux density. © [2018] IEEE. Reprinted, with permission, from [33]

Permeability can be assessed based on the B-H curve. As shown in Figure 6.12a and b (Ferrite N87 B-H curve at 20 kHz), the dashed curve (normal magnetization curve) consisting of all tips of the B-H curves clearly shows the important stages of the magnetization: first, the initial magnetization associated with low flux density less than 0.025 T, then magnetization with a higher incremental permeability, and finally rotation magnetization before saturation. The initial permeability and incremental permeability can be calculated based on this curve. Similarly, for the nanocrystalline Vitroperm 500 F material, the B-H curves and associated magnetization curve at different temperatures are summarized in Figure 6.12c and d, respectively.

The temperature dependence of the permeability of Ferrite N87 (both initial permeability and incremental permeability) is summarized in Figure 6.13a and b. At room temperature (298 K), the initial relative permeability is 2,510, which is consistent with the value reported in this core's datasheet, 2,200 ± 25 percent. At cryogenic temperature (93 K), the initial permeability is 350, which has decreased from the room temperature value by a factor of 7.2. Similarly, the incremental permeability also decreases as temperature decreases. The temperature dependence of permeability for Vitroperm 500F is displayed in Figure 6.13c and d. The initial permeability at 93 K is 42 percent smaller than the room temperature value. Also, the incremental permeability as a function of magnetic field strength shows a trend similar to that of the initial permeability when temperature decreases.

6.3 Power Stage Design Considerations at Cryogenic Temperature

A typical power stage with key functional blocks highlighted for motor drives in aircraft applications is illustrated in Figure 6.14. The power stage primarily consists of power devices and gate drivers. The special design and selection considerations for a power stage operating at cryogenic temperature are discussed in the following sections.

6.3.1 Power Device Selection

The performance of the power stage is strongly dependent on the quality of its power devices, whose key characteristics include on-state resistance, switching loss, and breakdown voltage. As discussed in Section 6.2.1, Si MOSFETs are preferred for cryogenic power electronics. The characterization results in Section 6.2.1.2 showed that, in general, a Si MOSFET displays reduced on-state resistance, faster switching speed, and lower breakdown voltage at cryogenic temperature than at room temperature.

Key considerations for selecting appropriate power devices for cryogenic power circuits are summarized as follows.

1. The reduced on-state resistance at cryogenic temperature facilitates increased current. Thus, a device with smaller die size can be selected for higher-power applications without the penalty of excessive conduction loss.

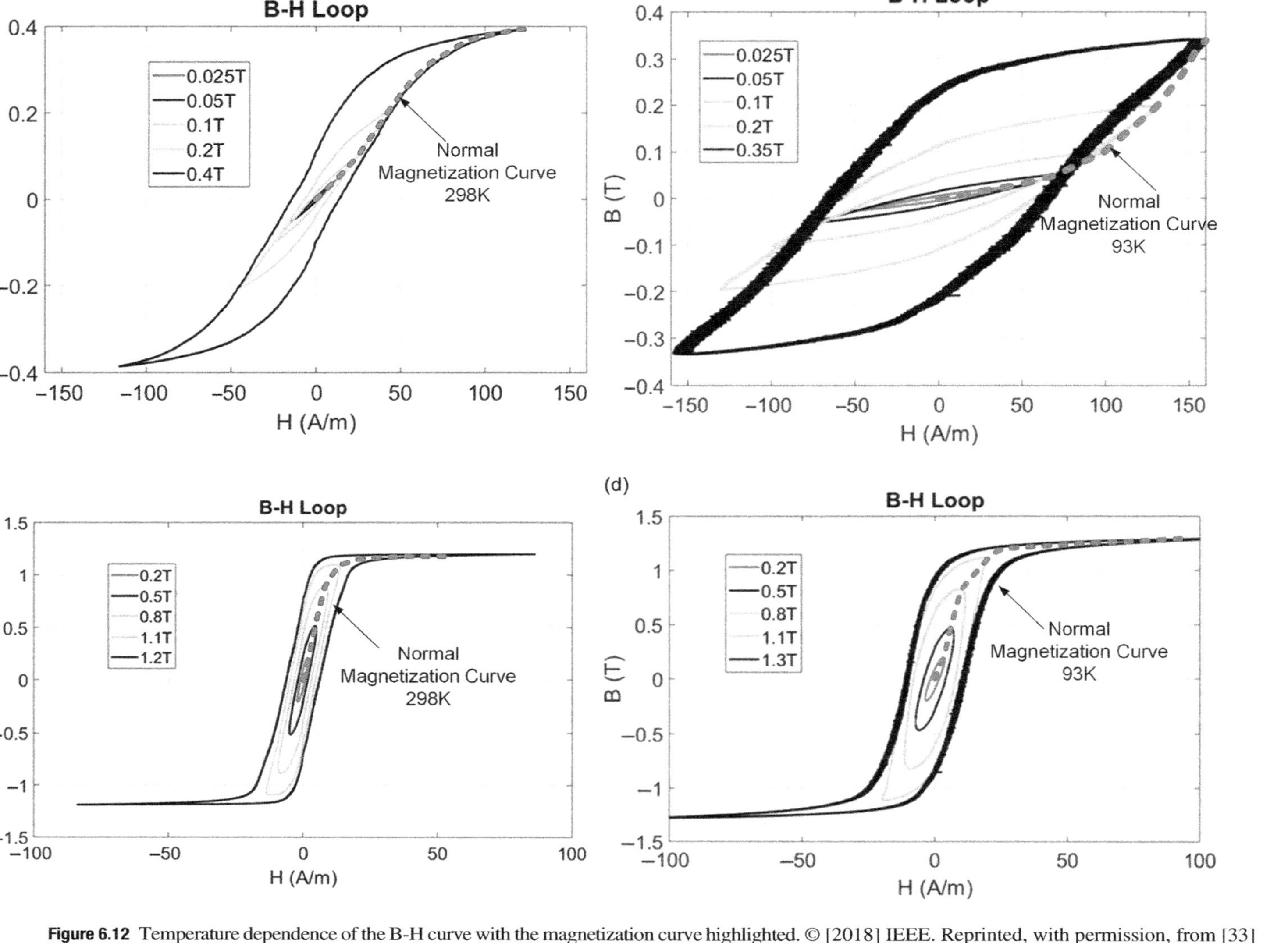

Figure 6.12 Temperature dependence of the B-H curve with the magnetization curve highlighted. © [2018] IEEE. Reprinted, with permission, from [33]

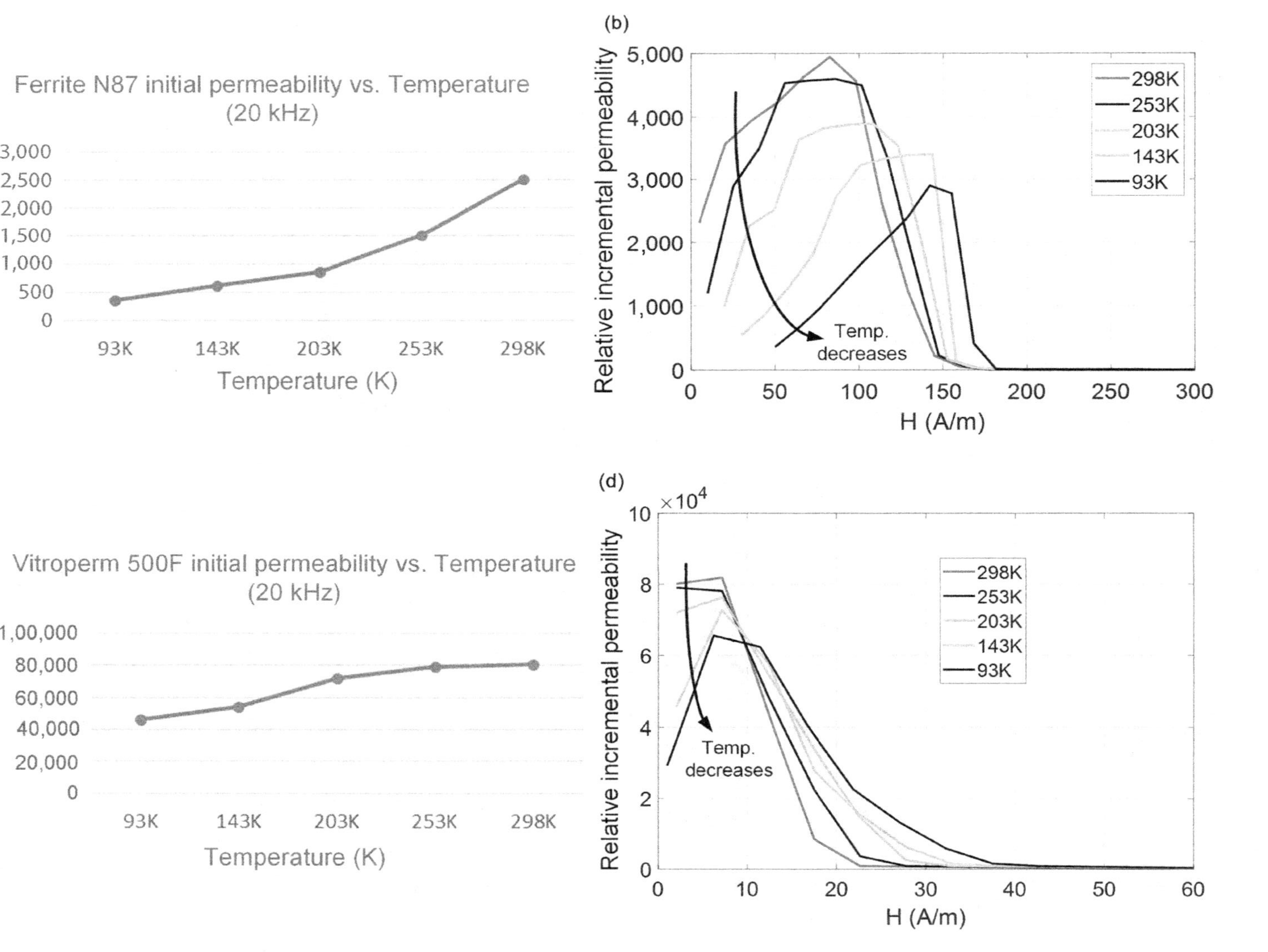

Figure 6.12 Temperature dependence of permeability at 20 kHz. © [2018] IEEE. Reprinted, with permission, from [33]

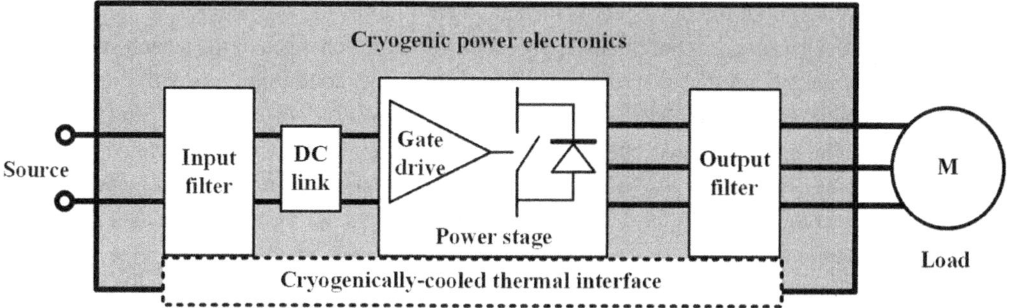

Figure 6.14 Diagram of power stage for cryogenic power electronics in dc-fed motor drives.

2. Because of the reduced switching losses enabled at cryogenic temperature, much faster switching frequencies can be used than are possible at room temperature. This may not directly affect selection of the switching device itself, but it will have large impacts on the design of passives. Although largely beneficial, fast switching speeds also pose challenges for converter design, including gate drive bandwidth and noise immunity, and layout design challenges due to the increased influence of parasitics on the converter performance.
3. Due to the reduced breakdown voltage of Si devices at cryogenic temperature, the voltage margin of device in the given operating condition should be chosen more conservatively than it would be at room temperature. This is especially true in aircraft applications, where the high-altitude, low-pressure environment imposes more severe voltage related issues, including radiation and corona. As a result, in some cases, a multi-level converter or series-connected device must be utilized to compensate for this drawback.
4. The latest-generation Si MOSFETs, although demonstrating significantly better performance at room temperature than their counterparts based on legacy technology, do not necessarily perform better at cryogenic temperature. As a result, device selection for cryogenic power electronics may not be the same as it would be in a traditional room temperature design. For instance, as illustrated in Figure 6.6, the on-state resistance of a 900 V C3 CoolMOS device is comparable to or even lower than that of a 600 V C7 CoolMOS at cryogenic temperature. For today's aircraft applications with a ±270 V dc bus, as compared to 600 V device, a 900 V device can simplify the power stage without requiring a complicated multi-level topology or one with switches in series. Therefore, 900 V C3 CoolMOS (an older technology) may be a preferred option.

6.3.2 Gate Drive Design Considerations

The gate drive is the main link between power devices and the logic-level control signals. A good gate drive is able to efficiently and reliably control power devices under static, dynamic, and fault conditions. Specifically, in the static state, a good gate drive can hold the device in its on state with minimized conduction loss and can keep the device safely in the off state, preventing spurious changes of the switch state due to external or internal disturbances. During transient conditions, a good gate drive is capable of driving the device between the on and off states quickly and with low switching losses, acceptable EMI, low voltage overshoot, and limited parasitic ringing. In case of any hazardous events, the gate drive must quickly and reliably protect the device from damaging shoot-through, overcurrent, overvoltage, and over-temperature faults.

Considering the extremely low-temperature environment along with the inherent properties of power devices operating at cryogenic temperature, special considerations are required for cryogenic gate drive design. These mainly include the following:

1. The gate drive electronics, such as the integrated circuit (IC) and buffer circuit, should be capable of operating properly at low temperature. This is because, unlike the gate drive isolation components (e.g., signal isolator and isolated power supply), it is preferred to locate the gate drive IC and buffer circuit close to the power switching devices for parasitic minimization and switching performance enhancement.
2. The on-state resistance of preferred device candidates greatly decreases at cryogenic temperature. This is beneficial for reducing conduction loss but detrimental for overcurrent protection due to the increased saturation current of power devices with the given gate voltage at low temperature and its resultant much larger shoot-through current. Hence, the short-circuit protection circuit has to be carefully designed.
3. As stated in Section 6.3.1, the switching speed of power switches increases at cryogenic temperature. Although a promising feature for switching loss reduction, this creates a challenge for the gate drive design by adding more stringent requirements for gate driving capability and noise immunity. Therefore, the design of a gate drive that can fully utilize the high-speed switching capability of cryogenic devices and overcome the fast-switching-induced issues is critical.

In the following discussion, low-temperature gate drive electronics and short-circuit protection are the focus. Gate drive design for fast switching devices has been studied extensively for emerging wide bandgap (WBG) devices, such as SiC and GaN [34]. These results can be leveraged for power devices operating at low temperature, but they will not be repeated here.

6.3.2.1 Low-Temperature Electronics

Generally speaking, ICs can be categorized into two groups: bipolar junction transistor (BJT) based and complementary metal oxide semiconductor (CMOS)-based. However, a third group – silicon-on-insulator (SOI) technology – is garnering increased attention for the development of ICs in harsh environments. A detailed discussion of each of these at cryogenic temperature is summarized in the following sections.

BJT-Based Electronics

Extensive literature has shown that the BJT performs poorly at low temperature [35–39], which implies that BJT-based ICs are not suitable for operation at cryogenic temperature.

The *dc current gain* β of a typical NP-N BJT [36] is plotted in Figure 6.15. It is clear that β drops significantly with the decrease of temperature. This is mainly because the bandgap in the emitter region drops at low temperature, which results in a severe decrease of emitter injection efficiency. Another contributor is the reduction of the base transport factor caused by the reduction of carrier lifetime.

Additionally, the base-collector breakdown voltage decreases with decreasing temperature due to the typical P-N diode behavior. The collector-emitter breakdown voltage increases due to the reduced current gain. The switching speed increases because of the increase of diffusion coefficient and decrease of carrier lifetime. However, the advantage is not enough to compensate for the declined current gain.

CMOS-Based Electronics

Prior studies have shown that the power MOSFET exhibits improved performance at low temperature. For a CMOS-based IC, the main improvement is the switching speed [41–44]. Due to the increase of carrier mobility and saturation velocity at low temperature, the transconductance of both N-channel and P-channel MOSFETs increases.

The absolute value of threshold voltage of both N-channel and P-channel MOSFETs increases as temperature decreases because of the increased surface potential caused by decreased intrinsic carrier concentration.

Despite the benefit of faster switching speed brought by low temperature, special attention should be paid to the decrease in reliability caused by hot carrier degradation [43–47]. With the increased carrier mobility and mean free path, the probability for

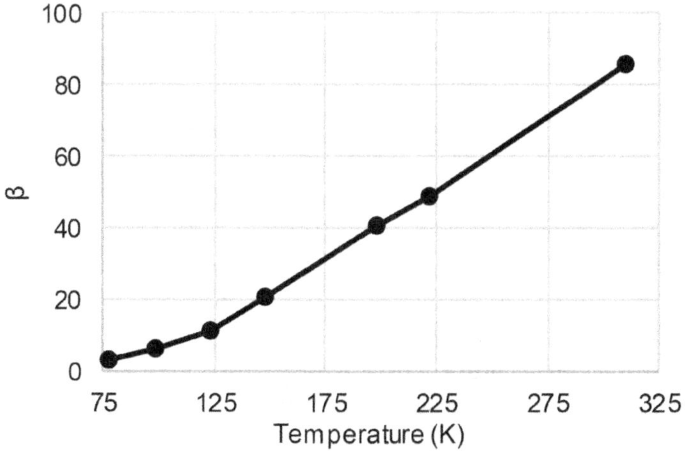

Figure 6.15 Temperature dependence of the dc current gain β of a typical N-P-N BJT. © [2020] IEEE. Reprinted, with permission, from [40]

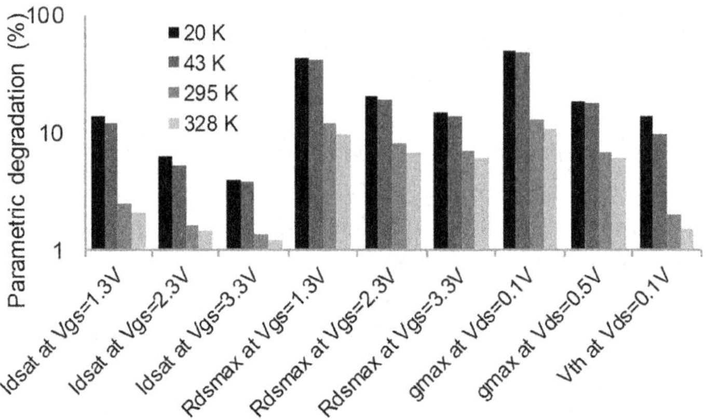

Figure 6.16 Transconductance degradation of a CMOS-based analog circuit. © [2020] IEEE. Reprinted, with permission, from [40]

carriers to gain enough energy to enter and get trapped in the gate oxide increases. Therefore, the interface states are easier to be formed at low temperature, which is the source of transconductance degradation and threshold voltage shift. It should be noted that this phenomenon does not damage the device directly, but it limits the long-term reliability and lifetime. The transconductance degradation of a CMOS-based analog circuit at different temperatures after the same operating time is shown in Figure 6.16 [46]. It is clearly observed that the degradation increases when the temperature drops. The potential solution to mitigate the impact of hot carrier effect is to reduce the supply voltage at cryogenic temperature so that the electric field is decreased, and the possibility for carriers to gain enough energy to flow into the gate oxide is reduced.

SOI-Based Electronics

SOI technology can provide faster speeds, higher efficiency, and a smaller package when compared to conventional CMOS-based ICs. Similar to CMOS technology, the speed of an SOI device improves at low temperature. However, SOI devices suffer from a *kink effect* at low temperature, especially partially depleted (PD) SOIs [45, 48–50]. This phenomenon is one of the main floating body effects. Due to the higher-impact ionization efficiency at low temperature, carriers are generated, but they are not able to flow through the Si substrate, and some of them are trapped to form a forward bias of the body region, which reduces the threshold voltage and consequently increases the channel current. The kink effect reduces gain, increases loss, induces low-frequency noise, and reduces the lifetime of the device. This issue can be suppressed by using a fully depleted (FD) SOI.

Discussion

While BJT-based circuits are not suitable, CMOS-based circuits are promising for cryogenic operation. Though SOI-based technology also shows some benefits at low temperature, it is not likely to be used for cryogenic power electronics in the near future due to its limited availability and high cost, especially when compared to CMOS. Consequently, CMOS-based circuits are the most likely candidates to be seen in cryogenic gate drive designs for the foreseeable future.

6.3.2.2 Short-Circuit Protection

Short-circuit protection is critical for voltage-source-based power electronics, which are among the most widely used converter types in many applications, including aircrafts. It is required when both the upper and lower device in the same phase-leg turn on simultaneously to short the dc energy source. As a result, large shoot-through current flows through the devices, causing thermal breakdown and long-term stability issues.

In comparison with room temperature, at cryogenic temperature, the unique challenge for short-circuit protection is due to the increased fault current caused by the reduced on-state resistance and increased saturation current of the power devices. As can be observed in Figure 6.17 under the same fault condition, the tested peak shoot-through current at 77 K is almost doubled when compared to the same test at room temperature.

Because of this larger shoot-through current, several criteria for short-circuit protection design should be taken into account: (1) fast protection response time to guarantee that power devices operate within safe operating area (SOA) margins; (2) enhanced noise immunity capability to avoid false trigger by noise during the normal fast switching transient; and (3) soft turn-off scheme to mitigate the voltage overshoot when larger shoot-through current is being cleared. Regarding the first two criteria, extensive studies have been conducted for SiC MOSFETs at room/high-temperature operation [51]. The developed techniques can be utilized for power devices operating at cryogenic temperature. The third criterion can be physically implemented to activate a large turn-off resistor under the fault condition to slowly

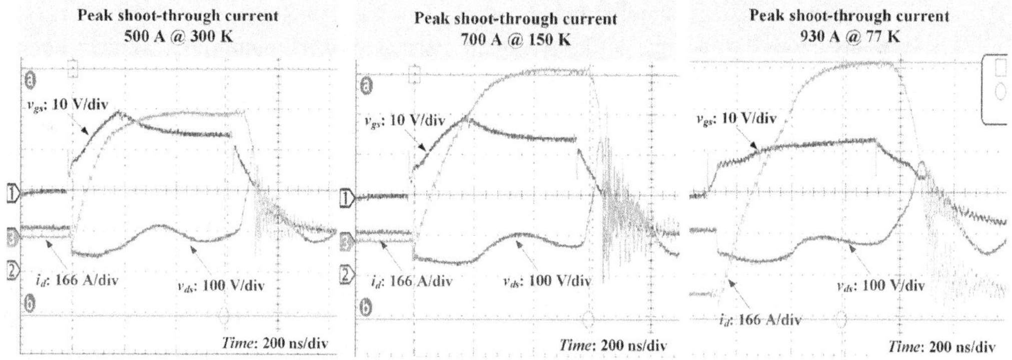

Figure 6.17 Shoot-through current dependence on temperature. © [2020] IEEE. Reprinted, with permission, from [26]

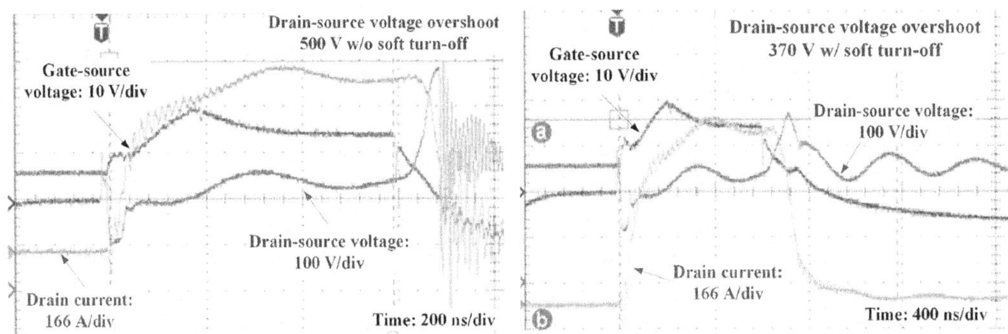

Figure 6.18 Soft turn-off for drain-source voltage overshoot mitigation when fault current is cleared.

shut down the device channel with limited *di/dt*. Hence, the voltage overshoot can be minimized. The effectiveness of the soft turn-off scheme for over-voltage mitigation is demonstrated in Figure 6.18.

6.4 Filter Design Considerations at Cryogenic Temperature

6.4.1 Inductor

An inductor in a power converter consists of two main parts: a magnetic core and a winding. For the magnetic core, as discussed in Section 6.2.2, core loss, saturation flux density, and permeability are temperature dependent. Taking Ferrite N87 as an example, as temperature decreases, core loss significantly increases, saturation flux density slightly decreases, and permeability reduces. All of these indicate the worst performance of a magnetic core is at low temperature.

On the other hand, regarding the winding, the electrical resistivity of commonly used materials, e.g., aluminum and copper, decreases greatly at low temperature. The temperature dependence of the electrical resistance of aluminum (1350) and copper (C101) are given Table 6.1. Nevertheless, considering the ac resistance – e.g., skin effect, which is critical for high-frequency applications – the skin depth is highly temperature dependent due to the reduced electrical resistivity. Under sufficiently low-temperature and high-frequency applications, the mean free path of the electrons becomes higher than the skin depth. In such cases, the classical skin effect theory is not valid and the anomalous skin effect occurs [52–57]. Figure 6.19 shows the calculated surface resistance of the copper with residual resistivity ratio (RRR) equal to 2,000 based on classical and anomalous skin effect at 100 kHz, 1 MHz, and 10 MHz, respectively. It is observed that at low temperature, the anomalous skin effect theory predicts higher resistance than the classical theory.

Considering the temperature-dependent characteristics of both the core and winding, the performance of inductors in cryogenic power electronics is not clear.

Table 6.1 Comparison of electrical resistance of copper and aluminum between room and cryogenic temperature.

Aluminum (1350)		**Copper (C101)**	
Electrical resistance at 298 K	Electrical resistance at 77 K	Electrical resistance at 298 K	Electrical resistance at 77 K
2.83×10^{-5} Ω/mm	3.84×10^{-6} Ω/mm	1.72×10^{-5} Ω/mm	2.72×10^{-6} Ω/mm

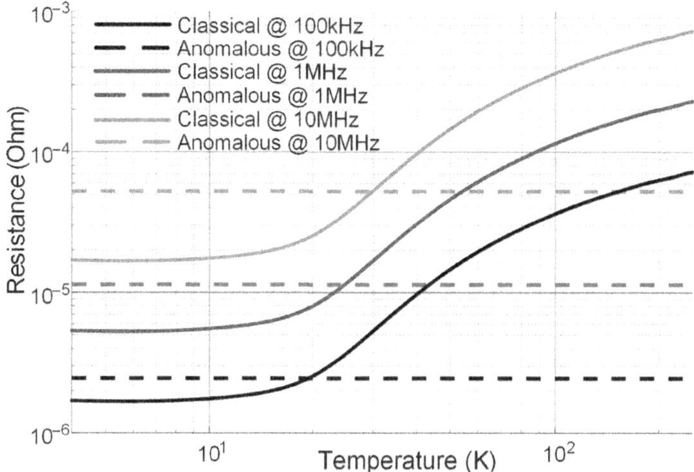

Figure 6.19 Surface resistance with classical and anomalous skin effect of RRR = 2,000 copper at different frequency. © [2020] IEEE. Reprinted, with permission, from [40]

Thus, a trade study is conducted below based on an example design of a common-mode (CM) inductor for a 200 kW aircraft motor drive system [33].

The circuit under investigation is illustrated in Figure 6.20 with its filter highlighted. Since both dc and ac sides of the motor drive need to meet DO-160 EMI standards [58], the CM inductors are designed for both sides. The inverter switching frequency is selected as 40 kHz. The designed inductance of the ac CM inductor is 600 μH and the dc CM inductor is 500 μH. Nanocrystalline Vitroperm 500F material is selected for the CM inductor design due to its relatively better performance at cryogenic temperature when compared to that of a Ferrite core, based on the test data summarized in Section 6.2.2.

Two comparison designs are conducted. One design assumes low operating temperature (liquid nitrogen temperature 77 K), and the other assumes room temperature (298 K). The cooling method illustrated in [59] is employed for the CM inductor design. The room temperature and low-temperature design results, respectively, are illustrated in Figure 6.21. Several findings can be observed: (1) the low-temperature design requires more winding turns to achieve the necessary inductance due to the decreased permeability at low temperature; (2) the low-temperature design has more core loss and less winding loss, which can be explained because the core loss density and copper

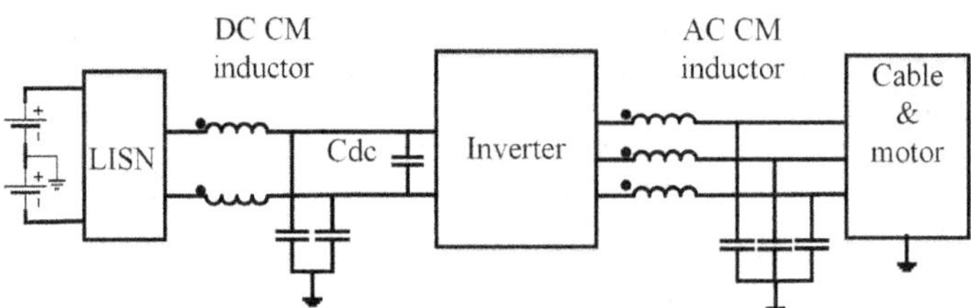

Figure 6.20 Case study: 200 kW motor drive system with CM filters in aircraft application. © [2018] IEEE. Reprinted, with permission, from [33]

resistivity show opposite trends as temperature varies; (3) the inductor loss is similar in the two designs since the benefit of winding loss reduction in the low-temperature design is canceled out by the increased core loss; (4) in this case study, a bulky core with few number of winding turns was used; therefore, the benefit obtained from the winding is not significant as compared to the penalty introduced by the core at cryogenic temperature. However, for other cases where a small core with a high number of winding turns is used, the cryogenic cooling system can bring more improvement with respect to filter efficiency and power density since the winding loss and weight are the dominant factors.

6.4.2 Capacitor

Capacitors made from different materials operating at cryogenic temperature have been investigated [2, 7, 60–66]. Due to the difference in dielectric constants across materials [61], capacitors show different temperature-dependent characteristics with respect to capacitance and dissipation factor. In Table 6.2 is summarized the variation of capacitance and dissipation factors with temperature. It is found that negative-positive 0 (NP0), polypropylene, polyphenylene sulfide (PPS), and mica perform well at cryogenic temperature. For applications where high capacitance is required, Tantalum is preferred, though its dissipation factor significantly increases. Electrolytic capacitors lose most of their capacitance at low temperature.

6.5 Example Cryogenic Power Electronic System

An example of a high-power cryogenically cooled inverter system for an aircraft application is presented in what follows to illustrate the design considerations when using a cryogenic cooling system. System integration details, a risk assessment and mitigation plan, a testing procedure with a protection scheme, and safety considerations are highlighted.

Table 6.3 lists the specifications of the inverter based on an actual National Aeronautics and Space Administration (NASA) requirement for future more electric aircraft (MEA) applications [1]. Based on their superior performance at cryogenic

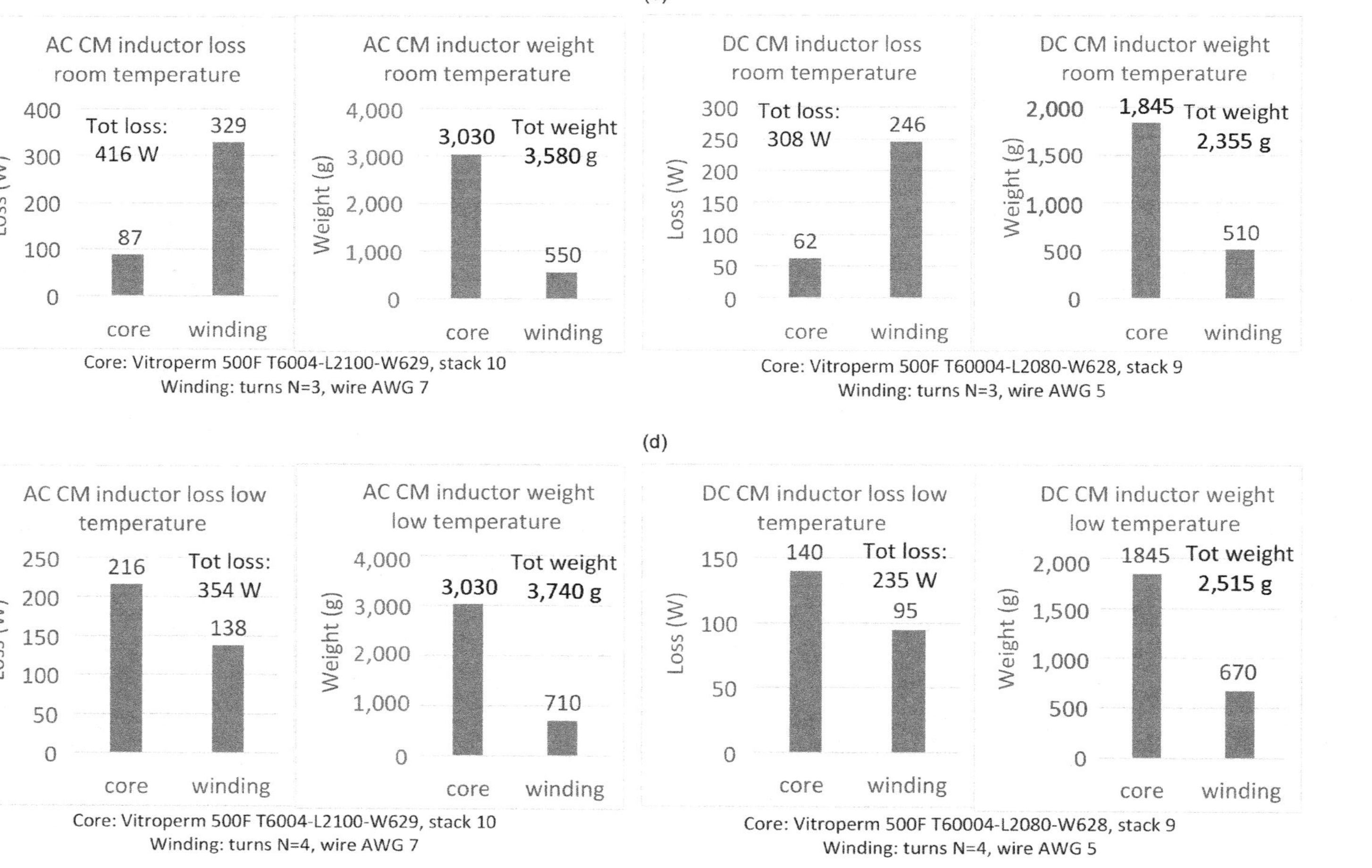

Figure 6.21 Temperature dependence of CM inductor design results. © [2018] IEEE. Reprinted, with permission, from [33].

Table 6.2 Performance changes of different capacitor technologies at cryogenic versus room temperature. © [2020] IEEE. Reprinted, with permission, from [40].

	Ceramic				Film						
Material	X7R	Y5V	Z5U	NP0	Polypropylene	PPS	Poly-ester	Poly-carbonate	**Mica**	**Electro-lytic**	**Tantalum**
Capacitance	↓↓	↓↓	↓↓	–	–	–	↓	↓	–	↓↓	↓
Dissipation factor	↑↑	↑↑	↑↑	–	↓	–	↓	↓	–	↓↓	↑↑

↑↑: increase greatly ↑: increase slightly –: keep constant ↓: decrease slightly ↓↓: decrease greatly

Table 6.3 Design specification of example cryogenic 40 kW inverter. © [2020] IEEE. Reprinted, with permission, from [26].

Input voltage	**1 kV_{dc}**
Output voltage	600 V_{ac} rms line-to-line
Output power	40 kVA
Fundamental frequency	0-3 kHz
Cryogenic coolant	Liquid nitrogen

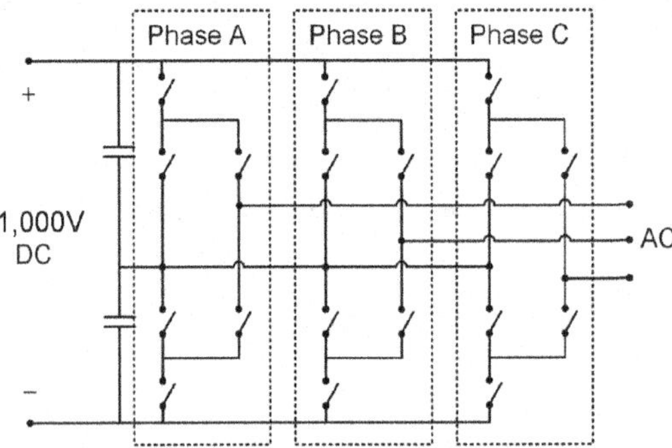

Figure 6.22 Three-level ANPC topology. © [2020] IEEE. Reprinted, with permission, from [26]

temperature, Si MOSFETs are selected as the switching device technology. Considering the voltage rating of available devices in today's market, a three-level active neutral point clamped (ANPC) topology is employed, as illustrated in Figure 6.22.

6.5.1 Cryogenic Cooling System and Integration

The cooling and system integration with special consideration for safety is critical. The requirements mainly include the following:

1. Ensure no leakage of the cryogenic coolant. Liquid nitrogen is more than 6,000 times as dense as gaseous nitrogen; thus, the explosion risk is large if there is a severe leak. Moreover, liquid nitrogen can cause cold injury to the operator.
2. Provide adequate thermal insulation between the cryogenic power electronics and the ambient environment. The temperature of the system should be thermally stable during the test with limited heat loss to the ambient environment.
3. Provide an inert environment for the equipment under test. Because the operating temperature is far below 0 °C, condensation due to the existence of air could cause short circuits within the power electronics converter.

To meet these requirements, a dedicated enclosure for the laboratory test is designed. As shown in Figure 6.23, multiple holes are drilled on the surface of the enclosure to provide an interface for liquid/gas nitrogen and electrical terminations. Also, slots are cut on the lid of the enclosure as the signal interface.

The conceptual diagram of the integrated system and the testing platform is shown in Figure 6.24. It should be noted that both liquid and gaseous nitrogen are utilized: liquid nitrogen flows through cold plates and lowers the temperature inside the enclosure while the gas nitrogen is used to create an inert environment inside the enclosure to avoid icing. Specifically, the gas nitrogen passes through first to push the air out of the enclosure before the liquid nitrogen is added. Inside the enclosure, pieces of Styrofoam with 1-inch thickness are closely attached to the inner walls for thermal insulation. For all the pipes used in the system, Yor-Lok tube fittings are adopted to guarantee proper sealing. Furthermore, cryogenic epoxy is adopted to fill the gap between tubes and the walls of the enclosure. In the end, the aforementioned requirements can be met. The inverter power stage, gate drives, analog controllers, auxiliary power supplies, and cold plate are all located inside the enclosure, while the interface board and digital signal processor (DSP) controller are located outside the

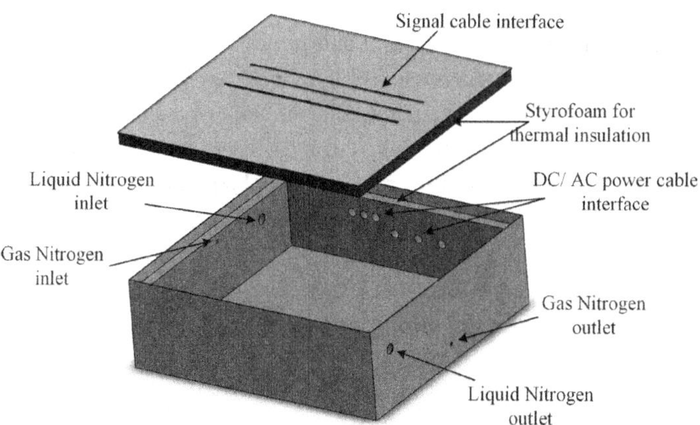

Figure 6.23 Dedicated enclosure designed for cryogenic power electronics. © [2020] IEEE. Reprinted, with permission, from [26]

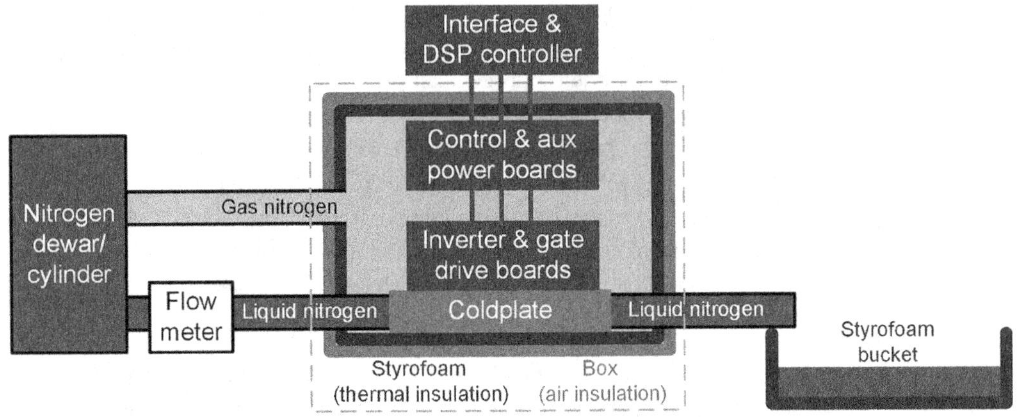

Figure 6.24 System diagram of cryogenic power electronics testing platform. © [2020] IEEE. Reprinted, with permission, from [26]

enclosure and operate at room temperature. Ribbon cables are used as the signal interface through the slots on the lid of the enclosure.

6.5.2 Safety Consideration and Mitigation Plan

In addition to the standard controls and protection methods for the power electronics converter, protection specific to the cryogenic system is critical. Before the implementation of prototyping and testing of cryogenic power electronics, a thorough safety assessment and mitigation plan should be devised, an example of which is summarized in Table 6.4. Accordingly, detailed control and protection are designed to mitigate failures and to protect both the equipment under test and (more importantly) the operator.

The basic architecture of the control and protection implemented for the testing of cryogenic power electronics is found in Figure 6.25. Special attention has been paid to cryogenic cooling-related safety issues. Specifically, the PC communicates with the temperature monitor, flow meter, and oxygen meter in real time. At the instant a fault is detected, such as under/over flow rate, abnormal temperature, or oxygen deficiency, the converter is controlled to be shut down and an emergency (E)-stop signal is triggered to close the liquid/gas nitrogen valves and alert the operators to evacuate.

Finally, as with all lab experiments, especially those with hazardous materials and/or test conditions, a dedicated testing procedure must be devised and followed. An example is summarized as follows.

- Step 1: Use personal protective equipment (PPE) at all the times during the test. A two-person rule must be followed – i.e., no one should work alone in the lab.
- Step 2: Start auxiliary power for the micro-controller, gate drive, and sensors (e.g., flow meter, temperature meter, oxygen meter) for protection.

Table 6.4 Example risk assessment and mitigation plan for testing of cryogenic power electronics.

Fault condition	Risk mitigation plan
Lack of liquid nitrogen during the test	• Monitor the remaining amount of liquid nitrogen in the Dewar. • Install a flow meter to monitor the flow rate at the outlet of the Dewar and enable protection if the flow rate is low.
Oxygen deficiency hazard	• Open all doors and run an exhaust fan in the lab during testing. • Install an oxygen deficiency monitor and alert operators in case of low oxygen level.
Liquid nitrogen leak	• Seal the outlet of the cryogenic cooling system and assess its tightness before the main test begins. • Add a lid screw with a spring for pressure relief.
Blocked liquid nitrogen path	• Gradually open the Dewar valve to ensure gas is observed at the end of the cryogenic cooling system. • Install a flow meter to monitor the flow rate at the outlet of the Dewar and enable protection if the flow rate is low.
Ice forming inside the enclosure	• Use room temperature nitrogen gas to force the air outside the enclosure. • Introduce short-circuit protection to shut down the circuit.

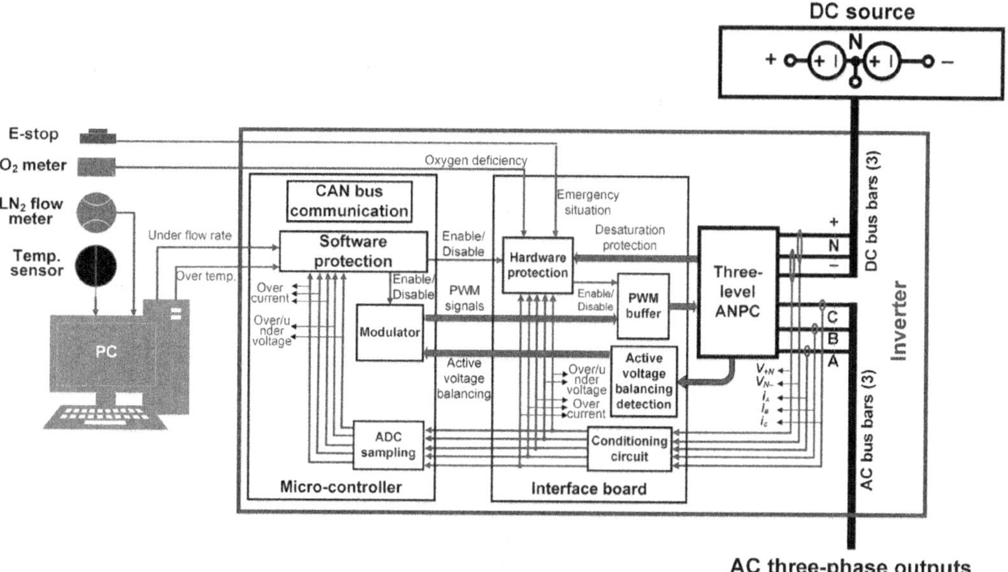

Figure 6.25 Control and protection architecture for testing of cryogenic power electronics. © [2020] IEEE. Reprinted, with permission, from [26]

- Step 3: Inject gas nitrogen to flush the air out of the enclosure to avoid condensation.
- Step 4: Open the liquid nitrogen valve and wait until the sensor indicates that the temperature of the cold plate is below 100 K.

- Step 5: Use caution to run the power test. Always carefully monitor the readout of all meters.
- Step 6: Shut down the main power when the test is completed.
- Step 7: Close the liquid nitrogen valve.
- Step 8: When the cold plate temperature is above 0 °C, turn off the gas nitrogen valve.
- Step 9: Shut down auxiliary power.

6.5.3 Experimental Demonstration

The testing platform and cryogenically cooled inverter prototype, respectively, are shown in Figure 6.26a and b. The key waveforms of the inverter operating at its rated condition, including output line-to-line voltage and three-phase currents, are displayed in Figure 6.27. The experiment demonstrates proper functioning of the developed power converter with a cryogenic cooling system.

6.6 Summary

Cryogenic power electronics can enable highly efficient ultra-dense power conversion systems which are critical for EAP. The benefits are achieved through improved power device performance, reduced conductor electrical resistivity, and increased temperature differential for heat transfer.

The main takeaways from this chapter, which provide a general framework with which to dive deeper into this important field, are as follows.

1. The Si MOSFET exhibits superior performance at cryogenic temperature, offering lower on-state resistance and faster switching speed than is possible at room temperature. However, its breakdown voltage decreases at low temperature. As a result, sufficient voltage safety margin must be maintained, especially for aircraft applications.
2. Among the emerging WBG devices, the SiC MOSFET is not suitable for cryogenic applications due to its significantly increased on-resistance at low temperatures. The GaN HEMT is more attractive, considering its improved static on-resistance and constant breakdown voltage at cryogenic temperature. However, the limited current ratings of commercially available GaN devices, additional switching losses (which might be attributed to dynamic on-resistance), and immature technology at cryogenic temperature are legitimate concerns that are yet to be fully addressed.
3. A nanocrystalline-type core performs relatively well at low temperature compared to a ferrite-based core and is the preferred option for cryogenic power electronics. However, when compared to its performance at room temperature, as the temperature decreases, the core loss increases and permeability decreases.
4. The electrical resistivity of commonly used conductors – e.g., copper and aluminum – significantly decreases as temperature drops. This essentially

(a)

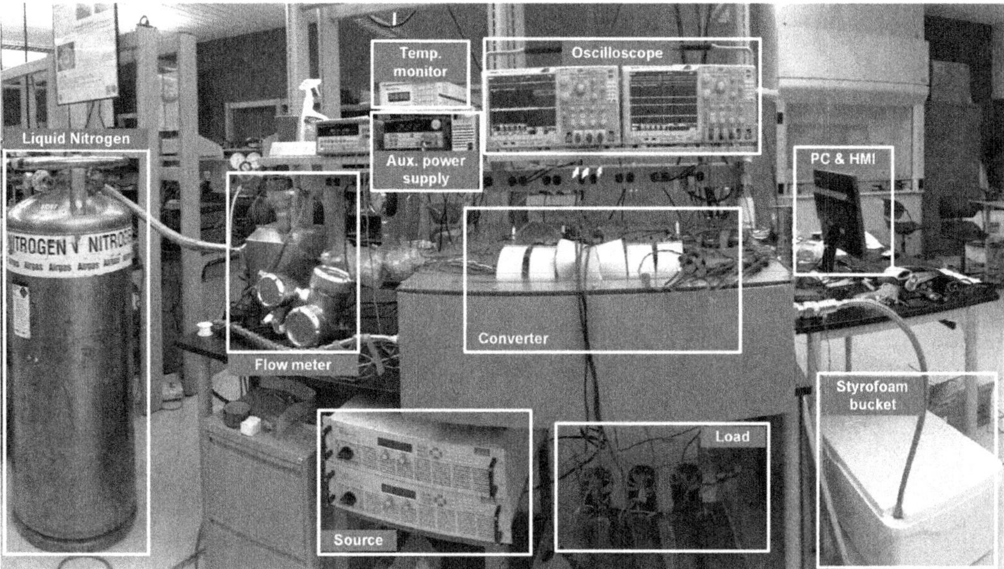

(b)

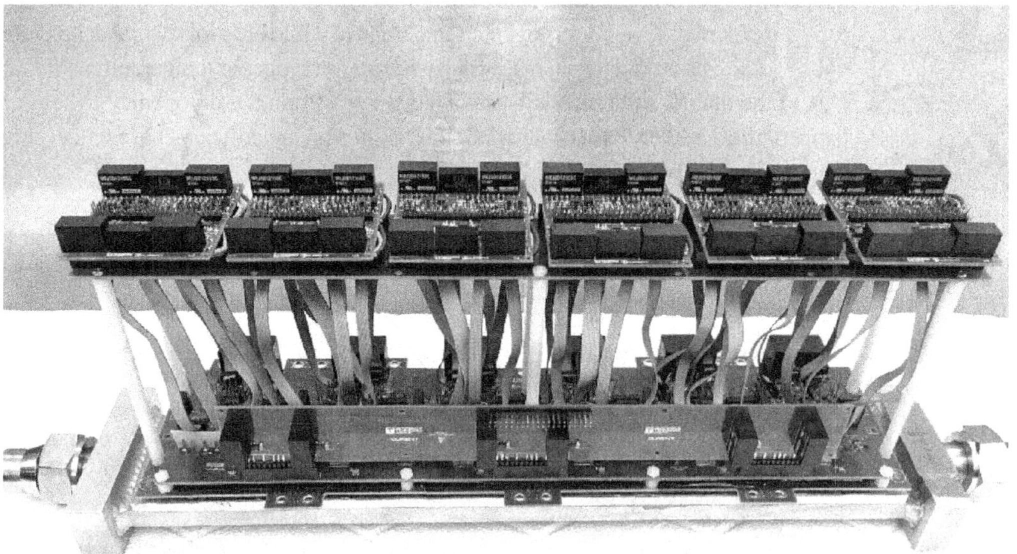

Figure 6.26 Cryogenically cooled power converter (inverter) testing platform. © [2020] IEEE. Reprinted, with permission, from [26]

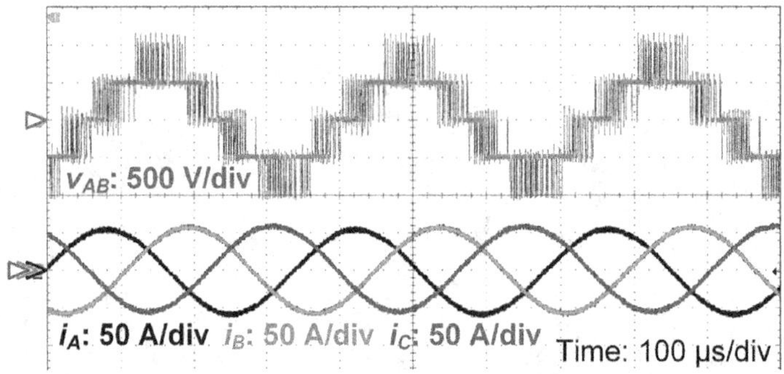

Figure 6.27 Key waveforms of the cryogenically cooled inverter operating at its rated condition. © [2020] IEEE. Reprinted, with permission, from [26]

eliminates their dc resistance. Unfortunately, for high-frequency applications, special attention must be paid to the ac resistance. Taking skin effect as an example, the skin depth decreases at lower temperature because of the reduced electrical resistivity. As a result, ac resistance can increase significantly, canceling out some (or all) of the benefits of lower dc resistance.

5. CMOS-based integrated circuits demonstrate better performance at low temperature than BJT-based versions. NP0, polypropylene, PPS, and mica capacitors perform well at cryogenic temperature, while electrolytic capacitors are not a viable option due to their low capacitance at cryogenic temperature.
6. Due to the unique safety hazards involved when working at cryogenic temperatures, a safety assessment and mitigation plan are critical. These will ensure the safety of all personnel and equipment.

Abbreviations

ac	alternating current
ANPC	active neutral point clamped
BJT	bipolar junction transistor
CM	common mode
CMOS	complementary metal oxide semiconductor
dc	direct current
DPT	double pulse test
DSP	digital signal processor
DUT	device under test
EAP	electrified aircraft propulsion
EMI	electromagnetic interference
FD	fully depleted
Ge	germanium
GaN	gallium nitride

HEMT	high electron mobility transistor
HSP	high specific power
IC	integrated circuit
IGBT	insulated gate bipolar transistor
MEA	more electric aircraft
NASA	National Aeronautics and Space Administration
NP0	negative-positive zero
PPE	personal protective equipment
PD	partially depleted
PPS	polyphenylene sulfide
RRR	residual resistivity ratio
Si	silicon
SiC	silicon carbide
SOA	safe operating area
SOI	silicon-on-insulator
WBG	wide bandgap

Variables

A_e	effective cross-sectional area, [m^2]
B	flux density, [T]
B_s	saturation flux density, [T]
β	dc current gain
E	core energy loss, [J]
H	magnetic field strength, [A-turns/m]
I_1	primary current, [A]
l_e	core mean magnetic path length, [m]
μ_r	relative permeability
n_1	primary turns
n_2	secondary turns
P_c	core loss, [W]
v_2	secondary voltage, [V]

References

[1] Amendment no. 3 to the NASA Research Announcement (NRA), "Research opportunities in aeronautics – 2015 (ROA-2015)," NNH14ZEA001N, 2015.

[2] J. Garrett, R. Schupbach, A. B. Lostetter, and H. A. Mantooth, "Development of a DC motor drive for extreme cold environments," presented at the IEEE Aerosp. Conf., Big Sky, MT, 2007, DOI: 10.1109/AERO.2007.352654.

[3] K. Rajashekara and B. Akin, "A review of cryogenic power electronics – status and applications," in *Proc. IEEE Int. Elect. Mach. & Drive Conf.*, Chicago, IL, May 2013, pp. 899–904.

[4] S. Yang, "Cryogenic characteristics of IGBTs," Ph.D. thesis, University of Birmingham, UK, 2005.
[5] R. Singh and B. J. Baliga, *Cryogenic Operation of Silicon Power Devices*. New York, NY: Springer Science and Business Media, 2012.
[6] C. Jia, "Experimental investigation of semiconductor losses in cryogenic DC-DC converters," Ph.D. thesis, University of Birmingham, UK, 2008.
[7] J. Bourne et al., "Ultra-wide temperature (–230 °C to 130 °C) DC-motor drive with SiGe asynchronous controller," presented at the IEEE Aerosp. Conf., Big Sky, MT, 2008, DOI: 10.1109/AERO.2008.4526594.
[8] S. Chen et al., "Cryogenic and high temperature performance of 4H-SiC power MOSFETs," in *Proc. IEEE Appl. Power Electron. Conf.*, Long Beach, CA, May 2013, pp. 207–210.
[9] T. Chailloux et al., "SiC power devices operation from cryogenic to high temperature: investigation of various 1.2 kV SiC power devices," in *Mater. Sci. Forum*, vol. 778, pp. 1122–1125, 2014.
[10] H. Chen et al., "Cryogenic characterization of commercial SiC Power MOSFETs," in *Mater. Sci. Forum*, vol. 821, pp. 777–780, 2015.
[11] H. Kim, J. Lim, and H. Cha, "DC characteristics of wide-bandgap semiconductor field-effect transistors at cryogenic temperatures," *J. Korean Phys. Soc.,* vol. 56, no. 5, pp. 1523–1526, 2010.
[12] H. Gui et al., "Characterization of 1.2 kV SiC power MOSFETs at cryogenic temperatures," presented at the 10th IEEE Energy Convers. Congr. & Expo., Portland, OR, 2018, DOI: 10.1109/ecce.2018.8557442.
[13] J. Colmenares et al., "Experimental characterization of enhancement mode gallium-nitride power field-effect transistors at cryogenic temperatures," in *Proc. IEEE Workshop on Wide Bandgap Power Devices and Appl.*, Fayetteville, AL, November 2016, pp. 129–134.
[14] X.-F. Zhang et al., "Electrical characteristics of AlInN/GaN HEMTs under cryogenic operation," *Chin. Phys. B,* vol. 22, no. 1, 017202, 2013.
[15] O. Katz, A. Horn, G. Bahir, and J. Salzman, "Electron mobility in an AlGaN/GaN two-dimensional electron gas. I. carrier concentration dependent mobility," *IEEE Trans. Electron Dev,* vol. 50, no. 10, pp. 2002–2008, 2003.
[16] S.-J. Chang et al., "Investigation of channel mobility in AlGaN/GaN high-electron-mobility transistors," *Jpn. J. Appl. Phys.,* vol. 55, no. 4, 044104, 2016.
[17] S. Nuttinck et al., "Cryogenic investigation of current collapse in AlGaN/GaN HFETS," in *Proc. Gallium Arsenide Appl. Symp.*, 2003, pp. 213–215.
[18] C.-H. Lin et al., "Transient pulsed analysis on GaN HEMTs at cryogenic temperatures," *IEEE Electron Device Lett.,* vol. 26, no. 10, pp. 710–712, 2005.
[19] R. Cuerdo et al., "The kink effect at cryogenic temperatures in deep submicron AlGaN/GaN HEMTs," *IEEE Electron Device Lett.,* vol. 30, no. 3, pp. 209–212, 2009.
[20] A. Caiafa et al., "IGBT operation at cryogenic temperatures: non-punch-through and punch-through comparison," in *Proc. IEEE Power Electron. Special. Conf.*, Aachen, Germany, November 2004, vol. 4, pp. 2960–2966.
[21] A. Caiafa *et al.*, "Cryogenic study and modeling of IGBTs," in *Proc. IEEE Power Electron. Special. Conf.*, Acapulco, Mexico, June 2003, vol. 4, pp. 1897–1903.
[22] Z. Zhang et al., "Characterization of high-voltage high-speed switching power semiconductors for high frequency cryogenically-cooled application," in *Proc. IEEE Appl. Power Electron. Conf.*, Tampa, FL, March 2017, pp. 1964–1969.

[23] K. K. Leong, B. T. Donnellan, A. T. Bryant, and P. A. Mawby, "An investigation into the utilisation of power MOSFETs at cryogenic temperatures to achieve ultra-low power losses," in *Proc. IEEE Energy Convers. Congr. and Expo.*, Atlanta, GA, September 2010, pp. 2214–2221.

[24] O. Mueller, "Properties of high-power Cryo-MOSFETs," in *Proc. IEEE Ind. Appl. Conf.*, San Diego, CA, October 1996, vol. 3, pp. 1443–1448.

[25] Y. Chen et al., "Experimental investigations of state-of-the-art 650-V class power MOSFETs for cryogenic power conversion at 77K," *IEEE J. Electron. Devices Soc.*, vol. 6, no. 1, pp. 8–18, 2018.

[26] H. Gui et al. "Development of high-power switching frequency cryogenically cooled inverter for aircraft applications," *IEEE Trans. Power Electr.*, vol. 35, no. 6, pp. 5670–5682, 2020.

[27] Z. Zhang et al., "Characterization of wide bandgap device for cryogenically-cooled power electronics in aircraft applications," presented at the 1st AIAA/IEEE Electric Aircraft Technol. Symp., Cincinnati, OH, July 2018, Paper AIAA 2018-5006.

[28] N. Ahmad, "Carrier freeze-out effects in semiconductor devices," *J. Appl. Phys.*, vol. 61, no. 5, pp. 1905–1909, 1987.

[29] M. Chen et al., "The magnetic properties of the ferromagnetic materials used for HTS transformers at 77 K," *IEEE Trans. Appl. Supercond.*, vol. 13, no. 2, pp. 2313–2316, 2003.

[30] J. Claassen, "Inductor design for cryogenic power electronics," *IEEE Trans. Appl. Supercond.*, vol. 15, no. 2, pp. 2385–2388, 2005.

[31] H. P. Quach and T. C. Chui, "Low temperature magnetic properties of Metglas 2714A and its potential use as core material for EMI filters," *Cryog.*, vol. 44, no. 6, pp. 445–449, 2004.

[32] S. S. Gerber, M. E. Elbuluk, A. Hammoud, and R. L. Patterson, "Performance of high-frequency high-flux magnetic cores at cryogenic temperatures," in *Proc. 37th Intersociety Energy Convers. Eng. Conf.*, Washington, DC, July 2002, pp. 249–254.

[33] R. Chen et al., "Core characterization and inductor design investigation at low temperature," presented at the. 10th IEEE Energy Convers. Congr. & Expo., Portland, OR, 2018, DOI: 10.1109/ecce.2018.8557779.

[34] Z. Zhang and F. Wang, "Driving and characterization of wide bandgap semiconductors for voltage source converter applications," presented at the IEEE Workshop on Wide Bandgap Power Devices and Appl., Knoxville, TN, October 2014, DOI: 10.1109/WiPDA.2014.6964609.

[35] J. D. Cressler, "Silicon bipolar transistor: a viable candidate for high speed applications at liquid nitrogen temperature," *Cryog.*, vol. 30, no. 12, pp. 1036–1047, 1990.

[36] W. P. Dumke, "The effect of base doping on the performance of Si bipolar transistors at low temperatures," *IEEE Trans. Electron Devices,* vol. 28, no. 5, pp. 494–500, 1981.

[37] R. Singh and B. Baliga, "Cryogenic operation of power bipolar transistors," *Solid State Electron.,* vol. 39, no. 1, pp. 101–108, 1996.

[38] J. Stork, D. L. Harame, B. Mayerson, and T. N. Nguyen, "Base profile design for high-performance operation of bipolar transistors at liquid-nitrogen temperature," *IEEE Trans. Electron Devices,* vol. 36, no. 8, pp. 1503–1509, 1989.

[39] J. Woo, J. D. Plummer, and J. Stork, "Non-ideal base current in bipolar transistors at low temperatures," *IEEE Trans. Electron Devices,* vol. 34, no. 1, pp. 130–138, 1987.

[40] H. Gui et al., "Review of power electronics components at cryogenic temperatures," *IEEE Trans. Power Electr.*, vol. 35, no. 5, pp. 5144–5156, 2020.

[41] W. F. Clark et al., "Low temperature CMOS-a brief review," *IEEE Trans. Compon. Packag. Manuf. Technol.*, vol. 15, no. 3, pp. 397–404, 1992.

[42] G. Ghibaudo and F. Balestra, "Low temperature characterization of silicon CMOS devices," in *Proc. Intl. Conf. on Microelectron.*, Nis, Serbia, September 1995, vol. 2, pp. 613–622.

[43] T. K. Makiniemi and P. J. Kosonen, "A low temperature pipelined analog-to-digital converter," in *Proc.8th IEEE Int. Conf. on Electron., Circuits and Syst.*, Malta, September 2001, vol. 2, pp. 849–852.

[44] B. Okcan, P. Merken, G. Gielen, and C. Van Hoof, "A cryogenic analog to digital converter operating from 300 K down to 4.4 K," *Rev. Sci. Instrum.*, vol. 81, no. 2, 024702, 2010.

[45] F. Balestra and G. Ghibaudo, *Device and Circuit Cryogenic Operation for Low Temperature Electronics*. New York, NY: Springer Science and Business Media, 2013.

[46] Y. Chen et al., "Design for ASIC reliability for low-temperature applications," NASA Jet Propulsion Lab., Pasadena, CA, Tech. Rep. 20060044221, 2005.

[47] A. Dejenfelt and O. Engström, "MOSFET mobility degradation due to interface-states, generated by Fowler-Nordheim electron injection," *Microelectron. Eng.*, vol. 15, no. 1–4, pp. 461–464, 1991.

[48] C. Claeys and E. Simoen, "The perspectives of silicon-on-insulator technologies for cryogenic applications," *J. Electrochem. Soc.*, vol. 141, no. 9, pp. 2522–2532, 1994.

[49] C. Claeys and E. Simoen, "Perspectives of silicon-on-insulator technologies for cryogenic electronics," in *Perspectives, Science and Technologies for Novel Silicon on Insulator Devices*, New York, NY: Springer, 2000, pp. 233–247.

[50] E. Simoen and C. Claeys, "The cryogenic operation of partially depleted silicon-on-insulator inverters," *IEEE Trans. Electron Devices,* vol. 42, no. 6, pp. 1100–1105, 1995.

[51] Z. Wang et al., "Design and performance evaluation of overcurrent protection schemes for silicon carbide (SiC) power MOSFETs," *IEEE Trans. Ind. Electron.*, vol. 61, no. 10, pp. 5570–5581, 2014.

[52] R. Chambers, "The anomalous skin effect," in *Proc. Royal Soc. London A: Math. Phys. Eng. Sci.*, 1952, vol. 215, no. 1123, pp. 481–497.

[53] M. Kaganov, G. Y. Lyubarskiy, and A. Mitina, "The theory and history of the anomalous skin effect in normal metals," *Phys. Rep.*, vol. 288, no. 1, pp. 291–304, 1997.

[54] H. London, "Alternating current losses in superconductors of the second kind," *Phys. Lett.*, vol. 6, no. 2, pp. 162–165, 1963.

[55] E. Maxwell, "Superconducting resonant cavities," in *Advances in Cryogenic Engineering*, New York, NY: Springer, 1961, pp. 154–165.

[56] A. Pippard, "The surface impedance of superconductors and normal metals at high frequencies: II. The anomalous skin effect in normal metals," in *Proc. Royal Soc. London A: Math. Phys. Eng. Sci.*, 1947, vol. 191, no. 1026, pp. 385–399.

[57] G. Reuter and E. Sondheimer, "The theory of the anomalous skin effect in metals," in *Proc. Royal Soc. London A: Math. Phys. Eng. Sci.*, 1948, vol. 195, no. 1042, pp. 336–364.

[58] RTCA: Environmental Conditions and Test Procedures for Airborne Electronic/Electrical Equipment and Instruments, DO-160G1, December 2014.

[59] J. Xue and F. Wang, "A practical liquid-cooling design method for magnetic components of EMI filter in high power motor drives," presented at the 8th IEEE Energy Convers. Congr. & Expo., Milwaukee, WI, September 2016, DOI: 10.1109/ECCE.2016.7854758.

[60] M. Elbuluk and A. Hammoud, "Power electronics in harsh environments," in *Proc. of the 40th IEEE Ind. Appl. Conf.*, Kowloon, Hong Kong, China, October 2005, vol. 2, pp. 1442–1448.

[61] L. Faria, A. Passaro, L. Nohra, and R. d'Amore, "Influence of the cryogenic temperature and the BIAS voltage on the spontaneous polarization effect of X5R dielectric capacitors," in *Proc. Int. Refereed J. Eng. Sci.,* vol. 1, no. 1, pp. 14–21.

[62] A. Hammoud, S. Gerber, R. L. Patterson, and T. L. MacDonald, "Performance of surface-mount ceramic and solid tantalum capacitors for cryogenic applications," in *Annual Report of the Conf. on Elect. Insul. and Dielectr. Phenom.*, October 1998, vol. 2, pp. 572–576.

[63] A. Hammoud and E. Overton, "Low temperature characterization of ceramic and film power capacitors," in *Proc. IEEE Conf. on Elect. Insul. and Dielectr. Phenom.*, Millbrae, CA, October 1996, vol. 2, pp. 701–704.

[64] M.-J. Pan, "Performance of capacitors under DC bias at liquid nitrogen temperature," *Cryogenics,* vol. 45, no. 6, pp. 463–467, 2005.

[65] R. L. Patterson, A. Hammond, and S. S. Gerber, "Evaluation of capacitors at cryogenic temperatures for space applications," in *Conf. Record of the IEEE Int. Symp. on Elect. Insul.*, Arlington, VA, June 1998, vol. 2, pp. 468–471.

[66] F. Teyssandier and D. Prêle, "Commercially available capacitors at cryogenic temperatures," presented at the 9th Int. Workshop on Low Temp. Electron., Guaruja, Brazil, June 2010, HAL: hal-00623399.

7 Electrochemical Energy Storage and Conversion for Electrified Aircraft

Ajay Misra

Introduction

There is significant worldwide interest in developing electrified aircraft propulsion (EAP) systems for various classes of aircraft, ranging from urban air mobility with two to four passengers to large single-aisle transport aircraft with hundreds of passengers. Both all-electric and hybrid-electric options are currently being considered for EAP. In the all-electric concept, an energy storage device or an energy conversion system (e.g., fuel cell, internal combustion or gas turbine engine) powers an electric motor to drive the propulsion fan. In the hybrid-electric concept, both an energy storage device and a gas turbine engine (or internal combustion engine) are used to power propulsion via some combination of turbofans and electric fans. The type of EAP used depends largely on the capability of the energy storage system, particularly its specific energy and power density. The pace of introduction for electrified aircraft propulsion in different market segments is also a strong function of the energy storage capability.

This chapter provides an overview of electrochemical energy storage and conversion systems for electrified aircraft. These systems include batteries, fuel cells, supercapacitors, and multifunctional structures with energy storage capability. The chapter begins with an overview of the state of the art (SOA) in battery technology and of the various electrified aircraft concepts and missions it enables. This is followed by a review of battery technology requirements for various classes of electrified aircraft. Battery technology advances are reviewed along with their applicability and limitations for expanding the electrified aircraft market segment beyond what is possible today. While batteries are the focus, alternative electrochemical energy storage and conversion systems (such as fuel cells and supercapacitors) are also addressed. The chapter concludes with a review of multifunctional structures with energy storage capability and their potential application for electrified aircraft.

7.1 Battery State of the Art (SOA)

The lithium-ion (Li-ion) battery, schematically shown in Figure 7.1, is the SOA battery used in today's electric vehicles. It consists of a graphite anode, organic liquid electrolyte (e.g., $LiPF_6$ salt dissolved in a mixture of organic solvents), and complex

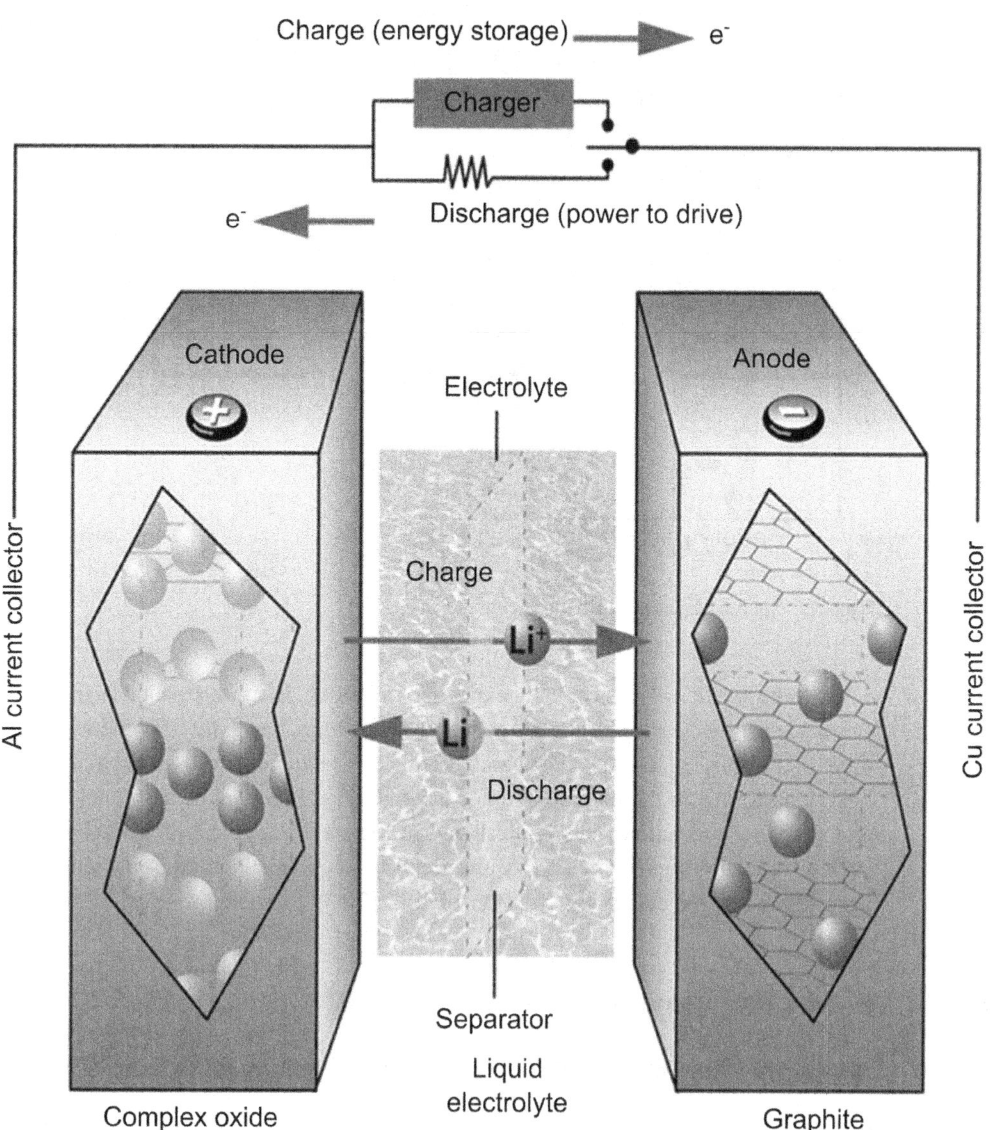

Figure 7.1 Li-ion battery.

oxide cathode such as $LiNiMnCoO_2$ (NMC) or $LiNiCoAlO_2$ (NCA). The anode and cathode are connected to the electric circuit through current collectors, typically aluminum for the cathode and copper for the anode. The separator – critical for cell safety – is a thin, porous membrane that physically separates the anode and cathode. As its name implies, its primary function is to prevent physical contact between the anode and cathode while also facilitating ion transport in the cell.

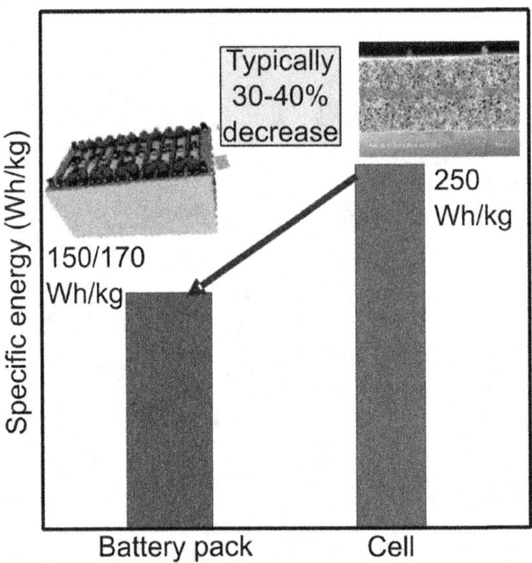

Figure 7.2 Specific energy of a battery cell and pack.

Battery *specific energy*, typically expressed in Wh/kg, is an important design parameter for electrified aircraft. The SOA Li-ion battery has a cell specific energy on the order of 250 Wh/kg. The specific energy is reduced as individual cells are combined in a battery pack, due to the incorporation of the battery and thermal management systems and structural protection. This is illustrated in Figure 7.2. Typically, there is a 30–40 percent reduction in specific energy from cell to pack. As such, SOA Li-ion battery packs have specific energy on the order of 150–170 Wh/kg.

7.2 Battery Requirements for Electrified Aircraft

7.2.1 All-Electric, Electrified Vertical Takeoff and Landing (eVTOL)

The term "eVTOL" is used to describe the family of vertical takeoff and landing (VTOL) aircraft that have partial or fully electric propulsion – i.e., "electrified" VTOL. There is considerable interest in developing all-electric eVTOL aircraft for urban air transportation. Uber has led the charge in the development of an eVTOL shared urban air taxi. In addition to Uber, many companies are developing eVTOL concepts. Both all-electric and hybrid-electric options are being considered. Many of the eVTOL concepts currently being developed are based on distributed EAP, which has multiple motors and fans driven from the same energy storage source. Battery capability, particularly specific energy, is one of the limiting factors for commercial introduction of eVTOL for urban air transportation.

Lilium GmbH, based in Germany, is currently developing an all-electric eVTOL with SOA battery technology [1]. Lilium successfully demonstrated a two-seat

Table 7.1 Gross weight and range as a function of battery pack specific energy for a four-passenger, all-electric eVTOL.

Spec. energy [Wh/kg]	Gross weight [lb]	Range [miles]	Number of passengers	Reference
400–600	3,000	75	4	McDonald and German [5]
250–350	5,000	75	4	McDonald and German [5]
200–300	7,000	75	4	McDonald and German [5]
400	4,800	90–140	4	Brown and Harris [6]
400	4,000	75–115	4	Brown and Harris [6]
400	3,000	45–70	4	Brown and Harris [6]
250–300	5,000	60	4	Brown and Harris [6]
375–425	3,000	60	4	Brown and Harris [6]
400	5,000	100	6	Johnson, Silva, Solis [7]
250	No design convergence	62	4	Finger et al. [8]
500	3,890	62	4	Finger et al. [8]
300	4,200	50	4	Duffy et al. [9]
400	4,500	100	4	Duffy et al. [9]
150	5,500	45	4	Fredericks et al. [10]
150	4,400	30	4	Fredericks et al. [10]

eVTOL aircraft concept in an unmanned flight test in 2017. This aircraft had a speed of 150 mph and a range of 150 miles on one battery charge. Joby Aviation is developing a five-seat eVTOL aircraft that can fly 150 miles on one charge using SOA batteries [2]. Karem Aircraft has been developing a butterfly quad-tiltrotor, all-electric eVTOL concept with five passengers with a range of 84 miles plus a six-mile reserve using SOA battery technology [3]. This satisfies Uber's Elevate vision for an urban air taxi, which calls for high vehicle utilization rates over rush hours in the morning and evening. Modeling by Karem Aircraft has shown that its eVTOL aircraft concept with SOA batteries can complete ten 25-mile missions over the three-hour peak period, with only eight minutes of charging between flights, before needing deeper recharging.

While SOA battery technology can enable the initial introduction of all-electric eVTOLs with limited capability for urban air transportation, many studies have indicated a need for batteries with higher specific energy in order to enable the introduction of all-electric eVTOLs with the desired capabilities. Uber has targeted a specific energy of 300 Wh/kg to enable a four-passenger, all-electric urban air taxi with a 60-mile range [4].

There have been very few system analyses that define battery requirements for eVTOL aircraft. In Table 7.1, the data from six such studies [5–10] are compiled. Although the results vary widely, some general trends can be inferred. The range for a four-passenger, all-electric eVTOL with 4–5 klbs of gross weight is approximately 30 miles with a SOA battery ($\approx$150 Wh/kg pack specific energy), 50–60 miles for a 300 Wh/kg battery pack, and 100 miles for a 400 Wh/kg battery pack. Sensitivity analyses by Brown and Harris [6] indicated that the benefits of higher specific energy diminish above 400 Wh/kg, which led them to conclude that battery pack specific energy of 400 Wh/kg is a critical enabling value for on-demand all-electric eVTOL aircraft. Similar conclusions were also reached by Duffy et al. [9] and Moore and Fredericks [11]. While the near-term goal of a 50–60-mile range for an all-electric, four-passenger eVTOL can be realized with a battery pack specific energy of 300 Wh/kg, the longer-term goal of 100-mile range will require a battery pack specific energy of 400 Wh/kg.

Besides the battery specific energy, *power density* is also important for eVTOL aircraft. Power density relates directly to the C-rating of the battery, which is a measure of the battery discharge rate. The C-rating is defined as the discharge current divided by the theoretical current draw under which the battery would deliver its nominal rated capacity in one hour. A 1C discharge rate would completely discharge the battery's rated capacity in 1 hour. A 2C discharge rate would deliver the battery's rated capacity in half that time (30 minutes), etc.

Analysis by Johnson et al. [7] indicates a 3C or higher discharge rate for an eVTOL during the hover phase for short-range (less than 30 miles) flight. Frederics et al. [10] have shown that for a 73-mile flight, the C-rate could be as high as 4C during takeoff and 5C during the landing phase of the flight, which would create severe thermal management challenges. Frequent operation at such high C ratings would reduce the life of batteries. Uber's all-electric eVTOL requirements specify a battery capable of a 3C discharge rate, but for many eVTOL missions, the required C-rating would be even higher than this. Thus, future batteries must have high power density in addition to high specific energy (HSE).

Developing batteries with sufficient lifetime is another challenge. Typically, SOA batteries have a cycle life of approximately 500, where the battery is 80 percent discharged after each cycle. For an electric vehicle with a 200-mile range for one charge, a 500-cycle life translates to 100,000 miles. The life would be longer if the battery is charged every day at a lower percentage of discharge. For eVTOL, if the range is 60 miles and assuming there are five flights every day, the battery goes through five charge–discharge cycles every day. If the battery has a 500-cycle life, the battery has to be replaced every 100 days. With a 1,000-cycle life, the battery must be replaced every 200 days. A recent trade study for Airbus Vahana [12] has shown a significant drop in direct operating cost as the battery cycle life increases from 500 to 2,000 cycles, after which the benefit of increasing cyclic life diminishes. Uber's cycle life goals for all-electric eVTOL [13] are 500 cycles by 2023, 1,000 cycles by 2028, and 2,000 cycles by 2032. Such high cycle life goals will be challenging as cycle life tends to decrease as specific energy increases.

The range of a four-passenger, all-electric eVTOL is expected to be on the order of 100 miles with a single battery charge as the battery pack specific energy increases from 150 Wh/kg to 400 Wh/kg, which is a significant challenge, as will be seen in later sections. For electrified eVTOL missions requiring a range of greater than 100 miles, hybrid electric options are being developed. In a hybrid electric option, the batteries are used for certain portions of the flight while a gas turbine or internal combustion engine is used for other portions of the flight. The primary benefit of using a hybrid concept is the reduction of operating costs and fuel burn, with the degree of hybridization being a function of the battery capability.

7.2.2 Conventional Takeoff and Landing (CTOL) Aircraft

In addition to eVTOL development, there has been considerable recent interest in developing all-electric CTOL aircraft for the two- and four-passenger general aviation markets. Table 7.2 shows the status of various commercial general aviation activities related to the introduction of all-electric CTOL aircraft using SOA batteries.

SOA Li-ion batteries are also creating new opportunities for the introduction of electrified aircraft for the *thin-haul* commuter market, which refers to a class of aircraft operating short flights (150–300 miles) with 10 passengers or fewer, using smaller airports. The relative quiet operation of electrified aircraft will enable the use of many smaller and regional airports for point-to-point travel. Besides low emissions and noise characteristics, lower operating cost is another major driver for the interest in electrified aircraft for the thin-haul market.

An Israeli start-up company, Eviation, has announced plans for the introduction of a nine-passenger, all-electric aircraft with a speed of 275 miles per hour (mph) and a range of 625 miles [14]. The aircraft will use a 900 kWh battery from Kokam with a pack specific energy of roughly 150 Wh/kg. The batteries will account for 60 percent of the aircraft takeoff weight. A Seattle-based start-up company, Zunum Aero, had targeted the introduction of a 12-passenger, hybrid electric aircraft with a cruising speed of 340 mph and a range of 700 miles before funding issues put their plans on hold in 2019 [15]. In the Zunum design, the battery weight would be 10–20 percent of the aircraft takeoff weight, and the placement of the batteries in the wing would enable quick swaps after each flight.

Table 7.2 Status of all-electric general aviation CTOL aircraft.

Company	Model	Number of passengers	Speed [mph]	Range	Comments
Pipistrel	Alpha Electro	2	100	100 miles	Commercialized as a trainer and for personal travel
Pipistrel	Panthera Electro	4	–	250 miles	Under development
Aero Electric Aircraft	Sun Flyer	2	63–138	3.5 hours	Commercialized as a trainer
Aero Electric Aircraft	Sun Flyer	4	63–150	4 hours	Under development
Airbus	E-fan	2	100	100	Discontinued development
Liaoning Ruixiang General Aviation	RX 1E-A	2	100	200	Flight demonstrated, commercial development in progress, plans for four passenger

For a CTOL (or any electric) aircraft, an increase in battery specific energy will enable longer range. The range R, in miles, of a CTOL aircraft is a function of several factors, including the *weight fraction* of batteries, speed, *lift-to-drag ratio* (L/D), and aircraft configuration. However, for basic quantitative comparisons, the range can be estimated as

$$R = 2.237 \times \eta E_{\text{batt}} \left(\frac{1}{g}\right) \left(\frac{L}{D}\right) \left(\frac{m_{\text{batt}}}{m}\right), \tag{7.1}$$

where E_{batt} is *battery pack specific energy*, in Wh/kg; η is *system efficiency* (i.e., how much electrical storage is converted to mechanical motion, assumed to be 0.8); g is the *acceleration due to gravity*, in m/s^2; m_{batt} is *battery mass*, in kg; and m is the *aircraft mass*, in kg. The term m_{batt}/m is the *battery mass fraction*.

Figure 7.3 shows R as a function of E_{batt} for different values of m_{batt}/m and L/D. For traditional, non-electrified aircraft, the *fuel weight fraction* is m_{fuel}/m, where m_{fuel} is the fuel mass at takeoff. For regional and 737-class aircraft, this ratio is on the order of 0.25, whereas larger, long-range aircraft have fuel weight fractions on the order of 0.4. Translating these numbers to their electric equivalents, one can choose $L/D = 18$ and $m_{\text{batt}}/m = 0.25$, which are the approximate values for small, regional, and 737-class traditional (non-electric) aircraft. The range for an all-electric aircraft with these values is 250 miles for a 300 Wh/kg battery, 330 miles for a 400 Wh/kg battery, and 500 miles for a 600 Wh/kg battery. Apparently, even when using batteries with exceptionally high specific energies, the range for all-electric aircraft is limited to 500 miles. According to (7.1) the range can be increased by increasing the battery weight fraction. There are many ways to do this; one promising method is to integrate the battery with the aircraft structure so that some structures are common to both the battery and the airframe, thereby reducing m and increasing m_{batt}/m. While Figure 7.3 gives some indication of

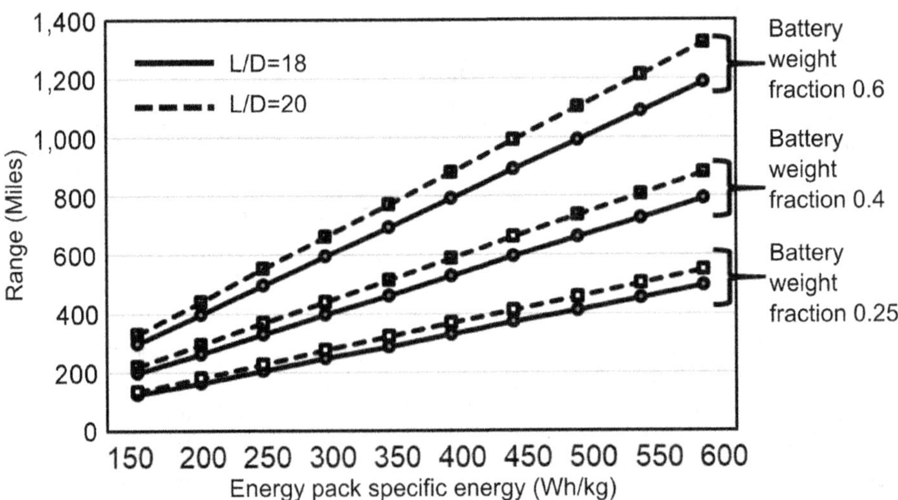

Figure 7.3 R vs. E_{batt} (pack level) for different L/D and m_{batt}/m in all-electric aircraft.

what is possible, the range is expected to be lower than what is shown if factors such as energy reserve requirements and power during takeoff and landing are considered.

There have been many studies analyzing battery requirements as a function of range and block fuel reduction for different classes of aircraft [16–25]. It is difficult to compare these results across studies, as they are based on different methodologies, technology assumptions, and aircraft configurations. More system studies must be undertaken to identify definitively the missions enabled by a specific value of E_{batt}. Despite this, some general trends in the capability of various CTOL aircraft as a function of E_{batt} can be identified based on these studies and the use of the simple range equation, (7.1). These will be summarized in the next section.

7.2.3 Summary of Electrified Aircraft Missions Enabled by Batteries

A notional summary of various eVTOL and CTOL missions that can be enabled by different values of E_{batt} is given in Table 7.3. The information is derived from various system studies and publicly available information. The results should be considered notional.

Table 7.3 eVTOL and CTOL missions enabled by battery packs with different values of E_{batt} (numbers shown should be considered as notional).

E_{batt} [Wh/kg]	eVTOL missions enabled	CTOL missions enabled
SOA (150–170)	• Urban air taxi, 4-passenger, all-electric, 25–30-mile range	• 10-passenger, all-electric and hybrid-electric thin-haul, range could be approximately 200 miles for all-electric and 500–600 miles for hybrid-electric
300	• Urban air taxi, 4-passenger, all-electric, 50-60-mile range	• 10-passenger, all-electric and hybrid-electric thin-haul, range could be approx. 300 miles for all-electric and more than 600 miles for hybrid-electric • 20-passenger commuter, approximately 200 miles all-electric and more than 300 miles hybrid-electric
400	• Urban air taxi, 4–5-passenger all-electric, 100-mile range • Multiple missions, 4–5-passenger, hybrid-electric, 200-mile range	• The same missions listed in the two preceding rows, but with longer range.
500		• 50–70-passenger regional, hybrid-electric, 300–500-mile range
600		• Large transport, narrow body, 180 passengers, 500–600 miles
750+		• 70-passenger regional, hybrid-electric, more than 1,000 miles • Large transport regional, hybrid-electric, 900 miles

The "sweet spot" for battery pack specific energy appears to be in the 300–400 Wh/kg range for a broad range of missions in the urban air mobility and thin-haul transportation markets. This level of specific energy will expand the market for electrified aircraft in these two segments. Increasing the specific energy to 500 Wh/kg will likely enable the introduction of electrified regional aircraft for short-haul travel within a 600-mile range, which accounts for a large percentage of daily flights worldwide [18]. Electrified regional aircraft and large transport aircraft with more than 900 miles of range will require battery specific energy greater than 750 Wh/kg.

7.3 Advances in Battery Technology for Lithium-Based Systems

7.3.1 Incremental Improvements to the SOA Li-Ion Battery

There is much research activity to improve the specific energy of SOA Li-ion batteries with graphite anodes, a good overview of which can be found in [26]. Most of the current research is focused on changing the cathode chemistry from NMC to high-nickel NMC, which has higher nickel content than SOA batteries. The other modification to the cathode includes the addition of a small amount of aluminum. It is forecasted that successful development of high-nickel NMC cathodes will increase the specific energy of cells from 250–270 Wh/kg to 300 Wh/kg. This will translate to pack-level specific energy of roughly 200 Wh/kg. The major advantage of this incremental approach is that new cells can be manufactured on the same production line that has been established for the SOA Li-ion batteries.

7.3.2 Battery Chemistries Beyond Li-Ion

Battery cells that utilize different anode and cathode materials than SOA Li-ion cells are currently being researched to drastically increase the specific energy beyond that of the SOA. Promising candidates include silicon metal anode, lithium metal cathode, sulfur cathode, and air or oxygen cathode. Theoretical specific energies for various anode and cathode combinations are shown in Table 7.4. When viewing these numbers, it is useful to keep in mind that theoretical specific energy is never realized

Table 7.4 Theoretical specific energy for various lithium-based battery options.

Anode	Cathode	Cell E_{batt} (theoretical) [Wh/kg]
Graphite (SOA)	NMC (SOA)	615
Silicon	NMC (SOA)	952
Lithium metal	NMC (SOA)	755
Silicon	Sulfur (Li_2S)	1,550
Lithium metal	Sulfur	2,500
Lithium metal	Air (oxygen)	3,500–5,200

in practice, and the actual specific energy of production batteries will be significantly lower. Details about the options in Table 7.4 are given in the sections that follow.

7.3.2.1 Silicon Anode with Layered Oxide Cathode and Liquid Electrolyte

Silicon is a promising anode material with a high theoretical specific energy. However, there is a large (400 percent) volume change in silicon associated with its charging and discharging reactions. Several approaches are being developed to accommodate the large volume change. One approach is to have an internally porous Si-C nanocomposite anode structure in which the internal nanoporous structure accommodates the large volume change [27]. The concept is being commercialized by Sila Nanotechnologies, which aims to increase the specific energy of Li-ion batteries by 40–45 percent [28]. An initial version of this silicon-anode battery is likely to exhibit a 20–25 percent increase in specific energy over the SOA, with the initial version from Sila Nanotechnology possessing cell specific energy on the order of 325 Wh/kg. Sila's ultimate cell specific energy target is 400 Wh/kg.

Another approach to accommodate silicon's volume expansion is to have three-dimensional (3-D) silicon nanowire anodes [29]. This technology, developed by Stanford University, is being commercialized by Amprius [30] and has thus far been used in small batteries (e.g., for cell phones and laptops). Recently, a battery using a 3-D silicon-nanowire anode powered the Airbus Zephyr unmanned aerial vehicle (UAV) high-altitude pseudo-satellite platform, which is intended to enable affordable, local satellite-like services. The goal of Amprius is to achieve specific energy of 435–450 Wh/kg for silicon-anode-based cells.

Despite the initial success demonstrated by nanostructured silicon anodes, there are challenges to manufacturing this technology at the scale required for electric vehicles and aircraft. It is expected that for large-scale applications, silicon-anode batteries might achieve cell specific energy on the order of 325–400 Wh/kg, which translates to pack specific energy of 225–280 Wh/kg. Studies by Gallagher et al. have indicated that the maximum pack-level specific energy that can be achieved using a silicon anode and NMC cathode is on the order of 260 Wh/kg [31].

7.3.2.2 Lithium-Metal Anode with Layered-Oxide Cathode and Liquid Electrolyte

An achievable cell specific energy when using a lithium-metal anode with a layered-oxide cathode (e.g., NMC) and liquid electrolyte is approximately 440 Wh/kg [32]. However, there are several challenges when using a lithium-metal anode with a liquid electrolyte [32, 33]. One of these is the cell short-circuiting, which is caused by the interfacial instability of the lithium metal–liquid electrolyte interface. The other challenge is increasing/overcoming the low cycle life caused by the continuous consumption of the liquid electrolyte and lithium. Several approaches are being pursued to address these challenges, which include interface engineering to modify the lithium metal–liquid electrolyte interface, alternate liquid electrolytes, and engineering of the anode architecture and morphology to accommodate volume expansion.

SolidEnergy is a start-up company developing a lithium-metal, semisolid battery with the potential for HSE [34]. The SolidEnergy approach consists of coating an ultrathin lithium metal foil anode with a mixed polymer-ceramic electrolyte and using a paste of solvent-in-salt liquid electrolyte dispersed in the high-nickel NMC cathode. The claimed cell specific energy reported by SolidEnergy is on the order of 500 Wh/kg [34]. The cell can only be recharged 200 times, compared to approximately 1,000 times for a SOA Li-ion battery. The SolidEnergy website [35] reports cell specific energy of 450 Wh/kg for 0.1 C rating, 408 Wh/kg for 1C rating, and 381 Wh/kg for 3C rating. While this HSE battery can find application in smaller vehicles (e.g., drones) today, its cycle life must be increased and its manufacturing process must be scaled up before it can find application in larger electrified aircraft for passenger transportation in the future.

Assuming cell specific energy in the range of 400–450 Wh/kg using a Li-metal anode, a liquid electrolyte, and an NMC cathode, the pack specific energy would be on the order of 280–315 Wh/kg, which is close to that predicted by Gallagher et al. [31]. It appears that a pack specific energy of 300 Wh/kg is achievable with a Li-metal anode and NMC cathode.

7.3.2.3 Lithium-Metal Anode with Sulfur Cathode and Liquid Electrolyte

The theoretical specific energy of a lithium-metal anode, sulfur-cathode cell is 2,560 Wh/kg, but practical and achievable values are on the order of 600–700 Wh/kg [36]. A number of companies and organizations claim to have achieved 400–450 Wh/kg with this technology in the laboratory [37–39]. These so-called lithium-sulfur (Li-S) HSE batteries have a low cycle life of less than 100 cycles, compared to the life of more than 1,000 cycles capable with Li-ion batteries [40]. Developmental Li-S batteries also have a lower C-rating. Major challenges for developing commercially viable Li-S batteries include (1) poor electronic and ionic conductivities for sulfur and its reaction products; (2) the transport of sulfur back and forth between positive and negative electrodes, called *polysulfide shuttle*; (3) large volume changes during the charging and discharging process (similar to silicon), which limits cycle life and results in capacity fade; and (4) lack of understanding of the various reactions involved [41]. Studies by Eroglu et al. show that the pack specific energy for Li-S batteries with liquid electrolytes is expected to be in the range of 150–450 Wh/kg [42]. The wide range reflects the uncertainties associated with the development of Li-S batteries with liquid electrolyte.

While the Li-S battery with liquid electrolyte has the potential to achieve cell specific energy of 500–700 Wh/kg, the performance of Li-S batteries currently under development makes them suitable for applications requiring low-mass batteries with low power and cycle life needs, such as high-altitude, long-endurance (HALE) UAVs. The cycle life and C-rating of Li-S batteries must be improved significantly before they become viable for electrified aircraft applications. Current research and development (R&D) activities on Li-S batteries include interface coatings for the lithium-metal anode, new electrolytes, interfacing the cathode architecture with the conductive network, and increasing the sulfur loading in the cathode.

7.3.2.4 Lithium-Metal Anode with Air/Oxygen Cathode and Liquid Electrolyte

Lithium-air (Li-air) or lithium-oxygen (Li-O) systems are attractive because of their very high theoretical specific energy. There has been a significant R&D effort to develop Li-air batteries during the last 15 years or so, but the major challenges are still the same as those identified in a 2010 paper by Girishkumar et al. [43]. The challenges include (1) very low power density, (2) low electrical energy efficiency, (3) low cycle life (less than 50–100 cycles), (4) the need to separate oxygen from ambient air to avoid contamination from moisture and carbon dioxide, (5) the instability of the electrolyte and the cathode interface under the stringent environment of both oxygen reduction and evolution reactions, and (5) a lack of understanding of various charge and discharge reactions. Current R&D activities are focused on the development of stable electrolytes, nanostructured and hierarchical porous cathode structures, and controlling the interfacial instability of the metallic lithium anode.

Even with the advances that are likely to come from new cell chemistries and design, the Li-air/Li-O system is expected to be very complex due to the need for additional mechanical components related to the introduction of oxygen and purification of air. Gallagher et al. have shown that the maximum pack specific energy achievable for a Li-O battery is on the order of 350 Wh/kg [31]. Such low specific energy at the pack level is due to the mechanical complexities of the Li-O system. Research on Li-air/Li-O batteries is in the early stages, and there are major technical barriers that need to be overcome before practical application.

7.3.2.5 All-Solid-State Battery with Lithium-Metal Anode

The discovery of new ceramic electrolytes with high ionic conductivity has generated significant interest in the development of all-solid-state batteries (ASSBs), which offer several advantages over batteries with liquid electrolytes. First, they do not contain any flammable liquid, which significantly enhances battery safety. Second, bipolar battery stacks can be manufactured with all solid components, increasing the specific energy at the battery pack level [44]. The thermal management system and battery management system can be simpler for ASSBs. Figure 7.4 shows the potential for achieving HSE at the pack level for ASSBs. It is anticipated the pack specific energy of ASSBs can be 90 percent of the cell specific energy, compared to 60–70 percent for batteries with liquid electrolyte.

There is significant ongoing research by industry, academia, national laboratories, and many start-up companies to develop solid-state batteries for electric vehicles. Although many industries are actively working to develop ASSBs, very few have reported any specific energy numbers or other performance numbers. One company, Solid Power, claims to have achieved cell specific energy of 350–400 Wh/kg [45].

Calculations on model ASSB cells have shown the potential for ASSBs to achieve cell specific energy of 425–480 Wh/kg using a lithium-metal anode and NMC cathode [46]. The cell specific energy is a function of several parameters, including electrolyte thickness, solid electrolyte volume fraction in the cathode, and lithium metal thickness. ASSBs with a lithium-metal anode and sulfur cathode offer the potential of

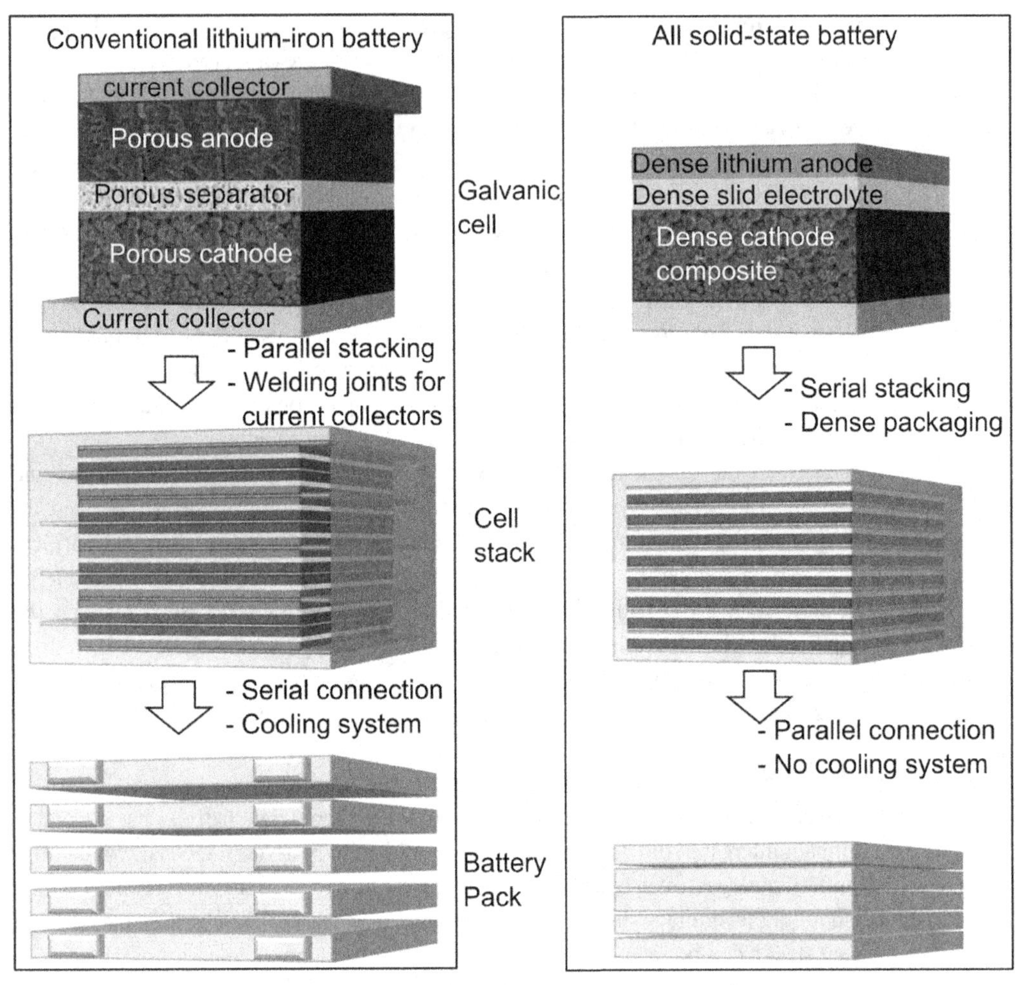

Figure 7.4 Potential for higher pack specific energy for ASSBs. © [2018] Elsevier. Reprinted, with permission, from [44]

achieving cell specific energies of more than 600 Wh/kg with the right combination of electrolyte type, electrolyte thickness, and sulfur loading in the cathode [47].

There is no consensus on the time frame for initial introduction of ASSBs in electric vehicles. Some believe that ASSBs could be introduced as early as 2022–2025, while others believe that commercialization of ASSBs is still a decade away. Irrespective of the forecast, there are significant challenges that need to be overcome before ASSBs can find application in electric vehicles. Some of these include the mechanical incompatibility of various layers leading to the movement of layers in each cycle, poor contact between the electrolyte and the cathode, manufacturing of batteries at a scale required for electric vehicles, low cycle life, low power density, and high cost. Some of the approaches to mitigate the challenges include 3-D electrode structures, interdigitated structures, hybrid polymer-ceramic-compliant electrolyte, innovative manufacturing processes, and new cathode chemistries [44, 47].

7.3.3 Progression of Battery Technology Development

A notional progression of battery technology development is shown in Figure 7.5. Both cell specific energy and pack specific energy are shown at different pack-specific-energy-to-cell-specific-energy ratios, with 0.7 corresponding to batteries with liquid electrolytes and 0.9 corresponding to ASSBs. The specific energy numbers shown for lithium-metal, high-nickel NMC and Li-S are for both liquid and solid-state batteries. The specific energy shown for the silicon-anode battery corresponds to a cell with a liquid electrolyte.

Any viable advanced cell technology must enable battery packs with at least 300 Wh/kg specific energy, which is required for all-electric urban air taxi and thin-haul aircrafts with a reasonable range. On the other hand, 400 Wh/kg, which is the "sweet spot" to enable a broad range of urban air and thin-haul transport, can currently only be achieved with ASSBs with a lithium-metal cathode and high-nickel NMC anode (at the upper limit of cell specific energy) or with Li-S batteries. It is likely that a pack specific energy of 500 Wh/kg, which is required to enable electrified regional aircraft, can only be achieved with Li-S ASSBs. A pack specific energy of 600 Wh/kg, which is required to electrify large narrow body transport aircraft with 500–600-mile range, can currently only be achieved at the upper limits of solid-state Li-S batteries. The path to achieve pack specific energy of greater than 600 Wh/kg is currently unclear. Clearly, expanding the electrified aircraft market for a broad range of applications will require the maturation of ASSBs and Li-S batteries.

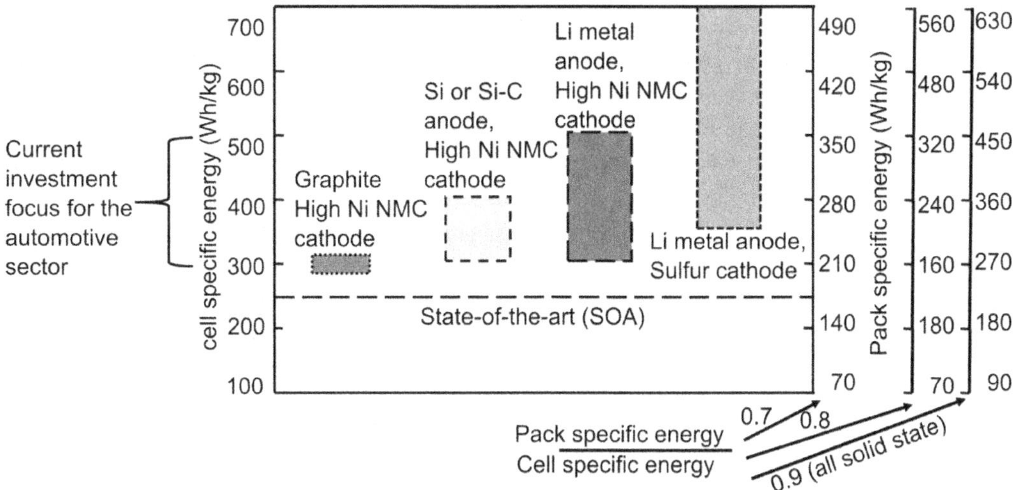

Figure 7.5 Notional evolution of battery technology.

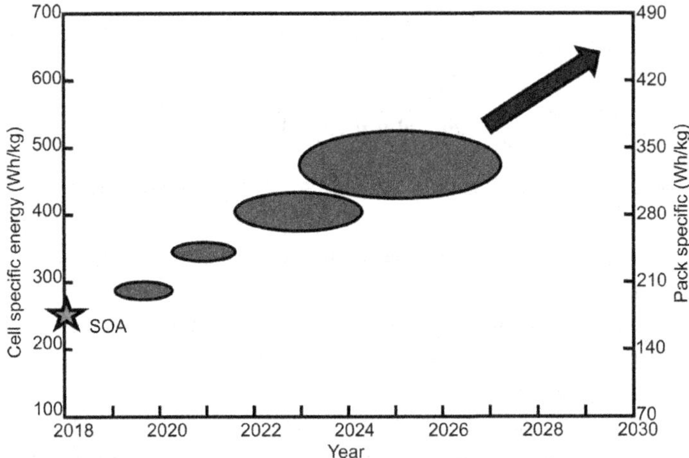

Figure 7.6 Notional increase in specific energy as a function of time for lithium batteries with liquid electrolyte.

Crafting a time frame for achieving any particular cell and pack specific energy would be highly speculative. However, there are some general trends that can be identified based on the significant ongoing worldwide research. For example, the US Department of Energy (DOE) is funding the Batt500 program, which aims to develop cell's with specific energy in the range of 400–500 Wh/kg, which translates to approximately 300 Wh/kg at the pack level for liquid electrolytes and 300–400 Wh/kg for ASSBs [48]. Based on the technical challenges related to the development of HSE batteries and the rate of progress being made, it is likely that batteries with pack specific energy of 300 Wh/kg will be available in the 2022–2025 time frame. Battery packs with specific energy of 400 Wh/kg are likely to be available in the 2030 time frame, assuming there are R&D investments to develop cells with specific energy greater than 500 Wh/kg. ASSBs with 300–350 Wh/kg pack specific energy might be available in the 2025 time frame and with 400 Wh/kg in the 2030 time frame. It is also likely that ASSBs for large power applications will not be available for another 10 years, considering the significant technical challenges. A notional projection of the increase in specific energy as a function of time is shown in Figure 7.6 for liquid electrolytes. These projections may be optimistic and will require on average an 8 percent increase in specific energy per year, which is higher than the historical value of 5 percent.

7.4 Other Battery Concepts

7.4.1 Flow Batteries

A *flow battery* is an electrochemical device that converts the chemical energy in the electroactive materials directly to electrical energy, similar to a conventional battery or

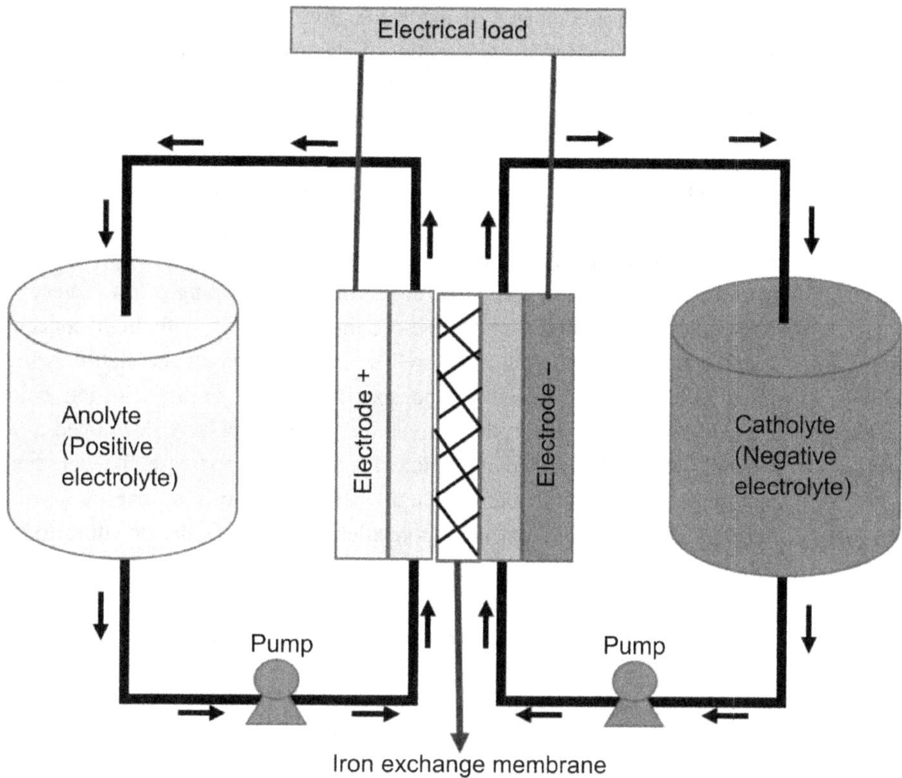

Figure 7.7 Flow battery schematic.

a fuel cell. The electroactive materials in a flow battery are stored externally in an electrolyte and introduced into the device only during operation. Systems in which all the electroactive materials are dissolved in a liquid electrolyte are called *redox flow batteries*. A schematic of a flow battery is shown in Figure 7.7.

A flow battery consists of electrochemical cells with porous electrodes separated by ion-selective, ionically conducting membranes. Energy storage tanks located outside of the electrochemical cell contain the electrolyte, which is comprised of the desired amount of dissolved redox species and soluble ions that transport across the membrane. A pumping system feeds the electrolyte to the electrochemical cell. Electrochemical reactions occur only on the surface of the porous electrodes in the reaction channel, which are typically comprised of porous carbon. Flow batteries can be recharged either by plugging into the grid or by swapping out spent with charged electrolyte. However, unlike conventional rechargeable batteries, the energy capacity and power are governed by separate components of a flow battery. The energy capacity is determined by the size of the electrolyte tanks, while the power is determined by the size of the electrochemical cell and stack. This allows power and energy density to be adjusted independently.

The major disadvantages of the redox flow battery are its extremely low energy storage density (limited by the amount of electroactive materials that can be dissolved in the electrolyte), the number of electrons involved in the electrochemical reaction, and low voltage. With the progress in computational design of organic molecules, molecular engineering approaches are being used to engineer the electrolyte and electroactive materials to increase energy density. One approach is to use insoluble solid particles as the electroactive materials that participate in the redox reactions. In this approach, the energy density is no longer limited by solubility limitations. A start-up company, Influit Energy, based in Illinois, is developing a flow battery with unique formulations of cathodic and anodic nanoelectrofuels with high concentrations of suspended active materials that are stably dispersed in the liquid, which has the potential to achieve 1.5 times the specific energy of SOA Li-ion batteries [49]. Recently, researchers from the University of Glasgow have developed a flow battery concept with concentrated nanomolecules that can increase the volumetric energy density to 225 Wh/L, which is roughly 10 times the SOA energy density for flow batteries. The concentrated nanomolecules concept has the potential to increase the volumetric energy density of flow batteries to 1,000 Wh/L, which is higher than the energy density of SOA Li-ion batteries (700 Wh/L) [50].

A novel concept that combines the best features of both a proton exchange membrane (PEM) fuel cell and a flow battery offers the potential for significant increases in energy density of flow batteries. In this approach, reversible partial electrochemical dehydrogenation of liquid organic fuels is combined with oxygen reduction at the cathode [51]. The concept, shown in Figure 7.8, uses two electrochemical reactions: (1) the anode reaction electro-dehydrogenates LH_n, a liquid organic hydrogen carrier (LOHC), to generate protons and electrons in the presence

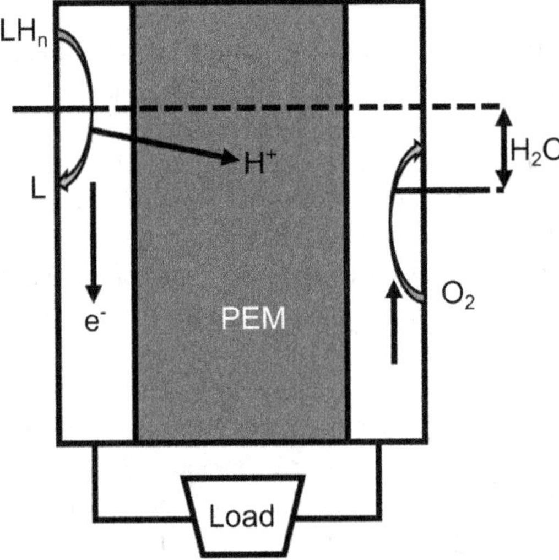

Figure 7.8 Fuel cell/flow battery based on dehydrogenation of liquid organic hydrogen carriers [51].

of an electrocatalyst, and (2) the protons that pass through the PEM react with oxygen at the cathode to generate water. The overall reaction is the partial oxidation of the LOHC. To recharge the flow battery, the reactions are reversed, and the organic liquid is either electrochemically rehydrogenated or rapidly replaced with the hydrogenated form at a refueling station. The system is a flow battery in which the high-energy hydrogenated fuel is stored separately from the electrochemical cell. The authors in [52] claim that the energy density can be as high as 1,350 Wh/kg, five to six times higher than the SOA Li-ion battery. Even so, there is significant opportunity to engineer the liquid organic hydrogen carrier and the solvent through molecular engineering using computational material design tools.

7.4.2 Aluminum-Air (Al-Air) Batteries

Aluminum-air (Al-air) batteries are gaining considerable attention because of their high theoretical specific energy (8,100 Wh/kg, second best to Li-air, which has a 11,000 Wh/kg theoretical specific energy), high theoretical voltage, inherent safety, material abundance, and low cost. The battery has several challenges, such as corrosion of the aluminum in the liquid electrolyte, a high self-discharge rate, a lack of rechargeability, and short shelf life [53]. The overall energy density is limited by any discharge or open-circuit corrosion that occurs during storage. This problem has prevented the widespread use of Al-air batteries. Approaches to mitigate these challenges include the use of aluminum alloys, the addition of corrosion inhibitors in the electrolyte, coatings on the aluminum anode, and electrolyte replacement.

The Al-air battery has found early application as a primary battery. Phinergy, an Israeli start-up company, is commercializing an Al-air battery targeting application as a range extender for electric vehicles, with the Al-air primary battery used to charge the onboard Li-ion battery. The Phinergy design uses an electrolyte consisting of deionized (drinkable) water mixed with an alkaline solution and a Teflon-based cathode [54]. The aluminum anode is consumed as it slowly dissolves in the electrolyte, resulting in electric current. The anode plates are swapped with new ones once the aluminum is fully consumed. Phinergy has demonstrated 1,850–2,500 miles of driving range with one charge when using 100 kg of aluminum. The anticipated battery life is one year in this application.

There is ongoing research to address the challenge of aluminum corrosion in Al-air batteries. Fuji Pigment Company has demonstrated that self-corrosion of the aluminum anode can be prevented by having alumina as a buffer between the aluminum anode and salt electrolyte and frequently refilling the salt solution [55]. The alumina buffer reduces corrosion by suppressing accumulation of the by-products on the aluminum anode, prolonging the battery life. The battery life is easily extended by simply refilling the salty water electrolyte occasionally. Researchers at MIT have developed a method to prevent corrosion of by introducing an oil barrier between the aluminum electrode and the electrolyte [56]. The oil is rapidly pumped away and replaced with electrolyte as soon as the battery is used, which has been demonstrated to decrease the energy loss by a factor of a thousand. The MIT researchers estimate that when an oil-protected Al-air battery pack is scaled up to the size needed for electric vehicles, it will have one-fifth the weight and half the size of a Li-ion battery pack.

Recently, an Al-air flow battery concept has been developed based on a novel silver nanoparticle seed-mediated silver manganate nanoplate architecture used for the oxygen reduction reaction [57]. This unique catalyst increased the discharge capacity by a factor of 17 compared to a conventional Al-air battery. It is predicted that this Al-air flow battery can achieve 2,500 Wh/kg cell specific energy.

Although in its infancy, the Al-air battery has potential for electrified aircraft applications. Vegh and Alonso have studied the design of short-range, Al-air-powered electric aircraft [58]. Their analyses have shown that the operating costs of electric aircraft can be substantially lower if Al-air primary batteries are used to power the aircraft. While primary Al-air batteries can be used for the electrified aircraft in the near term, rechargeable Al-air batteries will be attractive if commercially viable technologies can be developed to prevent aluminum corrosion in liquid electrolytes. ASSBs for the Al-air battery system can also be envisioned for the future.

7.4.3 Batteries Incorporating Molten Materials

There is increasing interest in the development of batteries with either molten electrolytes or molten electrodes that operate at temperatures above the melting point of the battery constituents. Molten salt electrolytes might be attractive for Li-O batteries for which degradation of the organic electrolytes and cathode materials by the reactive oxygen species limits the reversibility of the battery system. There have been several recent investigations on the use of molten salt electrolytes for Li-O batteries with demonstration of improved performance at laboratory scales [59–61]. The concept of a new class of molten air rechargeable batteries using highly conductive molten electrolytes and very-high-capacity multiple-electron compounds such as carbon, iron, and vanadium diboride show promise of intrinsic higher-energy storage capacity compared to the Li-O system [62]. Another molten battery concept is based on a solid electrolyte in combination with liquid metal electrodes [63].

Batteries incorporating molten materials are in an early stage of development and can potentially overcome the various challenges associated with metal-air batteries. However, for this to become a reality, molten batteries must be developed that can operate at lower temperatures in order to achieve the long-term durability that is required for electric aircraft. In addition, there are significant challenges related to the development of engineering systems incorporating molten salts and metals. Nonetheless, the immense potential warrants further R&D investment.

7.5 Fuel Cells

Fuel cells convert energy from chemical fuel into electricity by combining H_2 and CO in the fuel with oxygen, producing water and carbon dioxide as a by-product. Two types of fuel cells are of current interest for the electrified aircraft application. One is the *proton*

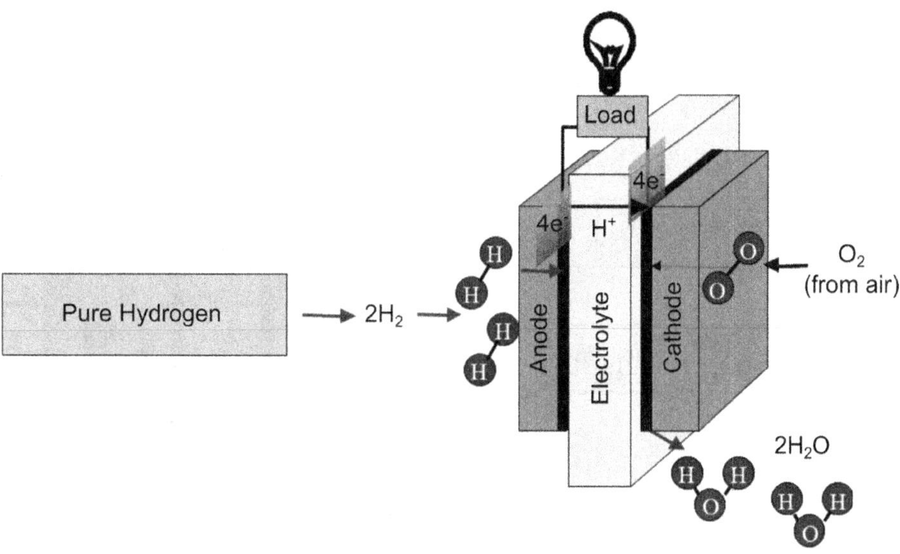

Figure 7.9 PEMFC schematic.

exchange membrane fuel cell (PEMFC), which has a polymer electrolyte that can transport H^+ ions. The principles of PEMFC operation are shown in Figure 7.9. The PEMFC operates at low temperature (approximately 80 °C) and requires pure hydrogen, because PEMFC performance is highly sensitive to impurities such as CO and S that are produced during the reforming of the hydrocarbon-based fuel to hydrogen.

The *solid oxide fuel cell* (SOFC), a schematic of which is shown in Figure 7.10, uses a ceramic electrolyte that transports oxygen ions. It operates at high temperature (600 °C–850 °C) and can use H_2 or $H_2 + CO$ generated by the reforming of hydrocarbon fuel. Direct utilization of hydrocarbon fuel is also possible with the SOFC system as the fuel cell has the capability for internal reforming.

The conversion efficiencies of PEMFC and SOFC are on the order of 50–60 percent. The fuel cell stack shown in Figure 7.11 consists of cells and bipolar plates stacked together.

7.5.1 Application of Proton Exchange Membrane Fuel Cell

In 2008, Boeing was the first to conduct a test flight with a fuel-cell-powered small plane, using pure hydrogen fuel [64]. The plane, a two-seat Dimona motor-glider, was modified by Boeing to include a hybrid PEMFC/Li-ion battery system to power an electric motor coupled to a conventional propeller. The batteries were used during the takeoff phase of the flight. During the three test flights, the plane's pilot climbed to 3,300 feet using a combination of battery power and fuel cell. After reaching cruising altitude and disconnecting the plane's batteries, the pilot was able to fly straight and level at 62 mph for 20 minutes using power generated by fuel cells alone.

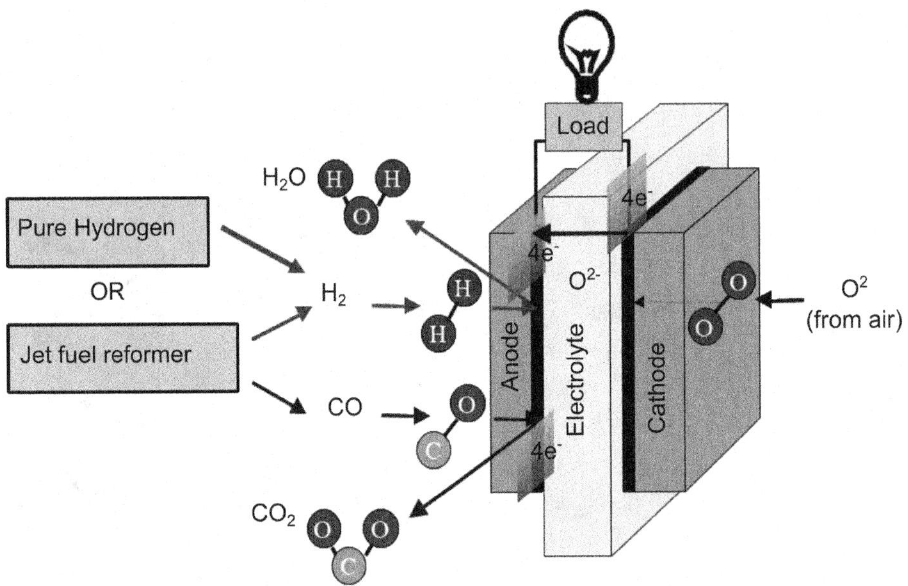

Figure 7.10 SOFC schematic.

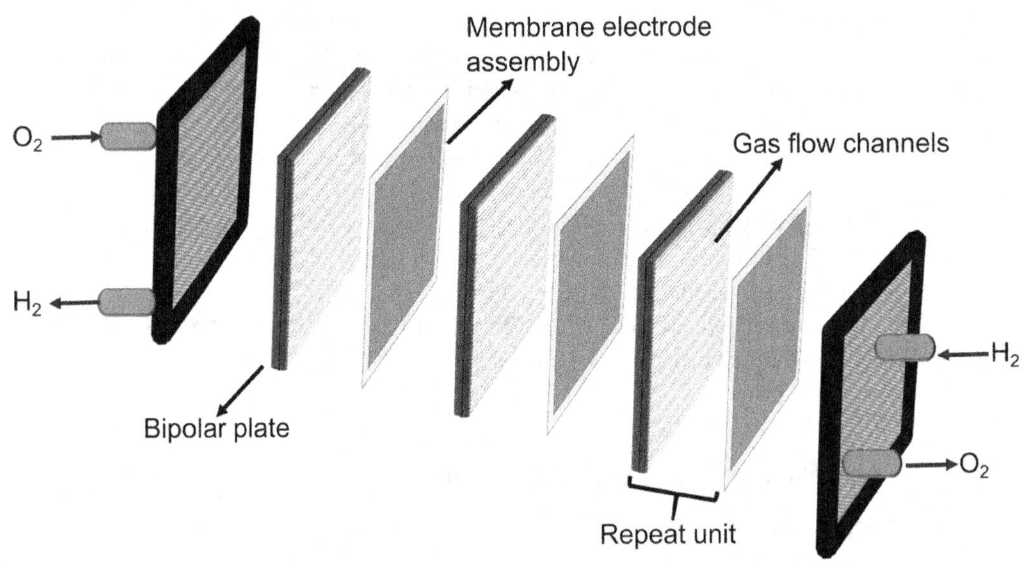

Figure 7.11 Fuel cell stack schematic.

The HY4 four-seater hydrogen-fuel-cell-powered passenger aircraft operated by DLR of Germany is the first of its kind in the world to be powered by a hybrid PEMFC-battery system [65–66]. The HY4 has a motor output of 80 kW and maximum and cruising speeds of approximately 125 and 90 mph, respectively. It can fly

Table 7.5 Aircraft PEMFC system-level power density requirements.

Aircraft Type	Power density required [kW/kg]
3-passenger general aviation	0.625
5-passenger business jet	0.84
8–9 passenger twin turboprop	0.88
37–40 passenger regional jet	1

470 miles on gaseous hydrogen and 930 miles on liquefied hydrogen. In comparison, when the aircraft is 100 percent battery powered, its range is only 200 miles. The experimental plane successfully completed a flight test in 2016. Small passenger aircraft, such as the HY4, can potentially be used in regional transport as electric air taxis for urban air mobility. For example, there are plans to develop 19-passenger regional hoppers using a hydrogen-fuel-based PEMFC energy storage system.

Recently, China has become the third nation in the world to successfully test an aircraft powered by hydrogen fuel. China's first manned fuel cell aircraft, powered by a 20 kW PEMFC, completed its maiden flight in 2017 [67]. The single-seat aircraft was powered by fuel cells and lithium batteries during its takeoff and climb, and entirely by fuel cells during the cruise phase. A charging time of 90 minutes enabled the aircraft to fly for 45–60 minutes.

The high energy density of SOA PEMFCs, on the order of 600–800 Wh/kg, is attractive for electrified aircraft applications. However, the low power density of the SOA PEMFC system limits its application. The power density of the SOA PEMFC stack is roughly 1 kW/kg, which translates to 0.5 kW/kg or lower at the system level. System studies have identified the PEMFC power density requirements at a system level for different types of aircraft [68], which is summarized in Table 7.5.

Analyses by Kadyk et al. have shown that for a 737-class large aircraft, the mass of the power system and fuel storage system using a 1.6 kW/kg PEMFC would be twice that of the two engines and fuel tank, which makes it impractical for using PEMFC for larger airplanes [69]. The PEMFC becomes a viable option for 737-class aircraft if the power density can be increased to 8–10 kW/kg.

The DOE PEMFC power density target for automotive applications is 0.65 kW/kg [70]. The PEMFC development effort for automotive and ground power applications is currently focused on increasing durability and reducing cost. Increasing the power density of the PEMFC system will require technology advancements in multiple fronts, such as a lightweight bipolar plate, low catalyst loading, thinner membranes, lightweight thermal management system, and balance-of-plant. Unfortunately, increasing power density is not the current focus of R&D for automobile and ground power applications.

7.5.2 Application of Solid Oxide Fuel Cell

The SOFC system has efficiency of approximately 60 percent when used in combination with a gas turbine, and it can directly utilize jet fuel with reforming and desulfurization. These two attributes render it attractive for powering electrified aircraft using fuel currently available at airports. Like the PEMFC, the low power density of the SOA SOFC systems (about 0.3 kW/kg) limits its application to smaller aircraft. Boeing has proposed a flight demonstrator concept to demonstrate a 50 percent reduction in fuel cost for a Cessna aircraft powered by the SOA SOFC during cruise and a combination of SOFC and battery power during takeoff and climb [71]. US National Aeronautics and Space Administration (NASA) studies using the Boeing SOFC system show more than a 40 percent reduction in fuel flow by mass at cruise, and more than a 50 percent reduction in fuel flow by volume [72]. NASA has recently completed an effort, named FUELEAP, the goal of which was to establish the feasibility of an integrated heavy fuel hybrid electric SOFC power system for a small electric airplane. The goal was to have a power system with greater than 75 kW capability, a power density greater than 300 W/kg for the entire system, and fuel-to-electricity efficiency greater than 60 percent.

While the power density of the SOA SOFC system is adequate for application in a small electric aircraft, significant increases in power density are required to power larger aircraft. Analysis by Borer et al. has shown that for the application of the SOFC in a large, single-aisle aircraft, the power-to-weight and power-to-volume ratios must improve by a factor of four relative to the current capability [73]. In addition, the fuel cell power output deterioration rates must be less than 2 percent per 10,000 hours of operation. Increasing the power density of the SOFC system will require decreasing the thickness of individual layers in the cell (cathode, anode, and electrolyte) and the bipolar plate through innovative manufacturing processes, engineering the pore structure for the anode and cathode, along with reduction in weight and complexity of the balance-of-plant.

7.5.3 Supercapacitors

Supercapacitors, also known as electrochemical double layer capacitors, store energy by creating a very thin double layer of charge between two plates made from highly porous materials soaked in an electrolyte. An example is shown in Figure 7.12.

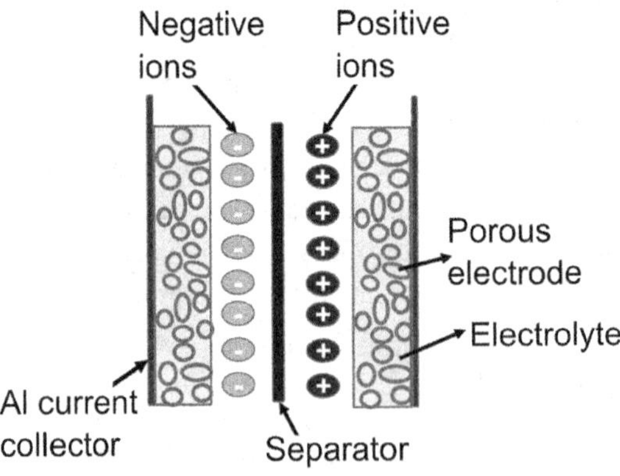

Figure 7.12 Schematic of a supercapacitor.

Compared to those of a standard capacitor, the supercapacitor plates have a larger surface area and less separation, which enables them to store more energy. Supercapacitors have higher power density but lower energy density than batteries and fuel cells. Supercapacitors are receiving considerable attention because of their high efficiency, rapid charge and discharge capability, simple support circuitry, long cycle life, and safety. The high power density of supercapacitors (up to 10 kW/kg) enables rapid charging and discharging, which is ideal if a sudden burst of power is required. The calendar life of supercapacitors can be greater than 15 years, compared to 5 years or less for batteries. The primary drawback of supercapacitors is their low energy density, which is less than 10 Wh/kg (compared to 150–180 Wh/kg for Li-ion batteries) for commercial SOA systems.

There is continuing research to increase the energy density of supercapacitors without sacrificing their high power density. Energy density numbers reported by various research groups range from 50 Wh/kg to 150 Wh/kg. These are based on laboratory scale experiments and have yet to be demonstrated at production scale. The advances in energy density are due to the use of nanotechnology, particularly technologies associated with graphene and carbon nanotubes. A hybrid battery/supercapacitor combination, where one electrode acts as a battery and the other acts as a supercapacitor, is another option that is being actively researched.

A breakthrough in supercapacitor technology with the potential to achieve specific energy of 180 Wh/kg (similar to that of Li-ion batteries) has been reported by researchers from the Universities of Bristol and Surrey [75]. The research team has been working with a start-up company, Superdielectrics, Ltd., to commercialize the product. Recently, Rolls-Royce signed a collaboration agreement with Superdielectrics to explore the potential of the recently discovered supercapacitor material to create next-generation high-energy density storage technology [76].

A comparison of supercapacitor capabilities with that of other energy storage systems is shown in Figure 7.13. With continued research in supercapacitors, it is expected that they will have similar specific energy as Li-ion batteries, but with higher specific power. Although the realization of a commercially viable, stand-alone supercapacitor energy storage system may be years away, supercapacitors with energy density in the range of 50–100 Wh/kg can find applications in combination with batteries or fuel cells. For example, a supercapacitor with high power density and reasonably high energy density can be used during the takeoff and climb portions of the flight, while the battery or fuel cell can be used during the cruise portion of the flight. This will have the benefit of using a smaller battery or fuel cell for the electrified aircraft.

7.5.4 Multifunctional Energy Storage Structure

In a conventional electrified aircraft, batteries are a stand-alone component of the overall aircraft structure, adding significant mass to the aircraft. However, if the aircraft structure can have some energy storage capability or the battery can carry some structural load, the overall system mass can be reduced, effectively increasing

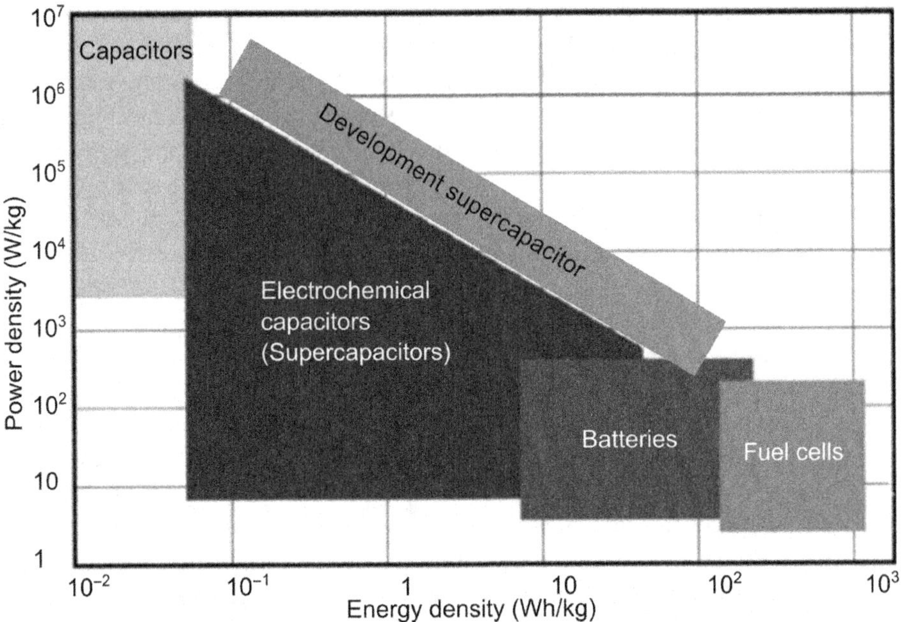

Figure 7.13 Supercapacitor compared to batteries and fuel cells.

the system-level specific energy. This concept forms the basis of multifunctional structures, or components with energy storage capability, which offer the potential for significant weight savings.

The weight saving potential of a multifunctional energy storage component is illustrated here with an example from the paper by Wetzel [77]. Assume that the system consists of 2,000 kg of polymer composite structure and 750 kg of battery. The *structural mass efficiency* is defined by σ_s, which represents the ratio of some structural property (e.g., stiffness or strength) relative to that of the polymer composite. For the polymer composite without any energy storage capability, σ_s is 1. For a battery with no load-bearing capability, σ_s is zero. Similarly, an *energy mass efficiency* (or energy storage capability) is defined as σ_c. For the original battery σ_c is 1. For the polymer composite without any energy storage capability, σ_c is zero. If the battery has full structural capability without degrading any energy storage capability, both its σ_s and σ_c are 1, which means that 750 kg of polymer composite structure is not required, resulting in weight savings of 750 kg for the system. Clearly, this situation is unrealistic, as a trade-off will exist, and the battery will lose energy storage capacity as it gains load-bearing capability. This trade-off is embodied in the following calculation of system mass:

$$\text{System mass} = \text{Polymer composite mass} + (1 - \sigma_s^{\text{batt}})(\text{Battery mass} \;/\; \sigma_c^{\text{batt}}). \tag{7.2}$$

If structural battery has 80 percent of its reference energy storage capability ($\sigma_c = 0.8$) and has 47 percent of the structural capability of the composite

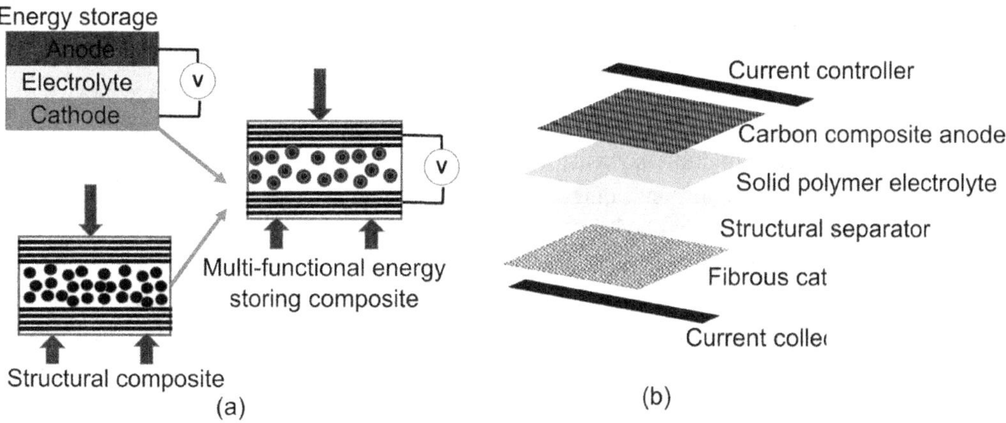

Figure 7.14 Concepts for multifunctional structures with energy storage capability: (a) composite with energy storage capability in fibers; (b) composite laminates with energy storage capability.

($\sigma_s = 0.47$), the system mass is 2,496 kg. Compared to 2,750 kg for the reference system (without load bearing capability for the battery), this combination reduces the mass by 9.2 percent. Other combinations with higher weight reduction can also be envisioned.

Multifunctionality in an energy storage structure can be achieved in two ways. First, two distinctive constituent components can be packaged together. An example is the incorporation of a battery skin on the surface of the wing of a UAV [78]. The other option is to have the constituents simultaneously and synergistically undertake two roles. Examples include multifunctionality for each layer of a composite laminate [77] and fibers with combined electrochemical and structural properties [79]. These concepts are schematically shown in Figure 7.14. These structures will require fibers having electrochemical properties in addition to their load bearing capabilities, plus solid-state electrolytes with desired structural properties. There is ongoing research to evaluate materials with both electrochemical and mechanical properties and to develop new materials with balanced properties [79–80]. The success of multifunctional energy storage structures with the constituents simultaneously and synergistically undertaking two roles hinges upon the development of multifunctional materials with desired properties and advanced manufacturing techniques to fabricate the structure.

Besides a reduction in system mass, multifunctional energy storage structures offer many other benefits, such as an increase aircraft range [79]. Such structures effectively increase the specific energy of the system without increasing the cell-level specific energy. For example, if an aircraft concept and mission require 500 Wh/kg cell specific energy but current technology only provides 400 Wh/kg, it might be possible to design the aircraft with the current battery technology if viable multifunctional energy storage structures can be developed. In other words, battery specific energy requirements can be relaxed to some degree by deploying multifunctional structures with energy storage capability.

7.6 Summary

It is expected that batteries will be the predominant energy storage device for electrified aircraft. Based on system studies over a wide range of aircraft and information from multiple sources, a general trend in battery requirements for different classes of aircraft and range has been established. SOA Li-ion batteries with pack specific energy of 150–170 Wh/kg can enable the initial introduction of four-passenger all-electric urban air taxi with limited range and all-electric/hybrid, 10-passenger thin-haul aircraft with limited range. However, the full potential of urban air taxi cannot be realized unless the battery pack specific energy increases to 300 Wh/kg. This specific energy will also enable more passengers and longer range for thin-haul aircraft. Many all-electric and hybrid eVTOL missions with 100–200-mile range will likely require further increases in battery pack specific energy to 400 Wh/kg. In general, a pack specific energy of 400 Wh/kg can be considered to be a "sweet spot" for electrified aircraft, as it is likely to enable a broad range of eVTOL and thin-haul missions along with short-haul commuter regional aircraft.

Increasing the battery pack specific energy to 500 Wh/kg can potentially enable 300–500-mile range for 50–70-passenger regional electrified aircraft. A battery pack specific energy of 600 Wh/kg can potentially enable the introduction of large, single-aisle, 180-passenger aircraft with 500–600-mile range. A range of 1,000 miles or more for large regional and single-aisle hybrid electric aircraft will require 750 Wh/kg or greater pack specific energy. These numbers should be considered as notional, and additional system analyses are required to more accurately quantify the trend.

Fifty percent of commercial aircraft fly 500 miles or less in each flight. Thus, half of the commercial air transportation market can be electrified with batteries if pack specific energy can be increased to 600 Wh/kg. This will have significant positive impact in decreasing greenhouse gas emissions and direct operating cost for the airlines.

Because of the need to increase the range of all-electric vehicles, there are significant R&D activities worldwide to increase the specific energy of Li-based batteries. Most of the effort to develop batteries with 400–500 Wh/kg cell specific energy fall into four categories, which are listed in the order of increasing difficulty.

1. Li-metal anode, liquid electrolyte, high-nickel NMC cathode (replacement of the graphite anode in SOA Li-ion batteries with Li metal)
2. ASSB with Li-metal anode, solid electrolyte, high-nickel NMC cathode
3. Li-metal anode, liquid electrolyte, sulfur cathode (Li-S)
4. ASSB with Li-metal anode, solid electrolyte, and sulfur cathode (solid-state Li-S)

A battery pack specific energy of 300 Wh/kg can be achieved with the first option, which is likely to materialize within the next few years. Achieving pack specific energy of 400 Wh/kg will require development of ASSBs with Li-metal or Li-metal-sulfur batteries with liquid electrolyte. Pack specific energies of 500 Wh/kg and 600 Wh/kg will require development of Li-S ASSBs, which are in a very early stage of development. A pack specific energy of 600 Wh/kg with a solid-state battery would require cell specific energy of roughly 700 Wh/kg, which is near the upper limit of practically achievable cell specific energy for a Li-S system. Thus, it is extremely challenging to achieve a pack specific energy of 600 Wh/kg with a Li-S system.

It is likely that batteries with pack specific energy of 300 Wh/kg could be available by 2025. A pack specific energy of 400 Wh/kg can probably be achieved by 2030 if various challenges related to ASSBs and Li-S batteries can be mitigated. There are no current investments in Li-based batteries specifically targeted to achieve pack specific energy greater than 400 Wh/kg (500 Wh/kg cell specific energy).

HSE batteries being developed for the automotive sector may not satisfy the high C-rating and power density requirements of eVTOL aircraft during takeoff and landing. These phases require high discharge currents, which impart a large thermal load; thus, battery pack thermal management systems must be designed to accommodate these.

Several new battery concepts – including flow batteries, Al-air, and batteries with molten materials – offer the potential for simultaneously achieving HSE and high power density. Many of these concepts are being explored for grid energy storage, so significant effort will be required to evaluate the applicability of these concepts for electrified aircraft.

Fuel cells used in combination with batteries show promise for electrified aircraft. The power density of SOA fuel cells limits their application to smaller aircraft. The power density needs to increase by a factor of two to four for application in larger aircraft (e.g., regional and single aisle). Today, SOA fuel cells are adequate for ground power applications, and most of the current development effort is focused on reducing costs and increasing durability for such applications. There are viable technology paths to increase the power density of fuel cells; however, there are no current investments of note in these areas.

Multifunctional structures with energy storage capability and/or multifunctional batteries with the structural capability can reduce the mass of the overall airframe-battery system, effectively increasing the specific energy of the overall system. Multifunctional structures are in early stages of development and may not mature for another 10 years or more. It is possible that this timeline can be accelerated if a particularly suitable application can be identified and an interdisciplinary approach is pursued to demonstrate the concept.

Abbreviations

3-D	three-dimensional
Al-air	aluminum-air
ASSB	all-solid-state battery
CTOL	conventional takeoff and landing
DOE	Department of Energy
EAP	electrified aircraft propulsion
eVTOL	all-electric vertical takeoff and landing
HALE	high altitude, long endurance
HSE	high specific energy
Li-air	lithium-air

Li-ion	lithium-ion
Li-O	lithium-oxygen
Li-S	lithium-sulfur
LOHC	liquid organic hydrogen carrier
mph	miles per hour
NASA	National Aeronautics and Space Administration
NCA	$LiNiCoAlO_2$
NMC	$LiNiMnCoO_2$
PEM	proton exchange membrane
PEMFC	proton exchange membrane fuel cell
R&D	research and development
SOA	state of the art
SOFC	solid oxide fuel cell
UAV	unmanned aerial vehicle

Variables

E_{batt}	battery pack specific energy, [Wh/kg]
g	acceleration due to gravity, [m/s^2]
η	efficiency
L/D	lift-to-drag ratio
m	aircraft mass, [kg]
m_{batt}	battery mass, [kg]
m_{batt}/m	battery mass fraction
R	range, [miles]
σ_c	energy mass efficiency
σ_s	structural mass efficiency

References

[1] The Vertical Flight Society, Fairfax, VA, USA. "Electric VTOL news: Lilium jet." [Online]. Available: http://evtol.news/aircraft/lilium/.

[2] The Vertical Flight Society, Fairfax, VA, USA. "Electric VTOL news: Advanced VTOL demonstrators accelerate full tilt." [Online]. Available: http://evtol.news/2018/02/26/joby-aviation-s4-evtol-tiltprop/.

[3] G. Warwick, "Karem's eVTOL shows way for Uber to meet goals," *Aviation Week and Space Technology*, May 24, 2018. [Online]. Available: http://aviationweek.com/future-aerospace/karem-s-evtol-shows-way-uber-meet-goals.

[4] Uber Elevate, San Francisco, CA, USA. "UberAir vehicle requirements and missions", June 6, 2018. [Online]. Available: https://s3.amazonaws.com/uber-static/elevate/Summary+Mission+and+Requirements.pdf.

[5] R. McDonald and B. German, "eVTOL stored energy overview," presented at the Uber Elevate Summit, Dallas, TX, 2017.

[6] A. Brown and W. L. Harris, "A vehicle design and optimization model for on-demand aviation," presented at the Joint AIAA SciTech Forum and AIAA/ASCE/AHS/ASC Structures, Structural Dynamics, and Materials Conf., Kissimmee, Florida, 2018, Paper AIAA 2018-0105.
[7] W. Johnson, C. Silva, and E. Solis, "Concept vehicles for VTOL air taxi operation," presented at the AHS Tech. Conf. on Aeromechanics Design for Transformative Vertical Flight, San Francisco, CA, 2018.
[8] D. F. Finger, F. Götten, C. Braun, and C. Bil, "Initial sizing for a family of hybrid-electric VTOL general aviation aircraft," presented at the 67th German Aerosp. Congress (DLRK), Friedrichshafen, Germany, 2018. DOI: 10.25967/480102.
[9] M. J. Duffy, S. R. Wakayama, and R. Hupp, "A study in reducing the cost of vertical flight with electric propulsion," presented at the 17th Joint AIAA Aviation Tech., Integration, and Operations Conf. and AIAA Aviation Forum, Denver, CO, 2017, Paper AIAA 2017-3442.
[10] W. L. Fredericks S. Sripad, G. C. Bower, and V. Viswanathan, "Performance metrics required for next- generation batteries to electrify vertical takeoff and landing (VTOL) aircraft," *ACS Energy Lett.*, vol. 3, pp. 2989–2994, 2018.
[11] M. D. Moore and B. Fredericks, "Misconceptions of electric propulsion aircraft and their emergent aviation markets," presented at the Joint 52nd AIAA Aerosp. Sci. Mtg. and SciTech Forum, National Harbor, MD, 2014, Paper AIAA 2014-0535.
[12] G. Bower, "Vahana configuration trade study: Part II." [Online]. Available: https://acubed.airbus.com/blog/vahana/vahana-configuration-trade-study-part-ii/.
[13] A. Rathi, "Uber will bring you flying taxis, if you can help build a magic battery," *Quartz*, April 11, 2018. [Online]. Available: https://qz.com/1243334/the-magical-battery-uber-needs-for-its-flying-cars/.
[14] C. Morris, "Eviation and Kokam announce electric aircraft battery supply deal," *Charged Electric Vehicles Magazine*, February 21, 2018. [Online]. Available: https://chargedevs.com/newswire/eviation-and-kokam-announce-electric-aircraft-battery-supply-deal/.
[15] Aerospace Technology, London, UK. "Zunum aero hybrid electric aircraft." [Online]. Available: www.aerospace-technology.com/projects/zunum-aero-hybrid-electric-aircraft/.
[16] M. Hepperle, "Electric flight: Potential and limitations," presented at the Energy Efficient Tech. and Concepts of Operation Workshop," Lisbon, Portugal, 2012, Paper STO-MP-AVT 209.
[17] Y. Fefermann et al., "Hybrid-electric motive power systems for commuter transport applications," presented at the 30th Congr. of the Intl. Council of the Aeronautical Sci., Daejeon, Korea, 2016.
[18] T. V. Marien et al., "Short-haul revitalization study final report," NASA, Hampton, VA, Tech. Rep. TM-2018-219833, 2018, Available: https://ntrs.nasa.gov/archive/nasa/casi.ntrs.nasa.gov/20180004393.pdf
[19] K. R. Antcliff et al., "Mission analysis and aircraft sizing of a hybrid-electric regional aircraft," presented at the 54th AIAA Aerosp. Sci. Mtg., San Diego, CA, 2016, Paper AIAA 2016-1028.
[20] K. R. Antcliff and F. M. Capristan, "Conceptual design of the parallel electric-gas architecture with synergistic utilization scheme (PEGASUS) concept," presented at the 18th AIAA/ISSMO Multidisciplinary Analysis and Optimization Conf., Denver, CO, 2017, Paper AIAA 2017-4001.

[21] A. T. Isikveren, C. Pornet, P. C. Vratny, and M. Schmidt, "Conceptual studies of future hybrid-electric regional aircraft," presented at the 22nd ISABE Intl. Symp. on Air Breathing Engines, Phoenix, AZ, 2015, Paper ISABE-2015-20285.

[22] M. Voskuijl, J. van Bogaert, and A. G. Rao, "Analysis and design of hybrid electric regional turboprop aircraft," *CEAS Aeronaut. J.*, vol. 9, pp. 15–25, 2018.

[23] C. Pornet, S. Kaiser, A. T. Isikveren, and M. Hornung, "Integrated fuel-battery hybrid for a narrow-body sized transport aircraft," *Aircraft Eng. Aerosp. Techn.*, vol. 86, no. 6, pp. 568–574, 2014.

[24] M. K. Bradley and C. K. Droney, "Subsonic ultra green aircraft research: phase I final report," NASA, Hampton, VA, Tech. Rep. NASA/CR 2011-216847, 2011, Available: https://ntrs.nasa.gov/archive/nasa/casi.ntrs.nasa.gov/20150017039.pdf.

[25] W. X. Ang, A. Rao, T. Kanakis, and W. Lammen, "Performance analysis of an electrically assisted propulsion system for a short-range civil aircraft," *J. Aerosp. Eng.*, vol. 233, no. 4, pp. 1490–1502, 2019.

[26] A. Manthiram, "An outlook on lithium ion battery technology," *ACS Cent. Sci.*, vol. 3, no. 10, pp. 1063–1069, 2017.

[27] B. J. Hertzberg, "Design of resilient silicon-carbon nanocomposite anodes," Ph.D. thesis, Georgia Institute of Technology, Atlanta, GA, 2011.

[28] C. Sealy, "Lithium-ion batteries charge to the next level," *Materials Today*, April 30, 2018. [Online]. Available: www.materialstoday.com/energy/news/lithium-ion-batteries-charge-to-the-next-level/.

[29] H. Wu and Y. Cui, "Designing nanostructured Si anodes for high energy lithium ion batteries," *Nano Today*, vol. 7, no. 5, pp. 414–429, 2012.

[30] Amprius, Inc., Fremont, CA, USA, "Amprius' silicon nanowire Li-ion batteries power airbus zephyr S HAPS solar aircraft," December 4, 2018. [Online]. Available: www.prnewswire.com/news-releases/amprius-silicon-nanowire-lithium-ion-batteries-power-airbus-zephyr-s-haps-solar-aircraft-300759406.html.

[31] K. Gallagher et al., "Quantifying the promise of Li-air batteries for electric vehicles," *Energy Environ. Sci.*, vol. 7, pp. 1555–1563, 2014.

[32] D. Lin, Y. Liu, and Y. Cui, "Reviving the lithium metal anode for high-energy batteries," *Nature Nanotechnol.*, vol. 12, pp. 194–206, 2017.

[33] X. Cheng, R. Zhang, C. Zhao, and Q. Zhang, "Toward safe lithium metal anode rechargeable batteries: a review," *Chem. Rev.*, vol. 117, no. 15, pp. 10403–10473, 2017.

[34] P. Patel, "New battery tech. launches in drones," *IEEE Spectrum*, vol. 55, no. 7 pp. 7–9, July 2018.

[35] SolidEnergy, Woburn, MA, USA, "How SolidEnergy is transforming the future of transportation and connectivity," 2017. [Online]. Available: http://assets.solidenergysystems.com/wp-content/uploads/2017/08/24022118/SES_WhitePaper.pdf.

[36] W. Xue et al., "Gravimetric and volumetric energy densities of lithium-sulfur batteries," *Curr. Opin. Electrochem.*, vol. 6, pp. 92–99, 2017.

[37] D. Sigler, "Oxis energy hits a lithium-sulfur battery high," *Sustainable Skies*, October 4, 2018. [Online]. Available: http://sustainableskies.org/oxis-hits-a-lithium-sulfur-battery-high-of-425-whkg/.

[38] M. Froese, "Sion power to produce licerion rechargeable batteries for EVs and drones," *Windpower Eng. Dev.*, January 31, 2018. [Online]. Available: https://www.windpowerengineering.com/sion-power-produce-licerion-rechargeable-batteries-evs-drones/.

[39] D. S. Kaskel, "Recent progress in lithium-sulfur-batteries," presented at the 7th Intl. Adv. Automotive Battery Conf., Mainz, Germany, 2017.

[40] A. Fotouhi et al., "Lithium-sulfur battery technology readiness and applications: A review," *Energies*, vol. 10, no. 12, p. 1937, 2017.

[41] Z. Lin and C. Liang, "Lithium-sulfur batteries: from liquid to solid cells," *J. Mater. Chem. A*, vol. 3, pp. 936–958, 2015.

[42] D. Eroglu, K. R. Zavedil, and K. G. Gallagher, "Critical link between materials chemistry and cell-level design for high energy density and low cost lithium-sulfur transportation battery," *J. Electrochem. Soc.*, vol. 162, no. 6, pp. A982–A990, 2015.

[43] G. Girishkumar et al., "Lithium-air battery: promise and challenges," *J. Phys. Chem. Lett.* vol. 1, pp. 2193–2203, 2010.

[44] J. Schnell et al., "All-solid-state lithium-ion and lithium metal batteries: Paving the way for large-scale production," *J. Power Sources*, vol. 382, pp. 160–175, 2018.

[45] J. Garvey, "Solid power," *Company Week*, January 1, 2018. [Online]. Available: https://companyweek.com/company-profile/solid-power.

[46] C. Li et al., "Estimation of energy density of Li-S batteries with liquid and solid electrolytes," *J. Power Sources*, vol. 326, pp. 1–5, 2016.

[47] D. Lin, Y. Liu, and Y. Cui, "Reviving the lithium metal anode for high-energy batteries," *Nat. Nanotech.*, vol. 12, pp. 194–206, 2017.

[48] U.S. Department of Energy Office of Technology Transitions, Washington, DC, USA. "Battery500 consortium to spark EV innovations: pacific northwest national laboratory-led, 5-year $50M effort seeks to almost triple energy stored in electric car batteries," July 28, 2016. [Online]. Available: www.energy.gov/technologytransitions/articles/battery500-consortium-spark-ev-innovations-pacific-northwest-national.

[49] D. J. Unger, "Illinois researchers seek a battery you can pump into your car's tank," *The Energy News Network*, March 1, 2017. [Online]. Available: https://energynews.us/2017/03/01/midwest/illinois-researchers-seek-a-battery-you-can-pump-into-your-cars-tank/.

[50] J. J. Chen, M. D. Symes, and L. Cronin, "Highly reduced and protonated aqueous solutions of $[P_2W_{18}O_{62}]^{6-}$ for on-demand hydrogen generation and energy storage," *Nat. Chem.*, vol. 10, pp. 1042–1047, 2018.

[51] C. M. Araujo et al., "Fuel selection for a regenerative organic fuel cell/flow battery: thermodynamic considerations," *Energy Environ. Sci.*, vol. 5, p. 9534, 2012.

[52] G. Soloveichik, "A novel concept for energy storage," presented at the Trans-Atlantic Workshop on Storage Tech. for Power Grids, Washington, DC, 2010, Available: www.energy.gov/sites/prod/files/piprod/documents/Session_D_Soloveichik.pdf.

[53] Y. Liu et al., "A comprehensive review on recent progress in aluminum-air batteries," *Green Energy Environ.*, vol. 2, no. 3, pp. 246–277, 2017.

[54] M. Rovito, "The alloyed powers," *Charged Electric Vehicles Magazine*, vol. 13, pp. 26–32, April 2014.

[55] R. Mori, "A novel aluminum-air rechargeable battery with Al_2O_3 as the buffer to suppress byproduct accumulation directly onto an aluminum anode and air cathode," *RSC Adv.*, vol. 4, pp. 30346–30351, 2014.

[56] B. J. Hopkins, Y. Shao-Horn, and D. P. Hart, "Suppressing corrosion in primary aluminum-air batteries via oil displacement," *Science*, vol. 362, no. 6415, pp. 658–661, 2018.

[57] J. Ryu et al., "Seed-mediated atomic-scale reconstruction of silver manganate nanoplates for oxygen reduction towards high-energy aluminum-air flow batteries," *Nat. Commun.*, vol. 9, 2018, Article 3715.

[58] J. M. Vegh and J. J. Alonso, "Design and optimization of short-range aluminum-air powered aircraft," presented at the 54th AIAA Aerosp. Sci. Mtg., San Diego, CA, 2016, Paper AIAA 2016-1026.

[59] V. Giordani et al., "A molten salt lithium-oxygen battery," *J. Am. Chem. Soc.*, vol. 138, no. 8, pp. 2656–2663, 2016.

[60] C. Xia, C. Y. Kwok, and L. F. Nazar, "A high-energy-density lithium-oxygen battery based on a reversible four-electron conversion to lithium oxide," *Science*, vol. 361, no. 6404, pp. 777–781, 2018.

[61] S. Feng, J. R. Lunger, J. A. Johnson, and Y. Shao-Horn, "Hot lithium-oxygen batteries charge ahead," *Science*, vol. 361, no. 6404, p. 758, 2018.

[62] S. Licht et al., "Molten air: A new, highest energy class of rechargeable batteries," *Energy Environ. Sci.*, vol. 6, no. 12, pp. 3646–3657, 2013.

[63] Y. Jin et al., "An intermediate temperature garnet-type solid electrolyte-based molten lithium battery for grid energy storage," *Nat. Energy*, vol. 3, pp. 732–738, 2018.

[64] N. Lapena-Rey, J. Mosquera, E. Bataller, and F. Orti, "First fuel-cell manned aircraft," *J. Aircraft*, vol. 47, no. 6, pp. 1825–1835, 2010.

[65] The German Aerosp. Center (DLR), Germany. "Zero-emission air transport: First flight of four-seat passenger aircraft HY4." [Online]. Available: www.dlr.de/dlr/en/desktopdefault.aspx/tabid-10081/151_read-19469/#/gallery/24480.

[66] Phys.org, "World's first 4-seater fuel-cell plane takes off in Germany." [Online]. Available: https://phys.org/news/2016-09-world-seater-fuel-cell-plane-germany.html.

[67] RT.com, "China becomes 3rd country to test hydrogen-powered plane – report," [Online]. Available: https://www.rt.com/news/373102-china-hydrogen-fuel-aircraft/.

[68] W. H. Wentz, R. Y. Myose, and A. S. Mohamed, "Hydrogen-fueled general aviation airplanes", presented at the 5th AIAA Aviation, Tech., Integration, and Operations Conf., Arlington, Virginia, 2005, Paper AIAA 2005-7324.

[69] T. Kadyk, C. Winnefeld, R. Hanke-Rauschenbach, and U. Krewer, "Analysis and design of fuel cell systems for aviation," *Energies,* vol. 11, no. 2, p. 375, 2018.

[70] US DRIVE Partnership, "Fuel cell technical team roadmap," [Online]. Available: www.energy.gov/sites/prod/files/2014/02/f8/fctt_roadmap_june2013.pdf.

[71] T. R. Stoia, S. Atreya, P. O'Neil, and C. Balan, "A highly efficient solid oxide fuel cell power system for an all-electric commuter airplane flight demonstrator," presented at the 54th AIAA Aerosp. Sci. Mtg., San Diego, CA, 2016, Paper AIAA 2016-1024.

[72] N. K. Borer et al., "Overcoming the adoption barrier to electric flight," presented at the 54th AIAA Aerosp. Sci. Mtg., San Diego, CA, 2016. Paper AIAA 2016-1022.

[73] B. Roth and R. Giffin III, "Fuel cell hybrid propulsion challenges and opportunities for commercial aviation," presented at the 46th AIAA/ASME/SAE/ASEE Joint Propulsion Conf. and Exhibit, Nashville, TN, 2010, Paper AIAA 2010-6537.

[74] W. Zuo et al., "Battery-supercapacitor hybrid devices: Recent progress and future prospects," *Adv. Sci.,* vol. 4, no. 7, 2017.

[75] University of Bristol, Bristol, UK. "Alternative to traditional batteries moves a step closer after exciting progress in supercapacitor technology," February 27, 2018. [Online]. Available: www.bris.ac.uk/news/2018/february/supercapacitor.html.

[76] Rolls-Royce, Derby, UK. "Rolls-Royce links up with UK-based Superdielectrics to explore potential of very high energy storage technology," March 19, 2018. [Online]. Available: www.rolls-royce.com/media/press-releases/2018/19-03-2018-rr-links-up-with-uk-based-superdielectrics.aspx.

[77] E. D. Wetzel, "Reducing weight: Multifunctional composites integrate power, communications, and structure," *The AMPTIAC Quarterly*, vol. 8, no. 4, pp. 91–95, 2004.
[78] J. P. Thomas and M. A. Qidwai, "The design and application of multifunctional structure-battery materials systems," *JOM*, vol. 57, no. 3, pp. 18–24, 2005.
[79] T. J. Adam et al., "Multifunctional composites for future energy storage in aerospace structures," *Energies*, vol. 11, no. 2, p. 335, 2018.
[80] G. Fredi et al., "Graphitic microstructure and performance of carbon fibre Li-ion structural battery electrodes," *Multifunct. Mater.*, vol. 1, Paper No. 015003, 2018.

8 Thermal Management of Electrified Propulsion Systems

Charles E. Lents

Introduction

A proper stand-alone introduction to the thermal management (TM) of hybrid electric (HE) propulsion systems (HEPS) – also referred to as electrified propulsion systems elsewhere in this book – would include the details of each HEPS architecture, including its component makeup and potential benefits. Luckily, these descriptions have already been given in Chapters 1 and 3 (Section 3.2 – Powertrains for Electrified Propulsion). Thus, this chapter limits its analysis to the TM challenges associated with these architectures.

A schematic of the two basic non-hybrid families of propulsion system – fueled-engine propulsion system (FEnPS) and battery-electric propulsion system (BEPS) – is shown in Figure 8.1. In FEnPS, the propulsion power comes from fans mechanically coupled to the fueled engine; in BEPS, it comes from fans driven by electric motors powered by electricity from batteries. Some authors include the BEPS as a subset of the hybrid electric propulsion system (HEPS) design space, but clearly a BEPS is not a hybrid. Thus, in this chapter FEnPS and BEPS architectures are treated separately.

Together, hybrid-fueled propulsion systems (HFPS) and HEPS make up the family of propulsion systems called hybrid propulsion (HP), and these are composed of various combinations of FEnPS and BEPS. Early work by the US National Aeronautics and Space Administration (NASA) defined a wide variety of potential HP system architectures, which were categorized into six basic architectures [1]. These should be familiar, as they were introduced and discussed in Chapter 1, illustrated in Figure 3.1 and discussed in Section 3.2, and touched on throughout the other chapters in this book. Here, they are presented with an eye toward analyzing their thermal performance and TM impacts on the aircraft. The analysis focuses on TM with gas turbine engines and ducted fans, but it is easily generalized to other engine types and propellers.

8.1 Electric Drivetrain Heat Sources and Aircraft Heat Sinks

Because each HP configuration discussed in Section 3.2 contains some percentage (possibly 100 percent) of electrified propulsion (EP), all require an electric drivetrain (EDT), which may contain any or all of the following: generators, rectifiers, feeders (wiring), electrical power system protection and control, motor drives, motors, batteries

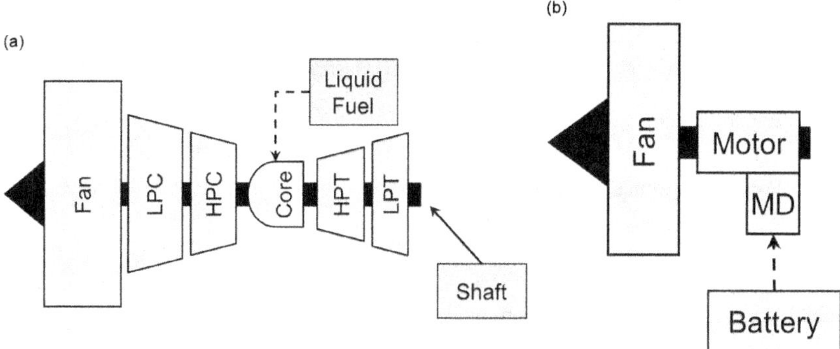

Figure 8.1 Non-hybrid propulsion systems: (a) FEnPS and (b) BEPS.

and battery management. None of these are present in a standard FEnP, so there are significant weight and loss penalties incurred when hybridizing the propulsion system. Therefore, there must be offsetting performance gains for the HP architecture to "buy its way" onto the aircraft (i.e., be economically viable; more on this concept in Chapter 9). Case in point: two ways the series HE architecture overcomes its EDT weight and loss penalties are by allowing the engine to operate at its peak efficiency throughout the mission profile and through improvements in propulsive efficiency by deep integration of the propulsion system into the aircraft. These are facilitated, respectively, by the battery's ability to act as a power buffer – supplying the peaks and absorbing the valleys in the load profile – and the distributed nature of the propulsion.

8.1.1 Heat Sources

EDT component material systems, including wire insulation, bearing oil, housing materials, and electronics, must be maintained at much lower temperatures than engine components, which can exceed 2,000 °F. Temperature requirements for various materials used in EDT components are listed in Table 8.1. Most limits shown are only allowed for a short duration, as component life can be compromised with extended time at high temperature. These limits can be used for component design if the design conditions only prevail for a short time. A short-duration design condition that is commonly used to size components is operation at maximum heat loss and "hot day ground ambient temperature" (40 °C, 104 °F). The International Standardization Organization (ISO) has established the *International Standard Atmosphere* (ISA), which provides atmospheric temperature and pressure versus altitude for standard day conditions [2]. The "hot day" condition is defined as the temperature at or below which the peak temperature at a given location on a given day will fall 95 percent of the time (i.e., there is only a 5 percent chance that the temperature will be higher). The hot day temperature is generally accepted as the ISA standard temperature at a given altitude plus 25 °C (called "ISA + 25 °C") [3]. When the components are operating under more moderate, lower temperature

Table 8.1 EDT component temperature limits.

Component	Sub-component or material	Temp. limit [°F]
Electric machine	Wire insulation	250
	Oil	300
Power electronics	Diodes and switches	200 (Si), 400 (SiC/GaN)
	Capacitors	160
	Inductor insulation and potting	300
	Gate drives	160
Wire	Insulation	250
Battery	Electrolyte	130

(i.e., normal day) longer-duration conditions, the TMS should cool them to a much lower temperature (between 10 °C and 30 °C lower).

8.1.2 Heat Sinks

EDT low-temperature heat, also known as *low-quality heat*, must be rejected to the ambient environment. The three possible heat sinks that facilitate this are ambient air, gas turbine engine fan duct air, and fuel; however, propeller-based aircraft do not have fan ducts, and fuel is not available as a heat sink in a BEPS.

Ambient air can be accessed either through skin cooling or by scooping air into a duct (i.e., *ram air*) and directing it through a heat exchanger (HX). For commercial applications, the ambient air temperature is highest on the ground and drops significantly as altitude increases. Thus, the TMS is typically designed to maintain the maximum allowable component temperatures as listed in Table 8.1 during hot day ground conditions. The temperature while cruising – almost always the longest flight segment – or on standard or cold days, is significantly lower than the hot day ground condition. Designing at the "hot day, takeoff condition" results in a TMS with the lowest weight and power that does not compromise component life, as the components operate with the lower temperature sink during most of their operating life.

The *fan air* of a high bypass ratio gas turbine engine is the air that flows through its fan duct. This air provides more than 95 percent of the engine's propulsive power, and it is relatively cool (temperatures do not exceed 170 °F). This air has been compressed by the fan, so it is hotter than ram air, but it is readily accessible to engine-mounted components like generators, motors, gearboxes, and the electric machines associated with parallel HEPS and turbo-electric propulsion systems. Fan air is also used to cool engine oil. Like ambient air, fan duct air is hottest at hot day ground conditions and cools as altitude increases. In commercial aircraft engines, fan air is "scooped" (i.e., directed) out of the fan duct through an *air oil cooler* (AOC) before being reinjected into the fan duct, as shown in Figure 8.2.

Fuel is used as a heat sink in both commercial and military applications. In commercial aircraft, only the "burn flow" is used as it flows from the fuel tank to the engine combustors. Fuel in excess of burn flow is not pumped out of a tank, circulated through

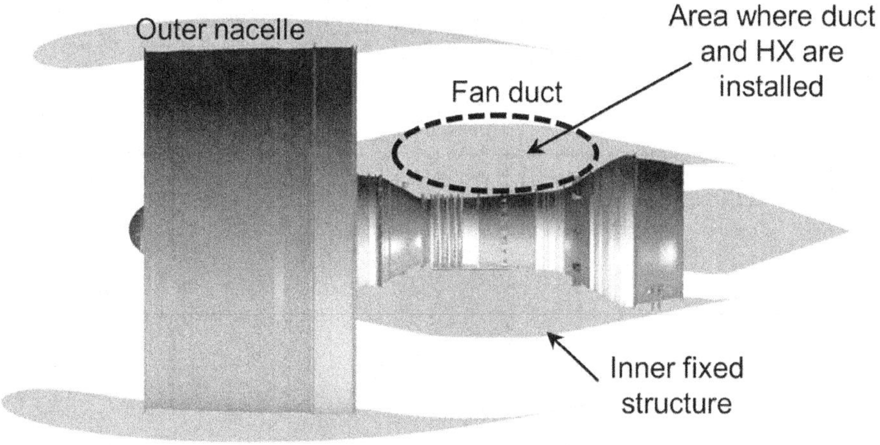

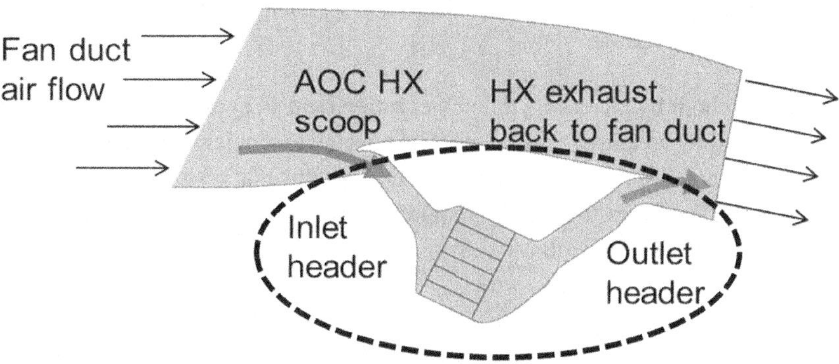

Figure 8.2 AOC using scooped fan duct air.

the TMS, and returned to tank, as it is in some military applications. Fuel flows through the cold side of the hydraulic-oil, generator-oil and engine-oil HXs.

There are two advantages to using fuel as a heat sink. First, raising the temperature of fuel before it reaches the engine combustor has a thermodynamic benefit. Hotter fuel requires less sensible heating (created by burning fuel) before vaporization, which increases engine fuel efficiency. Second, compared to air, liquid generally has a higher heat transfer coefficient, making it a better specific-heat transfer fluid. Thus, fuel-cooled HXs are generally smaller and lighter than their air-cooled counterparts. However, because only burn flow is used, fuel has a limited, variable capacity as a heat sink. Ram air has no such limitations.

8.2 Thermal Management of Electric Drivetrains

HP system EDT components add weight and insert losses between the engine or battery and the propulsor, as illustrated in Figure 8.3. The losses further impact aircraft

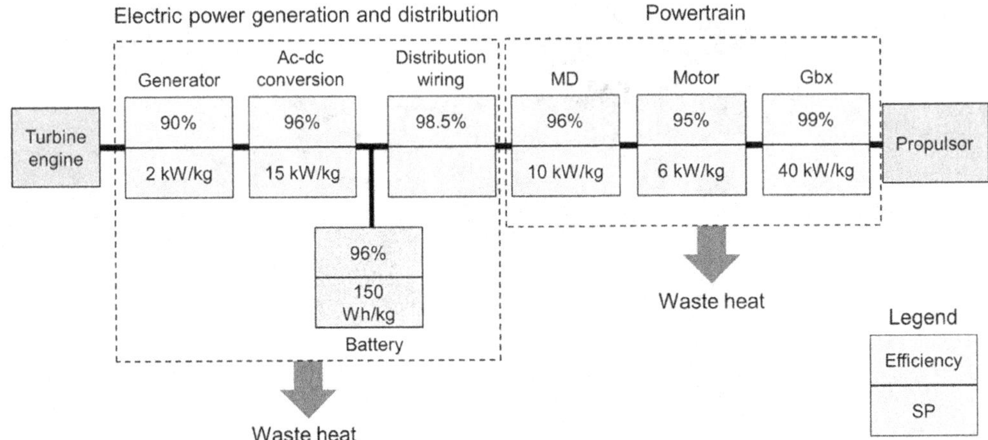

Figure 8.3 HP systems EDT power losses, heat rejection, and weight impact on aircraft performance.

weight both by increasing the power rating (and size) of upstream drivetrain components and by creating heat that must be rejected to ambient through the TMS. The TMS can include liquid pumps, fans, and heat pumps that also require power. This power demand increases the energy required to complete a mission, resulting in increased fuel and battery weight.

8.2.1 Electric Drivetrain Weight

Due to inefficiency, the power that must be supplied to each EDT component is greater than the power it processes. The chain of EDT components thus imposes a higher output power requirement on the engine than is needed to power the propulsor. This drives up the total energy required by the vehicle, which translates to added fuel and battery weight. In addition, the power rating and corresponding weight of each EDT component is driven up by the inefficiencies of every component after it (between it and the propulsor) in the chain.

The other parameter that impacts a component's weight is its power-to-weight ratio, or, *specific power* (SP), often expressed in units of kW/kg. The higher the SP, the lower is a component's weight for a given power rating. While SP has been touched on throughout this book, the extensive analysis of high SP electric machines in Chapter 3 is particularly germane here. Both SP and efficiency are functions of technology advancement. Thus, it is prudent to invest in technologies that increase both EDT component efficiencies and SP.

8.2.2 Thermal Management System Weight and Power

Component losses become heat, which must be rejected to the ambient air though a TMS to limit component temperature and ensure long life and good performance (high efficiency). The efficiencies shown in Figure 8.3 are for state-of-the-art (SOA) component performance. Even with these high efficiencies, at the megawatt levels of EDT power required for EP, heat is generated in the hundreds of kilowatts. To make matters worse, this is low-quality heat and must be processed by the TMS at relatively low temperature.

8.2.3 Thermal Management System Functions

EDT component heat transfer features facilitate the acquisition of waste heat by the TMS. Examples include rectifier and motor drive cold plates, spray-oil-cooled generator end-turns and back-iron cooling channels, exterior cooling fins of air-cooled motors, and air directed over wire bundles. The heat transfer features provide the component–TMS interface, and their weights are generally included in component weight. The TMS consists of components that transport the acquired heat to the aircraft heat sinks. Liquid cooling loops comprised of pumps and plumbing are an example. The TMS rejects heat to the available air and fuel heat sinks via HXs, surface coolers, direct-air, or fuel cooling. If the heat sink is air, air ducts and fans are used to direct the sink air through HXs. When components are air-cooled directly by the ambient sink air, the TMS heat transport and rejection functions happen simultaneously.

Thermodynamic principles dictate that heat only flows along a negative temperature gradient, from a hot source to a cool sink. Unfortunately, in aircraft (and many other) applications, the heat source is often cooler than the available heat sink. Cooling the aircraft passenger cabin (i.e., air conditioning) is a good example, where the ambient heat sink on a hot day can exceed 103 °F while the cabin temperature must be maintained at 70 °F. In order to provide the necessary cooling, a *heat pump* is used to pump heat from low to high temperature. Heat pumps require power to pump heat "uphill" against a negative temperature gradient, as shown in Figure 8.4. A heat pump can be located near the component it is cooling, at the heat acquisition section of the TMS, or near the ultimate sink at the heat rejection TMS section.

A TMS is comprised of plumbing, ducts, pumps, fans, heat pumps, and HXs. The weight of these components is directly proportional to the amount of heat to be transferred and the temperature difference between the heat sources and the available heat sinks at the design point. The greater the (positive) temperature difference, generally, the lower the weight required to reject the heat. If the temperature difference is negative at a design point, then a heat pump is required somewhere in the TMS.

8.2.4 Aircraft Performance Impacts

As is true with most, if not all, aerospace systems, the main way the EDT affects aircraft performance is through its impacts on vehicle weight. As was stated in Section

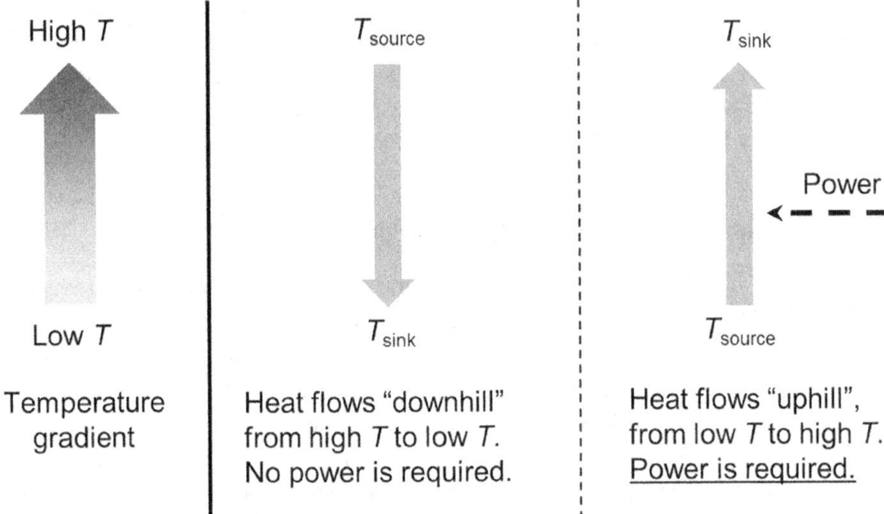

Figure 8.4 Power required to move heat from a cold source to a hot sink.

8.2.1, each EDT component has a characteristic SP. As the EDT power requirements go up, so do its component weights. The stacked-up inefficiencies of the chain of EDT components also result in higher-rated – and thus, larger – energy sources (engines and batteries). Technologies that increase EDT component efficiency (reduce waste heat) and temperature rating (i.e., increase the source-to-sink temperature gradient) reduce TMS weight. In addition, TMS weight is reduced with the development of advanced TMS hardware technology, such as additively manufactured HXs, lightweight materials, low-loss HX cores and manifolds, thermal storage, and new heat pump cycles.

8.3 Heat Acquisition: Component Cooling Approaches

Air- and liquid-cooled options exist for each EDT component. Common heat transfer features are listed in Table 8.2. Although they are listed here for reference, the design of these features is not the focus of this publication, as they are generally designed as part of the component and not in the control of the TMS system designer.

8.4 Thermal Management System Architectures

TMS architectures can take many forms, depending on the system they are cooling. Direct-air and liquid cooling are appropriate and most easily implemented when the ambient air (the final heat sink) is cooler than the components being cooled. Heat pumps are used to move heat against a thermal gradient and can be integrated with

Table 8.2 EDT component heat transfer features.

Component	Feature
Generator and motor	Air-cooled fin integrated windings Spray-oil-cooled end turns High conduction features Housing liquid channels for back-iron cooling Air-cooled housing external fins
Rectifier and motor drive	Liquid-cooled, finned cold-plates Conduction bars High conductivity potting Internal and external air-cooled finned heat sinks Fan-integrated heat sinks Immersion cooling
Wire	Thermally conductive electrical insulation
Battery	Liquid cooling channels Thermal mass and phase change materials Heat pipes and conduction bars

direct-air and liquid cooling features. Thermal storage and phase change materials can be used to smooth transient thermal loads, allowing the other TMS components to be smaller and more efficient. Each of these TMS architecture concepts is discussed in the following subsections.

8.4.1 Direct-Air Cooling

Perhaps the simplest approach to EDT thermal management is via direct cooling with ambient air, whereby the TMS simultaneously acquires, transports, and rejects heat. A direct-air-cooling TMS is comprised of ducts and fans to route air to various EDT components. If one or more components require sub-ambient cooling, heat must be transported against a thermal gradient to the ambient air. In this case, a heat pump can be integrated either directly at the component or in the air stream. If integrated at the component, the heat pump acquires the component heat, raises its temperature, and rejects it to the routed ambient air. If integrated into the air stream, the heat pump cools the air by removing heat from it and rejecting the heat to another ambient stream. The now sub-ambient cooled air is used for component cooling.

8.4.2 Liquid Cooling

The common alternative to direct air is liquid cooling. In this TM method, a liquid coolant loop acquires component heat, transports it to heat sinks (ambient or fan duct air or fuel) with plumbing and pumps, and rejects it through liquid-to-air or liquid-to-fuel HXs. Liquid-to-air HX air flow is generally supplied with ducts and fans. An example of a liquid-cooled subsystem is shown in Figure 8.5, in which a set of components – represented as circles – is cooled in parallel by an oil loop. The cooling

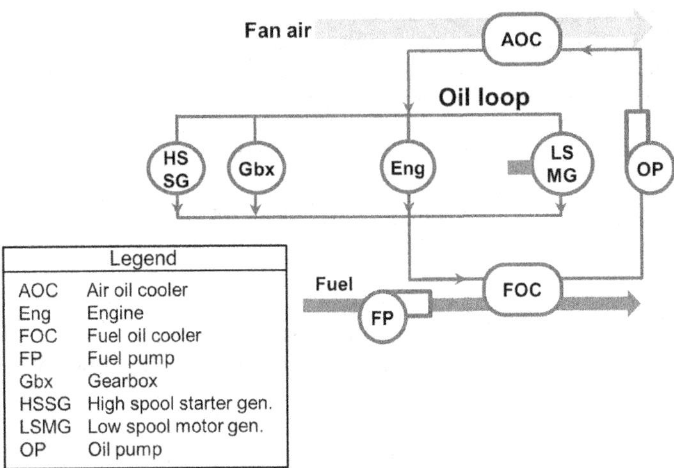

Figure 8.5 Liquid cooling loop subsystem.

oil flow is split among the four components, acquiring heat from each before being recombined. The heated oil first rejects heat to fuel via a fuel oil cooler (FOC) and then to fan air via an air oil cooler (AOC). The oil is then routed back to the components to complete the cooling loop. The process is driven by an oil pump (OP).

8.4.3 Liquid versus Air Cooling

Generators and motors and their associated power electronics (rectifiers and inverters) can be cooled by air or liquid. However, regardless of the cooling method, the heat sink for most components, ultimately, is ambient air. Liquid-cooled machines are generally smaller and lighter than air-cooled machines, but they require additional pumps and HXs. This is generally true of all EDT components, not just electric machines; thus, there clearly is a weight and volume trade involved in the decision to choose air or liquid cooling. Air cooling generally trades well with lower-rated machines (less than 30 kW) while liquid cooling is generally preferred for higher-rated machines (greater than 60 kW), although this is not a fixed rule.

8.4.4 Heat Pumps

A simple vapor cycle system (VCS), as shown on the left in Figure 8.6, is a good example of a heat pump system. Analogous to a residential air conditioner, in a VCS, heat is extracted from the component in the evaporator and rejected to the sink in the condenser. The powered compressor drives the process.

As shown in the diagram, $T_{\text{source, in}}$ is the temperature of a coolant stream that has been used to cool a low-temperature component. The coolant rejects its acquired heat and is cooled to $T_{\text{source, out}}$ by cold two-phase refrigerant that enters the evaporator (from point **a** in the diagram) at a temperature lower than $T_{\text{source, out}}$. As it acquires heat

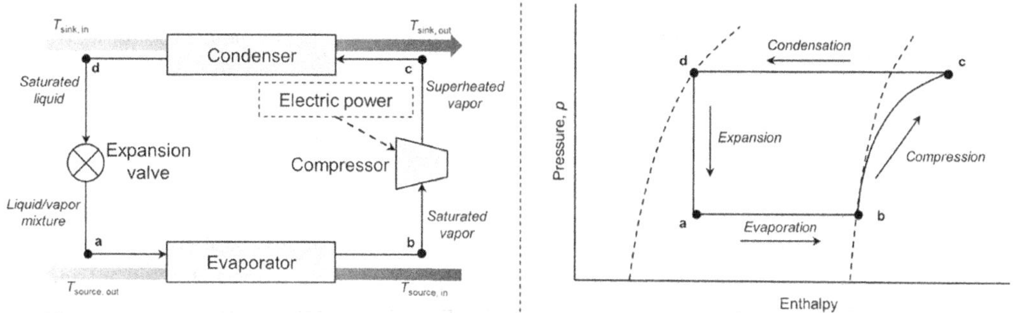

Figure 8.6 VCS (left) and pressure-enthalpy diagram (right).

from the coolant stream, the refrigerant evaporates, changing phase from liquid to gas at constant temperature and pressure. The coolant – now cooled to $T_{\text{source, out}}$ – is returned to cool the low-temperature component, increasing in temperature from $T_{\text{source, out}}$ to $T_{\text{source, in}}$. Note that the refrigerant itself could be supplied directly to the low-temperature component rather than using the intermediate cooling loop. However, the method shown here is usually preferred – especially if there are multiple components to be cooled – so that refrigerant does not have to be pumped throughout the aircraft. In general, the cooling loop fluid – which could be water, glycol, or other similar single-phase liquid – is easier to plumb and manage than two-phase refrigerant.

The refrigerant at the evaporator outlet (point **b**) is now a gas (i.e., vapor). It is compressed to high temperature and pressure by the compressor, whose pressure ratio is designed so that the outlet gas temperature is much higher than the heat sink inlet temperature, $T_{\text{sink, in}}$. The higher the compressor gas outlet temperature, the higher the quality of its heat and the less condenser area and weight required to reject the heat to the sink. The refrigerant then enters the condenser (point **c**) and changes phase at constant temperature and pressure back to liquid, rejecting its heat to the sink in the process. This raises the sink temperature from $T_{\text{sink, in}}$ to $T_{\text{sink, out}}$. The refrigerant (now at point **d**) is expanded at constant enthalpy to a low-pressure and low-temperature two-phase fluid by the expansion valve, after which the cycle repeats (back to point **a**). The VCS thus transfers source low-temperature heat to the higher-temperature sink.

The thermodynamic process just described is represented by the pressure-enthalpy diagram on the right side of Figure 8.6. The lower-left corner of the cycle (point **a**) represents the thermodynamic state of the VCS refrigerant as it enters the evaporator. The refrigerant's so-called vapor dome lies inside the boundary created by the two vertical arcs shown (the top of the dome is cut off in this illustration). Between these two lines, the refrigerant is a two-phase fluid; it is a mixture of liquid and vapor. In this state, its temperature remains constant as it is heated, but its enthalpy is increased as more liquid changes phase to gas (the reverse is also true as it is cooled). Outside the vapor dome (the right edge of the cycle, between points **b** and **c**), as the fluid, which is now a superheated vapor, is heated and its enthalpy is increased, the temperature also increases.

The cycle requires power at the compressor, and all energy flows must balance. Thus, the heat rejected to the sink at the condenser is equal to the heat acquired by the refrigerant at the evaporator, plus the power delivered by the compressor. The performance of a heat pump is characterized by its *coefficient of performance* (COP), defined as the rate of heat acquisition, or *heat duty*, of the evaporator divided by the power required to pump that heat to the sink. The higher the COP, the more efficient the heat pump. In the VCS, the denominator of the COP calculation is the compressor input power. This power is largely a function of the *temperature lift* ($T_{cond} - T_{evap}$) and, to a lesser extent, the thermodynamic properties of the refrigerant itself. The higher the lift for a given heat duty, the more pump power required and the lower the COP.

The weight of a HX is a function of the mean temperature difference between its hot- and cold-side fluids (more on this in Section 8.7). The greater the difference, the lighter the HX. However, in the case of the VCS condenser, a high temperature difference requires a greater temperature lift, which lowers the COP and likely increases the compressor weight. Thus, there is a trade between VCS weight and performance (COP).

Other heat pumps work in a similar way, using electric power to "pump" heat from a cold source to a hot sink. For example, an air cycle system (ACS) is like a VCS, but its cooling fluid (air) does not change phase. Thus, an ACS has a much lower COP than an equivalent VCS with the same lift. An ACS compresses air to high pressure and temperature with a compressor, rejects the heat to a sink fluid, dropping the air to medium temperature and high pressure, then expands the air to low pressure and temperature across a turbine, where it is used to cool a heat source. The turbine produces less power than is required by the compressor; thus, additional electric power is required to drive the cycle.

More exotic heat pumps also exist. For example, thermoelectric coolers use the Joule–Thompson effect to pump heat from a cold source to a hot sink by flowing current into the device being cooled. As these are rare on aircraft, especially when compared with VCS and ACS heat pumps, they will not be discussed further.

8.4.5 Thermal Storage

The ambient heat sink in commercial aircraft is highly variable: relatively speaking, it is hot on the ground but very cold at altitude. This presents a design challenge because components must be sized to handle the worst-case operating point. Because this occurs during the aircraft's relatively short time on the ground, components are oversized, inefficient, and heavy during most of their operating life. The use of thermal storage can help mitigate this. If heat is stored during taxi and takeoff, when the heat load and sink temperature are high, and processed (acquired, lifted, and rejected) during cruise when the load and sink temperature are lower, the TMS components can be rated for less heat and be much smaller and lighter. Both component thermal mass and phase-change materials can be used as thermal storage. Each is discussed in the following subsections.

Thermal Mass

Thermal mass describes a component's capacity to absorb heat. The higher the thermal mass, the more heat that can be absorbed. But heat absorption comes with a temperature rise; the higher the heat capacity, the less the temperature will rise for a given amount of absorbed heat. In general, more mass means more thermal mass; however, the ability of a material to store heat depends on its physical properties (e.g., *specific heat* and *heat capacity*) – water is much better at storing heat than plastic, for example. Most aircraft TMS components are made of metals and other materials that store heat effectively; thus, their mass translates to thermal mass.

A battery pack provides a good example of the importance of thermal mass. Even a pack with a cell-level energy density far exceeding 2,030 projections of 500 Wh/kg ([4], Section 7.2) will have significant mass from its battery cells and heat transfer features with high heat capacity. The corresponding high thermal mass can be used to absorb and store heat temporarily. For example, the NASA X-57 battery pack is precooled so that its thermal mass can absorb even more heat during flight [5]. Internal heat generated from battery current is not rejected to the ambient, but instead is absorbed by the battery pack's thermal mass, raising the battery temperature. A transient analysis for a battery pack with an optimistically low mass (and corresponding low thermal mass) with the details listed in Table 8.3 is illustrated in Figure 8.7. No cooling is provided while the ambient air is within 30 °C of the desired maximum battery cell temperature. The battery temperature, starting at 25 °C, slowly rises to 40 °C through taxi, takeoff, and climb to 20 kft. With a 12 °C ambient temperature at this altitude, cooling begins and the maximum cell temperature starts to drop.

Phase-Change Materials

Phase-change materials (PCMs), such as wax, absorb heat by using the energy to change phase (melt). The phase change occurs at roughly a constant temperature, so short periods of high heat load in a component can be absorbed without its temperature increasing. PCMs function like thermal mass, smoothing out the variability in heat load and sink temperature "seen" by the TMS. In their typical use, they are

Table 8.3 Battery pack transient analysis details.

Item	Assumption or parameter
1	500 Wh/kg cell energy density (SOA is approximately 250 Wh/kg)
2	Metal heat conduction features used to acquire cell heat at a *pack burden* (ratio of heat acquisition hardware weight to battery cell weight) of 15% (SOA is approximately 35%)
3	Max power is pulled during takeoff at a C-rate of 4 and then gradually drops to 0 kW after about 30 minutes. A 500 Wh/kg battery can provide 500 W of power for one hour (although margin of 20 percent charge should be left in the battery for long life). At a C-rate of 4, the battery provides 2 kW, but only for 15 minutes.
4	Recovered ram air starts at 40 °C, gradually rises to 46 °C by the end of climb out at 1,500 ft, and then begins to drop steadily to –18 °C at the top of climb.

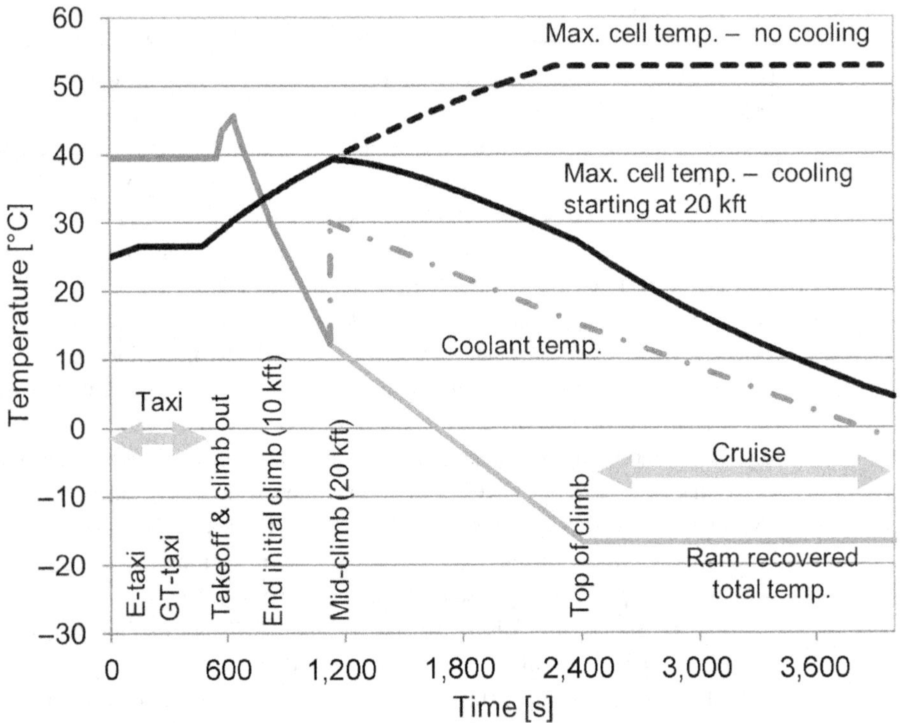

Figure 8.7 Transient battery temperature analysis.

installed as part of a phase-change HX, which has alternating layers of wax with fins and finned liquid channels, coupled with a pumped liquid loop that circulates coolant. The coolant is pumped to the sink, where the heat acquired in the HX is rejected. When the heat sink is cold (e.g., ambient air at cruise), the PCM can be solid. As the heat sink temperature and/or the heat loads increase, the coolant temperature rises until it starts melting the wax. In this way, much of the heat being transported by the liquid loop is stored in the PCM as it melts, rather than overloading the loop. The PCM is chosen based on its melt temperature to maintain the component generating the heat load at an acceptable level.

The use of thermal mass or PCMs has the potential to drive down the weight of TMS sink HXs, as both can smooth peak loads and temperatures. This drives down the HX weight, but the overall system weight can increase if the additional thermal storage mass is greater than the HX weight reduction.

8.5 A Reference Thermal Management System

Consider the notional TMS shown in Figure 8.8. Although developed for a parallel HEPS [6], it contains all the components needed to manage the heat loads of an EDT

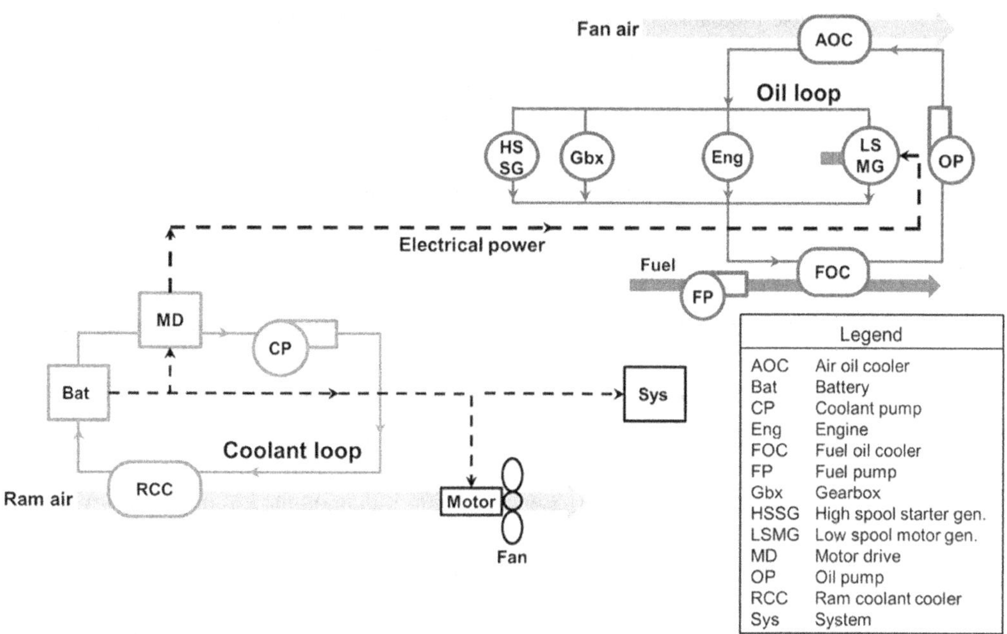

Figure 8.8 Reference EDT TMS.

for any HP system. The major components that require cooling are the propulsion motor, motor drive (MD), and battery (Bat). In the parallel HEPS, the motor is integrated in some manner with the engine low spool; thus, it is labeled low spool motor generator (LSMG) in the diagram. This, coupled with a cooling requirement of approximately 250 °F, makes the existing engine oil loop a good choice for transporting the motor heat. The dashed line indicates electrical power flow from the battery, which is routed through the MD to the propulsion LSMG, to the fan required to move cooling air through the TMS on the ground, and to other electrical systems (Sys).

The TMS is composed of two separate cooling loops, one to service the higher-temperature loads integrated on the engine (top half of the diagram in Figure 8.8) and one to service the lower-temperature loads (lower half of the diagram). The high-temperature (225 °F–300 °F) loop utilizes the existing engine oil circuit to service both conventional engine loads (Eng), accessories gearbox (Gbx), and high-spool starter/generator [HSSG]) as well as the HP system LSMG. The parallel oil streams are recombined and transport the acquired heat to the fuel and air heat sinks. First, heat is rejected to fuel at the FOC, but this sink can only accept approximately 20 percent of the oil heat because it has a relatively low flow rate and maximum allowable temperature. Thus, a second heat sink – fan duct air interfaced with the AOC HX – is required. The oil rejects the remaining 80 percent of its heat to the AOC before returning to the five parallel loads to complete the loop. The oil is circulated by a gearbox-mounted OP.

Because the battery and MD require cooling at temperatures far below that of the engine oil, they are cooled with a low-temperature ethylene-glycol-water loop. The coolant is first supplied to the battery, because it requires lower-temperature cooling, then the MD. The acquired heat is then rejected to ram air at the ram coolant cooler (RCC). At low speed, a fan is required to draw ram air through the RCC. The coolant is circulated by an electrically driven coolant pump (CP).

8.5.1 System Design Point

For the remainder of this chapter, the reference TMS shown in Figure 8.8 will be analyzed, and its components are designed to meet the requirements of a "worst-case" operating point. The design point chosen is the thermally stressing "hot day, end of climb-out," which occurs at an altitude of 1,500 ft. The operating conditions for this point are given in Table 8.4 [2, 3]. This point was chosen because, although the ambient temperature is slightly lower at 1,500 ft than on the ground, at end of climb-out, the aircraft is moving at Mach 0.4, so the recovered ram air is at its highest temperature. Also, the propulsion system is providing maximum power, so every EDT heat load is at its peak.

The reference TMS steady-state operation at this worst-case design point is illustrated in Figure 8.9. This is the same system schematic shown in Figure 8.8 but with operating temperatures (standard font), pressures (underlined), fluid mass flow rates (*italic*), electric power (***bold italic***), and heat flow rates (**bold**) superimposed. Note that the design values shown, although they are in units in common use in TMS design, are not SI units. Thus, unit conversions are necessary when using the design equations given in the remainder of the chapter. The author believes that describing TMS design in this realistic scenario with the engineering units most likely to be encountered will provide more utility to the reader.

As illustrated, each propulsion fan (one per engine) is powered by an LSMG that requires 2,101 kW (2.1 MW) of electric power during takeoff and climb-out to provide the needed boost power for the parallel HEPS [6]. Following the efficiency chain back to the battery (motor: 97 percent, MD: 96 percent, and transmission: 99 percent) results in 2,256 kW (2.26 MW) of required battery power. The battery also supplies 74 kW to meet the subsystem electric demand and 33 kW to power the

Table 8.4 Hot day, end of climb-out design point operating conditions.

Variable	Value [SI]	Value [other]
Altitude	457 m	1,500 ft
Ambient air temperature	36.3 °C	97.3 °F
Ambient air density	1.27 kg/m^3	N/A
Ambient air pressure	96,259 Pa	0.95 atm
Ambient air humidity	50%	N/A
Aircraft speed	137.2 m/s	Mach 0.4
Recovered ram air temperature	45.6 °C	114 °F

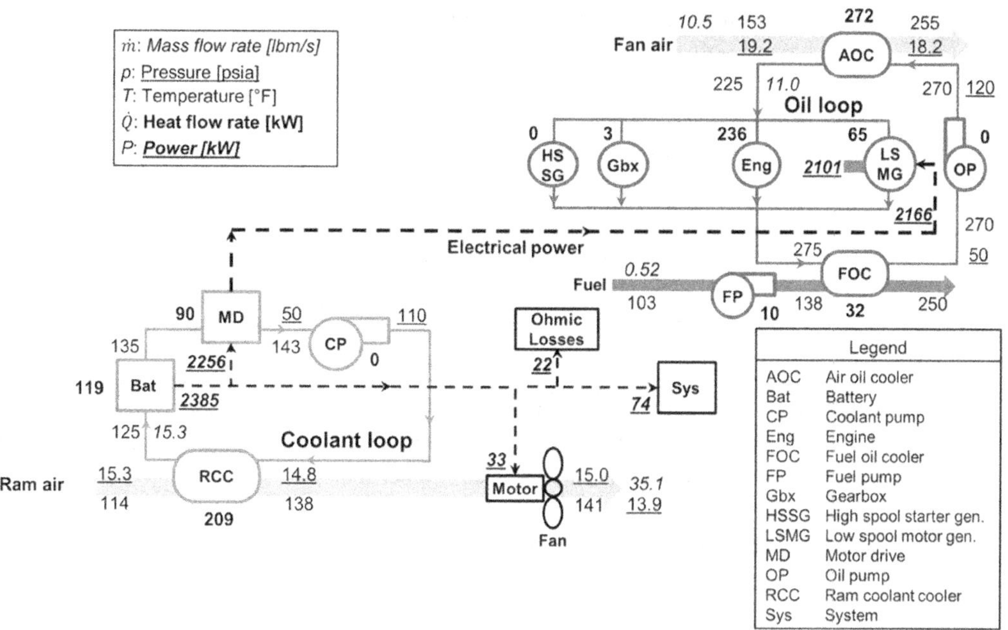

Figure 8.9 Steady-state operation of the reference EDT TMS at "hot day, end of climb-out."

coolant loop fan. There are 22 kW of Ohmic losses in the electrical distribution system. The heat rejection for each EDT component is derived from its input power and assumed component efficiency. Although the numbers in Figure 8.9 are given here without justification, the process for deriving and calculating them will be described in the remainder of the chapter.

8.6 Heat Transport: Pumps, Fans, and Plumbing

Pumps and fans, which consume power and add system mass, are required to move cooling fluids in a TMS. When designing these components, the power required to supply a desired flow rate across a given pressure rise is derived from fundamental physics-based models. In contrast, the efficiency and mass almost always are predicted and derived based on manufacturer historical data and learned expertise. As such, publicly available data and/or calculations for these are not readily available. An introduction to the design of heat transport components is given in the following subsections.

8.6.1 Pumps

While liquid cooling can substantially decrease the weight and volume of the heat-producing components, it adds liquid cooling pipes and pumps to the TMS. Of course, these additional components can also reduce the size of or eliminate fans and ducts, so there is a trade-off.

Pump Power

The power required to move an incompressible flow – the *fluid power* P_{fl}, in W, – is the product of *volumetric flow rate* $\dot{V}$, in m^3/s, and *pump pressure rise* Δp_{pump}, in Pa. The *pump input power* $P_{e,\text{pump}}$, in W, is the fluid power divided by the pump and drive motor *efficiencies* η_{pump} and η_{motor}, respectively:

$$P_{e,\text{pump}} = \frac{\dot{V}\Delta p_{\text{pump}}}{\eta_{\text{pump}}\eta_{\text{motor}}} = \frac{P_{fl}}{\eta_{\text{pump}}\eta_{\text{motor}}} \tag{8.1}$$

In (8.1), the variables are expressed in SI units. Often, however, flow rates and pressures are given in other units, in which case unit conversions are necessary. For example, pressures are often given in atm rather than Pa.

It is also useful to express (8.1) in terms of the fluid *mass flow rate* $\dot{m}$, in kg/s, which is $\dot{V}$ multiplied by the *fluid density* ρ_{fl}, in kg/m^3:

$$\dot{m} = \dot{V}\rho_{fl}, \tag{8.2}$$

which gives

$$P_{e,\text{pump}} = \frac{(\dot{m}/\rho_{fl})\Delta p_{\text{pump}}}{\eta_{\text{pump}}\eta_{\text{motor}}}. \tag{8.3}$$

Compared to other TMS components, liquid pumps typically consume relatively low power. As an example, using (8.3) an ethylene glycol pump delivering a fairly high mass flow rate of 5 kg/s at standard temperature and pressure (22 °C, 1 atm; $\rho_{fl} \approx 1{,}055\ \text{kg/m}^3$) across a large pressure rise of 1 atm (≈101 kPa) requires approximately 0.8 kW of pump power, assuming $\eta_{\text{pump}}\eta_{\text{motor}} = 60$ percent.

Pump Efficiency

The operating efficiency of any rotating machine is generally a function of its design efficiency and operating speed and power, as percentages of their rated values (i.e., their *part values*). Typically, operating efficiency drops with part power because there are fixed losses that are not a function of load. The relationship between efficiency and speed is not as easy to generalize. Design, part load, and part speed efficiency all depend on the manufacturer's design practices and know-how (engineering quality, machining tolerances, materials, etc.).

Normalized pump efficiency (the ratio of operating to design efficiency) versus normalized flow (the ratio of operating to design flow), commonly used to represent *part load performance*, is plotted in Figure 8.10. Without specific manufacturer's data, an assumption of 70 percent full-power, full-speed efficiency is reasonable, with a range of 40–85 percent possible. If the pump is electrically driven, the motor and drive efficiency should also be applied (assumptions of 90 and 95 percent, respectively, are typical).

Table 8.5 Notional centrifugal pump weight estimation parameters.

Parameter	High estimate	Low estimate
Diameter [in]	9	8
Height of full cone [in]	12	10
Height of pump [in]	5	4
Height of frustum [in]	7	6
Weight [lbm]	20	12

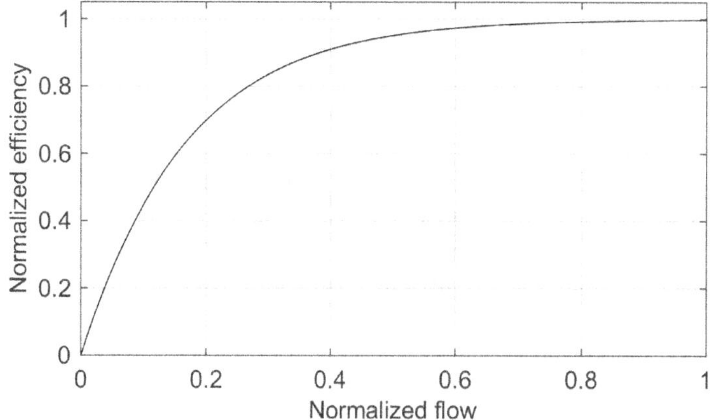

Figure 8.10 Normalized pump efficiency vs. normalized flow.

Pump Weight

Like its efficiency, a pump's weight is highly dependent on manufacturer design practices and know-how. Further, no good public data is available for an aerospace CP or OP. A reasonable weight estimate for a centrifugal pump, which is a common topology for a CP, can be derived using basic geometry. Assuming the pump has the shape of a right cone with its top cut off (a *frustum*), the parameters in Table 8.5 can be estimated. Estimating the weight as 70 percent of an aluminum frustrum gives the weight estimates. Given the relatively low weight, a fixed weight of 12–20 lbm is a reasonable estimate. If electrically driven, motor and MD weight should be added.

8.6.2 Plumbing

The fluid pressure drop and pipe weight for TMS plumbing can be derived from geometry and internal flow physics. Here, the term "pipe" is used to represent either an actual pipe or a duct. A fluid dynamics or heat transfer textbook (e.g., White [7]) provides the physics-based equations used to determine pressure drop as a function of mass flow rate in a round pipe. The *pipe inner diameter* D_{in}, in m, is calculated based on the desired *pipe pressure drop* Δp_{pipe}, in Pa. It is an iterative process, summarized as follows:

(1) Choose an initial guess for D_{in}, then calculate the resulting *pipe cross-sectional area* A_{in}, in m^2.

(2) Obtain the fluid transport properties (ρ_{fl}, *specific heat at constant pressure [volume]* C_p [C_v], both in J/kg-K, *thermal conductivity* κ, in W/m-K, and *fluid viscosity* μ_{fl}, in kg/m-s) from tabular data. These vary with temperature, for which it is standard to use the average cooling loop temperature.

(3) Determine the *fluid velocity* v_{fl}, in m/s, using (8.4) and the *Reynolds number* Re using (8.5), where L_{pipe}, in m, is the *pipe length*:

$$v_{fl} = \dot{m}/(\rho_{fl}A_{\text{in}}) \tag{8.4}$$

$$\text{Re} = \frac{\rho_{fl}v_{fl}L_{\text{pipe}}}{\mu_{fl}} \tag{8.5}$$

(4) Calculate the *friction factor* f based on the values obtained from the previous steps and round-pipe correlations found in tabulated data.

(5) The *K-factor* of the jth component K_j accounts for the pressure drop caused by bends and fittings, the so-called *minor losses*. It is common to assume a value of 0.5 for each component in the cooling circuit unless there is more detailed information.

(6) Calculate pipe pressure drop using (8.6):

$$\Delta p_{\text{pipe}} = \frac{\rho_{fl}v_{fl}^2}{2}\left(\frac{L_{\text{pipe}}}{D_{\text{in}}}f + \sum_j K_j\right) \tag{8.6}$$

(7) Iterate on D_{in} until desired pressure drop is reached. Generally, the pressure drop will decrease, all else equal, as D_{in} increases.

To estimate the plumbing weight, a stainless-steel pipe (remember, "pipe" can also mean "duct" here) can be assumed. Manufacturer websites can be used to provide a reasonable estimate of the wall thickness necessary to contain the required pressure. Alternatively, a hoop stress calculation can be used, but this generally results in the minimum allowable wall thickness, which is much thinner than what is used in practice. The metal thickness is added to D_{in} to obtain the *pipe outer diameter* D_{out}, in m. The *metal cross-sectional area* A_{metal}, in m^2, is the area of an annulus with inner and outer diameters D_{in} and D_{out}, respectively. This is multiplied by L_{pipe} to calculate the pipe *metal volume* V_{metal}, in m^3, and 20–50 percent is added to account for fittings and bends. The metal volume and *metal density* ρ_{metal}, in kg/m^3, are multiplied to get the *pipe mass*, m_{pipe}, in kg.

Finally, the plumbing may be insulated. This is not likely for coolant or oil plumbing, as the desire is to reject acquired load heat, and some of this can be rejected through uninsulated plumbing. If there is insulation, its weight can be calculated by using an assumed insulation thickness and the calculations described in the previous paragraph.

8.6.3 Fans

As discussed in Section 8.1.2, air is the major heat sink available for heat rejection. Fans are required to move air, either across/through air-cooled components or through HXs and ducts.

Fan Power

The *fan power* P_{fan}, in W, is derived from the equations for isentropic compression of an ideal gas (air) and is calculated as

$$P_{\text{fan}} = \dot{m} C_p T_{\text{inlet}} \frac{\text{Pr}^{(\gamma-1/\gamma)} - 1}{\eta_{\text{fan}}}, \tag{8.7}$$

where $\gamma \triangleq C_p/C_v$ is the *ratio of specific heats*, T_{inlet}, in K, is the *inlet temperature*, Pr is the fan *pressure ratio* (ratio of fan outlet to inlet pressure), and η_{fan} is the *fan efficiency*. P_{fan} is the power required at the fan shaft. The fan efficiency is the ratio of the actual shaft power required to the ideal power required, assuming isentropic compression. If the fan is electrically driven, then a motor and MD efficiency must be lumped into η_{fan}.

Fan Efficiency and Weight

Fan design is beyond the scope of this book, and fan performance and weight are highly dependent on manufacturing quality. However, some general guidance is provided. A high-end fan operating at its design point is roughly 70 percent efficient. When operating at off-design points, the efficiency can drop below 50 percent. Fans of interest to TMS designers are generally motor driven, and in them, most of the mass is contained in the motor and motor drives. In relation to these, the fan blades themselves are very light.

8.7 Heat Rejection: Heat Exchangers

Compact Heat Exchangers, by Kays and London, is the definitive source for information on the design and performance prediction of compact, plate-fin HXs [8]. The following discussion follows the Kays and London method to size the AOC HX for the reference TMS introduced in Section 8.5 and shown in Figure 8.8. The process can be applied to the FOC and other HXs as well.

8.7.1 Air Oil Cooler Design

As a design example, consider the AOC, which is a liquid-to-air HX transferring heat from the oil loop to the fan air. Later, in Section 8.8, the process is described for deriving the design conditions for this HX. Here, however, those conditions are assumed known and used for sizing. The reader should be aware of this and feel free to skip around as appropriate to best understand the design processes that follow.

Referring to Figure 8.9, at the design point the AOC heat load (272 kW), oil inlet temperature (270 °F), and fan air inlet temperature (153 °F) are boundary conditions imposed by environmental conditions and component constraints. The other necessary AOC HX design parameters can be derived after the following four free parameters have been selected: (1) the temperature difference ΔT, in K (or, equivalently, °C), across the oil system heat-rejecting components; (2) ΔT between the AOC outlet air and the AOC inlet oil (the *pinch temperature*, ΔT_{pinch}, in K); and (3) the oil-side and (4) air-side pressure drops. Once 50 °F is selected for (1), the AOC oil outlet temperature (225 °F = 275 °F – 50 °F) and oil flow rate (11 lbm/s) are determined. Likewise, once the pinch temperature (15 °F) is chosen for (2), the AOC air outlet temperature (255 °F = 270 °F – 15 °F) and air mass flow rate (10.5 lbm/s) are determined. Both the air and oil mass flow rates are calculated as

$$\dot{m} = \frac{\dot{Q}}{C_p \Delta T}, \tag{8.8}$$

where $\dot{Q}$ is the HX heat duty, in W, and C_p and ΔT are chosen for the fluid (air or oil) whose mass flow is being calculated. This provides all the necessary information to size the AOC, whose performance within the reference TMS is summarized in Figure 8.11. As will be discussed in Section 8.8, the free parameters can be optimized (as opposed to selected by engineering judgment) to minimize an objective function.

In convection heat transfer, the rate of heat transfer $\dot{Q}$ from a hot source to a cooling fluid is directly proportional to the sink fluid *heat transfer coefficient*, the *heat transfer surface area* A, in m^2, and the temperature difference between the hot source and the cool sink ΔT_{h-c}, in K. The same is true in a HX, where the *overall heat transfer coefficient* U, in W/m^2-K, includes the two fluid heat transfer coefficients and the conductance of the HX walls separating the two fluids:

$$\dot{Q} = UA\Delta T_{h-c}. \tag{8.9}$$

The relationship expressed in (8.9) indicates that high values of U, A, or ΔT_{h-c} result in high $\dot{Q}$, which makes intuitive sense. TMS designers commonly refer to the *UA*

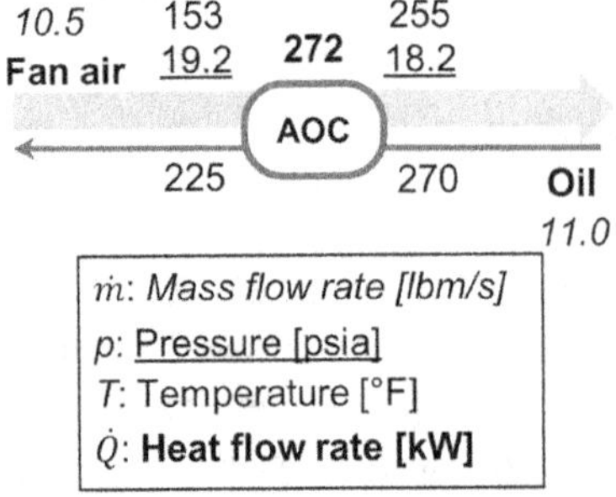

Figure 8.11 AOC design conditions.

product as "UA," and the UA required to achieve a given heat duty is easily found by rearranging (8.9) as

$$UA = \frac{\dot{Q}}{\Delta T_{h-c}}. \tag{8.10}$$

It is desirable to maximize U so that less A is required, as A essentially translates into weight.

In practice, HX designers commonly use the *Effectiveness-NTU Method.* HX *effectiveness* ε, which ranges from zero to one, is defined as the actual heat transfer rate divided by its maximum possible value. In theory, the maximum $\dot{Q}$ is achieved when either the temperatures of the cold-side outlet and hot-side inlet or the hot-side outlet and cold-side inlet are equal. However, in practice, it is most often the case that one fluid stream has a higher *heat flow capacity* $\dot{C}$, in W/K, than the other, where

$$\dot{C} \triangleq \dot{m} C_p \tag{8.11}$$

and

$$\dot{Q} = \dot{C} \Delta T. \tag{8.12}$$

Thus, the ΔT of the higher-$\dot{C}$ fluid at a given $\dot{Q}$ is lower than that of the other fluid, and the theoretical maximum $\dot{Q}$ is not possible. In this case, the maximum heat transfer rate, $\dot{Q}_{\max}$, is found from (8.12) as the product of the smaller $\dot{C}$, $\dot{C}_{\min}$, and the ΔT between the hot- and cold-side inlets $\Delta T_{(h,\mathrm{in}-c,\mathrm{in})}$. The actual $\dot{Q}$ is calculated from (8.12) using either hot or cold fluid values (both fluids will have the same absolute value of $\dot{Q}$ with opposite signs). Effectiveness ε is, then,

$$\varepsilon = \frac{\dot{Q}}{\dot{Q}_{\max}} = \frac{\dot{C} \Delta T}{\dot{C}_{\min} \Delta T_{(h,\mathrm{in}-c,\mathrm{in})}}. \tag{8.13}$$

The HX fluid temperatures along the length of the AOC heat exchanger are shown in Figure 8.12.

NTU is the *number of transfer units*, defined as the UA divided by the minimum HX heat flow capacity:

$$NTU = \frac{UA}{\dot{C}_{\min}}. \tag{8.14}$$

Using the HX effectiveness and the heat flow capacity of each fluid, NTU can be derived using the method provided by Kays and London in [8]. In [8] the ε-NTU relationships for different HX configurations are derived, and the relationship for a cross-flow HX is provided in Figure 8.13. The relationship in (8.14) then provides the UA needed to transfer the required heat flow. Note that, unlike component efficiency, effectiveness is not inherent in any component and is not a measure of "goodness." Rather, it is a requirement derived from the system analysis.

Finally, in Figure 8.12, observe the effect of the selected $\Delta T_{\mathrm{pinch}}$ on the overall system performance. Low $\Delta T_{\mathrm{pinch}}$ results in high ΔT across the air side; thus, a lower

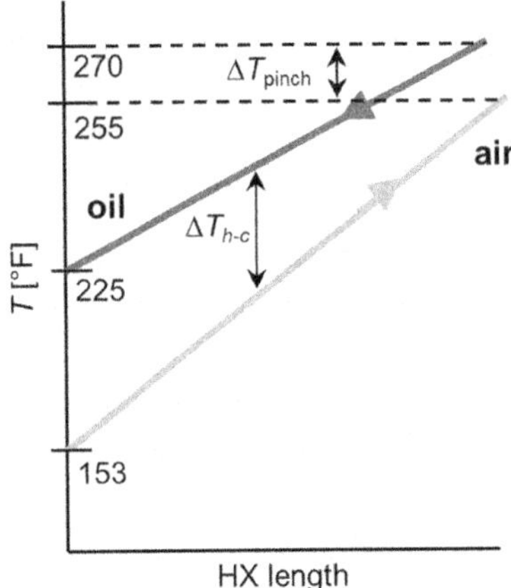

Figure 8.12 HX hot and cold fluid temperatures along the AOC HX length.

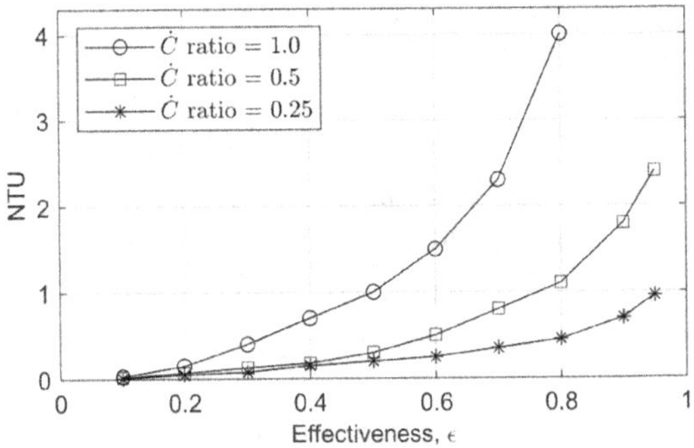

Figure 8.13 Cross-flow HX effectiveness vs. NTU at different heat flow capacity ratios.

air mass flow rate is required, and lower power is required to move the air. However, the effectiveness necessary to achieve this is large, which, according to Figure 8.13, increases UA and HX weight.

8.7.2 Geometry Design: The Plate-Fin Cross-Flow Heat Exchanger

As the process thus far described has made evident, the required UA can be determined with no knowledge of the HX geometry. Therefore, once the required UA has been established, the TMS designer must iterate on the HX geometry to achieve the required performance, constrained by the maximum allowable pressure drops on each side of the HX.

The basic geometry of a plate-fin cross-flow HX is shown in Figure 8.14. Alternating layers of hot and cold fluid flow channels are stacked so that the fluids flow across each other at right angles, separated by thin parting plates. Although only

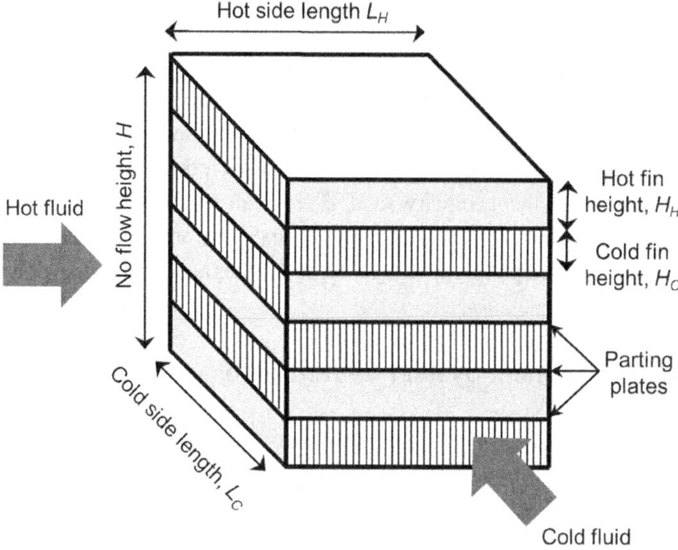

Figure 8.14 Plate-fin HX geometry.

three hot and cold layers are shown, the total number of layers is a design variable. The cold (hot) *flow height* (not labeled) is the product of the number of cold (hot) flow layers and the cold (hot) *fin height* (H_C, H_H, respectively), and the *total HX height* (labeled "no flow height H") is the sum of the cold and hot flow heights and the total thickness of all the parting plates. The cold- (hot-) side *flow area* is the product of cold (hot) flow height and hot- (cold-) side length (L_H, L_C, respectively; notice the cold flow height is multiplied by the hot-side length, and vice versa) less the area blocked by the fins, which is the product of fin thickness, fin density (*fins per inch* [FPI]) and cold- (hot-) side flow height. This geometry also determines the *hydraulic diameter*, a generalized version of diameter, which applies to non-round (e.g., rectangular) pipes and ducts.

With the fluid transport properties (ρ_{fl}, C_p, κ, and μ_{fl}), hydraulic diameters, and flow areas now known, the Prandlt and Nusselt numbers, v_{fl}, Re, f, and heat transfer coefficient for each fluid stream can be calculated. The pressure drop for each fluid stream, fin efficiency, and U are then calculated. These calculations are detailed in [8] and are not explained further here. The fin density and height, number of hot and cold layers, and L_H and L_C are used to compute the heat transfer area A. Thus, UA and pressure drop can then be determined for the selected geometry.

Using the design process described, L_H and L_C, the number of flow layers , and the fin height, density (FPI) and type on each of the hot and cold sides and are all iterated until the required UA is achieved. The result is constrained such that the pressure drops on each side do not exceed the requirements. There are many more free parameters than constraints, so an optimization is executed to minimize weight while meeting the UA and pressure drop requirements.

8.7.3 Performance Prediction

The HX has been designed to meet the worst-case operating conditions, but it is also necessary to compute its expected performance across the operating envelope as the heat duty and heat sink temperatures change. Once the HX geometry is known, its performance can be predicted across all TMS operating points. In performance prediction, U and UA are calculated based on the geometry, and ε is given by the NTU-ε relationship (e.g., Figure 8.13). With the known heat sink inlet temperature and calculated ε, the heat source inlet temperature can be predicted. The outlet conditions are derived knowing the heat duty.

8.8 Thermal Management System Optimization

In this section, the reference TMS in Figure 8.9 is revisited. Thus far, all the operating conditions shown (mass flow rates, temperatures, pressures, and heat flow rates) have been assumed given and used for illustrative purposes. Now, details are given on how those numbers have been calculated or chosen, either through engineering calculations and judgments or optimization. For the following discussion, the parameters in Table 8.6 have been assumed.

The operating conditions in Figure 8.9 result from ambient conditions and heat loads at the design point (hot day, end of climb-out), component temperature constraints, and free parameters selected by engineering decisions. As noted in Section 8.7.2, the free parameters can be optimized to minimize some cost function, but it is useful to understand the trade-offs that these parameters influence through an example. In the discussion that follows, constraints are underlined, *imposed conditions are italicized*, calculated values are in standard font, and **free parameters are bolded**. Do not confuse these style choices with the labels for the various physical parameters in Figure 8.9! They are summarized for easy reference in Table 8.7.

Table 8.6 EDT TMS system parameters.

Parameter	Value
Pump efficiency (combined pump and motor) (OP, FP, CP)	N/A*
C_p fan air	1,003 J/kg-K
C_p oil (Mobil Jet Oil II)	2,195 J/kg-K
C_p coolant (ethylene-glycol-water)	3,094 J/kg-K
C_p ram air	1,003 J/kg-K
ρ_{fl} coolant (ethylene-glycol-water)	N/A*
ρ_{fl} oil (Mobil Jet Oil II)	N/A*

* The efficiency and fluid densities are used in the pipe and pump loss calculations. In our example, these losses – and their associated fluid temperature increases – are assumed to be negligible compared to the others. Thus, these parameters are not used in the example. As an example, a relatively inefficient water pump operating at 60% efficiency with a 50 psid pressure rise only increases the water temperature by 0.55 °C.

Table 8.7 Legend for TMS optimization description.

Legend
Constraint
Imposed condition
Calculated value
Free parameter

8.8.1 Oil Loop

The design starts with the oil loop, also called the oil system, which is the top cooling loop in Figure 8.9. It includes heat sources (the components being cooled), fuel pump (FP), FOC, OP, and AOC. Each of these is addressed in the following paragraphs.

Heat Sources/Cooled Components

The HSSG, Gbx, Eng, and LSMG heat sources – the components that must be cooled by the oil loop – dictate the oil temperatures. Without these temperature constraints, it would be desirable to run the oil system very hot, as this would result in a high ΔT between the hot oil and the fan air and fuel heat sinks, resulting, respectively, in lightweight FOC and AOC HXs. The components require cooling at a maximum oil temperature of 275 °F. A **temperature rise of 50 °F across the oil loop component loads** is selected, resulting in a component inlet temperature of 225 °F. Given the known *total component heat of 304 kW (3 + 236 + 65)*, (8.11) and (8.12) are used to compute the oil mass flow rate (with a unit conversion) of 11.0 lbm/s.

Fuel System and Fuel Oil Cooler

The fuel system contains the flowing fuel and the FP. At the design point, the *fuel mass flow rate 0.52 lbm/s* is given by the engine demand, and the *hot day ambient temperature is 103 °F*. FP losses heat the fuel flowing to the FOC slightly from ambient to 138 °F. The *250 °F maximum fuel temperature* is a fixed constraint, which sets the maximum temperature rise across the FOC (250 °F – 138 °F = 112 °F). This ΔT and the fuel mass flow rate are used in (8.11) and (8.12) to calculate the FOC heat rejection rate from oil to fuel of 32 kW. With the FOC heat flow rate known, these equations are now used with the oil mass flow rate to calculate the FOC oil exit temperature of 270 °F.

Oil Pump

A **70 psid OP pressure rise** (from 50 psia to 120 psia) is chosen. This is used with the oil mass flow rate, oil fluid density, and motor and pump efficiencies, to compute pump power using (8.3). The pump pressure provides the pressure head needed to overcome pressure drops in the oil loop HXs, plumbing, and loads. The higher the pump pressure rise, the greater the pump power and weight but the lower the plumbing and HX weight.

The pump efficiency and power are used to compute the power loss (heat rate) from the pump into the oil, which from (8.8) gives the temperature rise across the pump. In this design example, the pump power loss is negligible, and the temperature of the oil supplied at the AOC oil inlet is equal to the temperature at the FOC oil outlet, 270 °F.

Air Oil Cooler

The heat duty of the AOC is known: it is the total oil system heat rate less the rate of heat removed at the FOC (304 kW – 32 kW = 272 kW). The *fan duct air temperature* of *153 °F* is known for this operating condition. The **AOC air exit temperature of 255 °F** is a free parameter. It can either be specified directly or determined by either (1) specifying the HX cold-side ε using (8.13) and noting that $\dot{C} = \dot{C}_{\text{min}}$ on the air side, so that $\varepsilon = \Delta T_{(c,\text{out}-c,\text{in})}/\Delta T_{(h,\text{in}-c,\text{in})}$, or (2) specifying ΔT_{pinch}. In order to cool the oil, ΔT_{pinch} must be positive, and it specifies how far the AOC air outlet temperature must be below the AOC oil inlet temperature. For this exercise, **a** ΔT_{pinch} **of 15 °F** is used, resulting in an AOC air exit temperature of 255 °F (270 °F – 15 °F). With the air temperatures now specified and the heat duty known, the AOC air mass flow rate of 10.5 lbm/s is calculated using (8.8). Air mass flow rate could be used as a free parameter instead of the AOC air exit temperature, but using the pinch temperature as a free parameter guarantees that the air exit temperature will be below the oil inlet temperature. Low ΔT_{pinch} results in high air ΔT and low air mass flow rate (which causes fan duct drag) but high AOC weight. The other free parameter associated with the AOC is the **AOC air-side pressure drop of 1.0 psid**. High Δp results in high fan duct drag but low AOC weight.

This completes the oil loop, with temperatures, pressures, and mass flow rates at each component specified. With this information, components can be sized and system weight, power draw, and fan drag can be determined. These additional design calculations are omitted from this text.

8.8.2 Coolant Loop

The design continues with the coolant loop, which is shown in the bottom left of Figure 8.9. Like the oil loop, it includes heat sources – the Bat and MD – and it also contains a CP and RCC. Design details for each of these are now addressed.

Heat Sources/Cooled Components

The maximum allowable battery temperature of 135 °F is set by the battery pack limits, and this sets the battery coolant exit temperature. The **battery coolant temperature rise of 10 °F** is a free parameter, but it is constrained to less than 21 °F in this case to ensure that the battery coolant inlet temperature is greater than the RCC air inlet temperature of 114 °F. This is because the coolant rejects heat to ram air through the RCC, so it can never be at a lower temperature than the ram air sink. Given the known *119 kW battery heat rejection*, the coolant mass flow rate of 15.3 lbm/s is calculated using (8.8). The *90 kW MD heat rejection* is also known, so it is used with the coolant mass flow rate in (8.8) to compute the 143 °F MD exit temperature.

Coolant Pump

The **CP pressure rise is chosen as 60 psid** (from 50 psia to 110 psia). This, along with the coolant mass flow rate (and its known density), yields pump power from (8.3). The pump pressure provides the necessary pressure head to overcome the pressure drops in the RCC, plumbing, and heat loads. The higher the pump pressure rise, the greater the pump power and weight but the lower the plumbing and RCC weight.

The CP efficiency and power are used to compute the power loss (heat rate) from the pump into the coolant. In this case, the pump heat is negligible, and from (8.8), the temperature at the RCC coolant inlet is equal to the temperature at the MD outlet (143 °F).

Ram Coolant Cooler

The *RCC heat duty of 209 kW* is an imposed condition: it is the total coolant system heat rate due to the battery (119 kW), MD (90 kW), and coolant pump (≈0 kW). The *ram air temperature of 114 °F* is known for this operating condition. The RCC air exit temperature of 138 °F is a free parameter. The narrative in Section 8.8.1 for the AOC, regarding cold-side ε, ΔT_{pinch}, and air mass flow rate applies to the RCC as well. An **RCC ΔT_{pinch} of 5 °F** is used as the free parameter, resulting in an RCC air exit temperature of 138 °F (143 °F – 5 °F). With the RCC air temperatures specified and the heat duty known, air mass flow rate of 35.1 lbm/s is calculated using (8.8). As with the AOC, a low RCC pinch temperature results in high air ΔT, low air mass flow rate (which causes ram drag), and high RCC weight. The other free parameter associated with the RCC is **RCC air-side pressure drop (0.5 psid)**. High Δp results in high ram drag but low RCC weight.

On the ground, ram air is generated by the ram fan, which is shown in the bottom center in Figure 8.9. Once the aircraft is moving, ram recovery provides the recovery pressure that drives air flow through the ram duct and RCC. Running the ram fan during flight is also an option. The **ram fan pressure ratio** becomes a free parameter in this case. The higher the pressure ratio, the lower the ram drag but the higher the fan power. If ram air is supplied to the ram exit nozzle at the same pressure that was recovered in the ram inlet diffuser, there is no drag. For ground operation, the fan pressure ratio is determined by the RCC Δp, because the fan must boost the pressure back to the ground ambient pressure. Thus, the RCC air-side Δp affects both ram drag during flight and fan power (and size) required for ground operation.

8.8.3 Optimization

The narratives in Sections 8.8.1 and 8.8.2 identified several TMS free parameters (in bold) and described how their values affect the overall TMS performance in terms of weight, volume, etc. These parameters and their chosen values are listed in Table 8.8.

In a "real" design (i.e., for a commercial or military aircraft), the free parameters are typically optimized to minimize a system cost function that is based on specific design goals.

Table 8.8 Free parameters for optimization of reference EDT TMS.

Free parameter	Value
ΔT across oil loop component loads	50 °F
Δp_{pump}, OP	70 psid
Δp, HSSG, Gbx, Eng, LSMG	Unspecified
ΔT_{pinch}, AOC	15 °F
Δp, AOC air side	1.0 psid
ΔT, Bat coolant	10 °F
Δp_{pump}, CP	60 psid
ΔT_{pinch}, RCC	5 °F
Δp, RCC air side	0.5 psid
Ram fan pressure ratio	Unspecified

For commercial aircraft, fuel burn rate or total energy consumption provide good cost functions. The TMS free parameters in Table 8.8 impact the vehicle performance in three key ways: weight, power draw, and aerodynamic drag. Fuel is required to lift the TMS weight, to power TMS pumps and fans, and to overcome aerodynamic drag caused by using ram and engine fan air as TMS heat sinks. An engine performance model can provide the sensitivity of fuel burn rate to each of these three influences in the form of lbm/hr of fuel burned per "x." For example, if 0.3 lbm/hr of fuel is burned for every horsepower of power extracted from the engine high spool to power hydraulic pumps, electric generators, etc., the power draw sensitivity factor would be 0.3 lbm/hr/hp. These fuel burn penalties vary over the mission, but an example calculation using typical cruise values is provided in (8.15) for reference.

$$\begin{aligned} \text{Fuel Burn Rate} = 0.025 &\left[\frac{\text{lbm (fuel)}}{\text{h}}\right]\left[\frac{1}{\text{lbm (weight)}}\right] \times \text{Weight [lbm]} \\ &+ 0.3\left[\frac{\text{lbm (fuel)}}{\text{h}}\right]\left[\frac{1}{\text{hp}}\right] \times \text{Power [hp]} \\ &+ 0.5\left[\frac{\text{lbm (fuel)}}{\text{h}}\right]\left[\frac{1}{\text{lbf}}\right] \times \text{Ram and Fan Drag [lbf]} \end{aligned} \tag{8.15}$$

With this formulation, the free parameters can be optimized to minimize the fuel burn objective function.

8.9 Summary

This chapter has provided an introduction to the TM of HEPS. EDT heat sources and aircraft heat sinks were discussed in Section 8.1, followed by a summary of the TM challenges that are germane to EDTs in Section 8.2. TMS heat acquisition and component cooling approaches were described in Section 8.3, followed by a detailed look at TMS architectures in Section 8.4. A reference TMS that was used to introduce analysis concepts in the remainder of the chapter was presented in Section 8.5. Heat transport and heat rejection components, including the physics-based equations necessary for their analysis, were detailed in Sections 8.6 and 8.7. The chapter concluded with a detailed step-by-step design of the reference TMS, listing constraints, imposed conditions, calculated values, and free parameters.

This chapter has been structured and written to provide the reader with a detailed introduction to the unique challenges of managing the heat produced by an EDT. Armed with an understanding of this material, the reader will be well prepared to pursue advanced study in this area and will possess a more well-rounded knowledge of the design challenges of electrified propulsion.

Abbreviations

ACS	air cycle system
AOC	air oil cooler
Bat	battery
BEPS	battery-electric propulsion system
CONOPS	concept-of-operations
COP	coefficient of performance
CP	coolant pump
Eng	engine
EP	electric/electrified propulsion
FEnPS	fueled-engine propulsion system
FOC	fuel oil cooler
FP	fuel pump
FPI	fins per inch
GaN	gallium nitride
Gbx	gearbox
HE, HEPS	hybrid-electric, hybrid-electric propulsion system
HFPS	hybrid-fueled propulsion system
HP	hybrid propulsion
HSSG	high-spool starter/generator
HX	heat exchanger
ISA	international standard atmosphere
ISO	international standard organization
LPT	low-pressure turbine
LSMG	low-spool motor generator
MD	motor drive
NASA	National Aeronautics and Space Administration
NTU	number of transfer units

OP	oil pump
PCM	phase-change material
RCC	ram coolant cooler
SiC	silicon carbide
SOA	state of the art
SP	specific power
Sys	system
TMS	thermal management system
UA	heat exchanger *UA* product

Variables

A	heat transfer surface area, [m^2]
A_{in}	pipe inner cross-sectional area, [m^2]
A_{metal}	metal cross-sectional area, [m^2]
$\dot{C}$, $\dot{C}_{\text{min}}$	heat flow capacity, minimum value of two fluids in a HX, [W/K]
C_p, C_v	specific heat at constant pressure/volume, [J/kg-K]
D_{in}, D_{out}	pipe inner/outer diameter, [m]
Δp_{pipe}	pipe pressure drop, [Pa]
Δp_{pump}	pump pressure rise, [Pa]
ΔT, ΔT_{h-c}	temperature difference, from hot source to cold sink, [K]
$\Delta T_{(h,\text{in}-c,\text{in})}$	temperature difference between hot inlet and cold inlet, [K]
ΔT_{pinch}	temperature difference between the AOC outlet air and inlet oil, [K]
ε	effectiveness
f	friction factor
γ	ratio of specific heats
η_{fan}, η_{motor}	fan/motor efficiency
η_{pump}	pump efficiency
H	heat exchanger "no flow height," [m]
H_C, H_H	heat exchanger cold/hot fin height, [m]
κ	thermal conductivity, [W/m-K]
K_j	*K*-factor of *j*th component
L_C, L_H	heat exchanger cold/hot side length, [m]
L_{pipe}	pipe length, [m]
μ_{fl}	fluid viscosity, [kg/m-s]
$\dot{m}$	mass flow rate, [kg/s]
m_{pipe}	pipe mass, [kg]
$P_{e,\text{pump}}$	pump input electrical power, [W]
P_{fan}, P_{fl}	fan/fluid power, [W]
Pr	pressure ratio
$\dot{Q}$, $\dot{Q}_{\text{max}}$	heat transfer rate (heat duty), maximum possible value, [W]
ρ_{fl}, ρ_{metal}	fluid/metal density, [kg/m^3]
Re	Reynolds number

T_{cond}, T_{evap}	VCS fluid condensation/evaporation temperature, [K]
T_{inlet}	inlet temperature, [K]
$T_{sink,\ in/out}$	VCS sink temperature at condenser input/output, [K]
$T_{source,\ in/out}$	VCS component heat source temperature at evaporator input/output, [K]
U	overall heat transfer coefficient, [W/m^2-K]
v_{fl}	fluid velocity, [m/s]
$\dot{V}$	volumetric flow rate, [m^3/s]
V_{metal}	metal volume, [m^3]

References

[1] J. L. Felder, "NASA electric propulsion system studies," NASA, Cleveland, OH, Tech. Rep. GRC-EDAA-TN28410, 2015.

[2] E. Torenbeek, *Advanced Aircraft Design: Conceptual Design, Analysis and Optimization of Subsonic Civil Airplanes*, 1st ed., West Sussex, UK: John Wiley and Sons, 2013.

[3] I. Moir and A. Seabridge, *Aircraft Systems*, 3rd ed., West Sussex, UK: John Wiley and Sons, 2008.

[4] J. Liu et al., "Pathways for practical high-energy long-cycling lithium metal batteries," *Nat. Energy*, vol. 4, pp. 180–186, 2019.

[5] N. K. Borer et al., "Design and performance of the NASA SCEPTOR distributed electric propulsion flight demonstrator," presented at the AIAA Aviation Techn., Integr., and Operations Conf., Washington, D.C., 2016, Paper AIAA 2016-3920.

[6] C. E. Lents, L. W. Hardin, J. M. Rheaume and L. Kohlman, "Parallel hybrid gas-electric geared turbofan engine conceptual design and benefits analysis," presented at the 52nd AIAA/SAE/ASEE Joint Propulsion Conf., Salt Lake City, Utah, 2016, Paper AIAA 2016-4610.

[7] F. M. White, *Fluid Mechanics*, 8th ed., New York, NY: McGraw-Hill, 2016.

[8] W. M. Kays and A. L. London, *Compact Heat Exchangers*, 3rd ed., New York, NY: McGraw-Hill, 1984.

9 Performance Assessment of Electrified Aircraft

Jonathan C. Gladin

Introduction

Ever since electrification began to garner serious interest within the aerospace community many years ago, there has been a steady stream of literature forecasting the performance gains of various concept hybrid- and all-electric vehicle (HEV and AEV, respectively) configurations. Most studies have defined a specific HEV or AEV concept and computed its potential benefits and key dependencies while attempting to answer the question: Can this powertrain "buy its way" onto an aircraft? In other words, can this new concept equal or outperform traditional ones (e.g., gas turbine or internal combustion based) on cost and other key factors? An important subtext to this question – often not directly addressed – is *when*, specifically, will the concept architecture become physically and economically viable? In short: "Is it as good or better than what we have today, and when can we get it?" Answering these two questions is the goal of the fundamental *performance assessment process*, a systematic method to analyze the trade-offs involved when choosing an electric system over a traditional one (and the hydrocarbons that fuel it).

A systematic performance assessment process for deriving system-level figures of merit (FOMs) for electrified aircraft (EA) concept architectures is described in this chapter. Key steps in the process are identified, and details are given on how they might be reasonably performed. Concepts and conclusions from earlier chapters are called upon as needed.

9.1 Electric Aircraft Architecture Assessment Overview

Although numerous variations exist, the assessment process that is the focus of this chapter is shown in Figure 9.1. It includes several elements that are required to answer the fundamental assessment questions. The first step is to define a baseline, non-electric airplane against which to compare the electrified concept. The baseline serves two purposes: first, it defines a set of minimum requirements for the concept plane to achieve (design range, payload, thrust, etc.); second, it establishes a reference point for current aircraft state-of-the-art (SOA) technology. The baseline is projected to a so-called *entry-into-service* (EIS) date, which is the earliest date at which the "new baseline" aircraft would be available for service. The "new baseline" is essentially

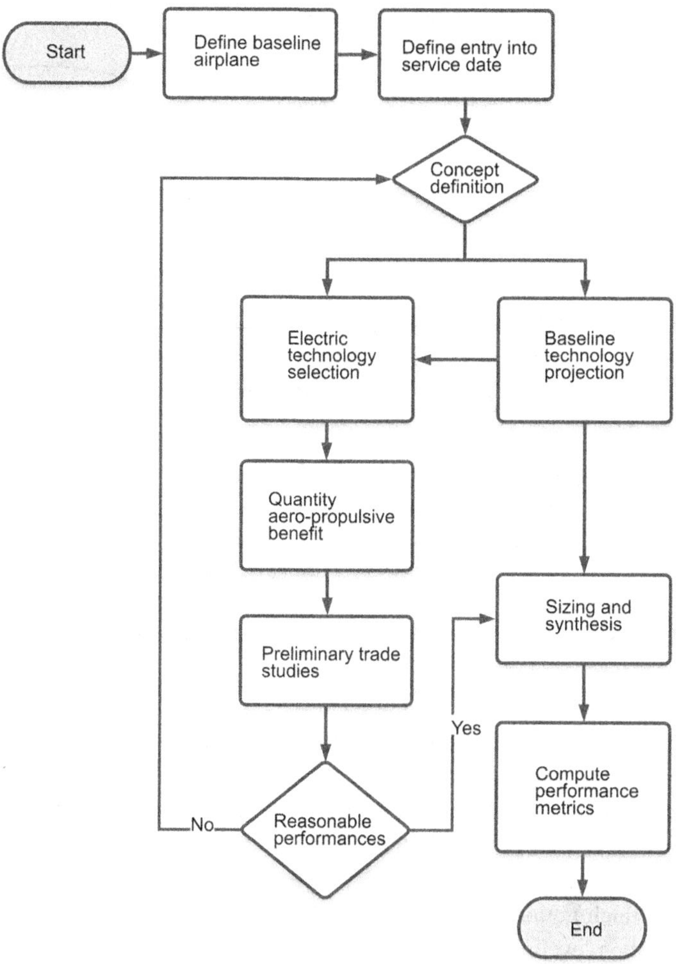

Figure 9.1 EA architecture performance assessment process.

the current SOA projected forward in time to the EIS for the to-be-defined concept architecture, assuming typical improvements. The EIS date may depend on various factors, including the intended time frame of interest for introducing the technology, market dependencies, or other business-related factors.

Next is arguably the most important step in the assessment process: defining the concept EA. In this step, various HEV and/or AEV options are considered, and a final concept (or multiple concepts) for evaluation is/are selected. The aircraft propulsion system powertrain is considered, along with various options for the airplane configuration, propulsor layout, and other configurational options. All these factors that affect the FOMs are then compared, and a final solution is converged upon. This then becomes the concept EA that is carried forward through the rest of the assessment process.

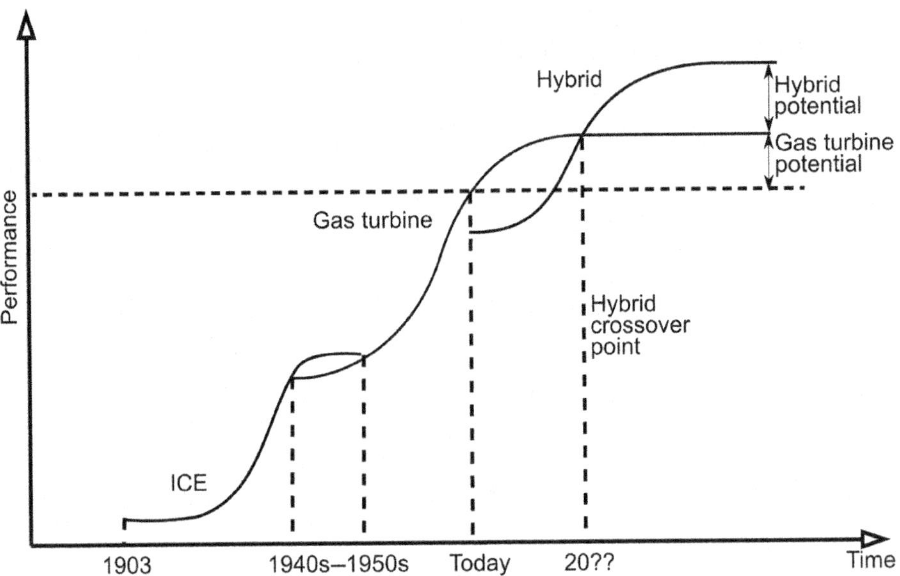

Figure 9.2 Notional propulsion system S-curve charts for large transport airplanes.

With the concept architecture identified, the baseline is projected forward in time to allow for improvements to its conventional propulsion system that can reasonably be expected while the concept is being developed. An illustration of such improvements is shown in Figure 9.2, which starts with the original internal combustion engines used on some of the very first airplanes. The curves shown are known as technology *S-curves*, which notionally illustrate the slow start and development, rapid improvement, and then leveling-off of propulsion technologies over time.

The gas turbine engine, which is the prime mover for most large modern airplanes, has undergone a long progression of technological improvements. Its cycle efficiency has improved primarily due to increasing metal temperatures and pressure ratios and improved cooling technologies and turbomachinery design procedures. The last few decades have seen significant improvement in the gas turbine, and it is expected that this progression will continue into the near future, with some debate about the extent to which the gas turbine engine cycle can be improved up to the limits of the Brayton cycle on which it is based. Various hybrid configurations have the potential to move beyond the gas turbine S-curve, but substantial investment in developing electric technologies is required to achieve the power- and energy-to- weight ratios that are comparable to the gas turbine engine. In order to properly assess the potential of a given concept at an assumed EIS, it is necessary to project both technology S-curves (concept and traditional technology) forward in time to calculate the difference between the two.

Many hybrid aircraft architectures have been designed to enable benefits that may or may not be captured in standard assessment metrics – for example, a so-called *aero-*

propulsive benefit, which is a beneficial interaction between the airframe and the propulsors. These types of benefits must be accurately quantified for their enabling architectures to be competitive in the final assessment of the airplane metrics.

To assess a concept architecture, it is necessary to perform preliminary trade studies to determine if it has reasonable potential for performance improvement. These trades are conducted using basic first principles of flight, thermodynamics, and laws of motion while assuming certain parameters for the technologies of the baseline and concept electrified propulsion (EP) system. If, after the initial trade studies, the performance of the concept system is insufficient to continue with the exercise, it may need to be revised or eliminated from the design space. If, however, the initial trade studies indicate a potential improvement, then a more detailed airplane sizing and synthesis study must be carried out to fully flesh out the electric aircraft (EA) definition. Detailed modeling and simulation is required to compute the aerodynamic, propulsive, and structural properties of the concept, and any additional degrees of freedom (DOFs) related to the electrical system must be investigated in order to determine the best operational usage of the additional energy and thrust systems on the airplane. Doing this gives the electrified concept the fairest comparison to the conventional system when projected forward to the EIS date.

Finally, with the concept airplane fully defined and sized, it is necessary to compute the various metrics of interest that will determine its viability in comparison to the conventional. This is essential to answer the fundamental assessment question: "Does it buy its way on?" These metrics could be the block fuel burn or energy usage of the airplane, the cash operating cost, or other various environmental aspects such as CO_2 or other emissions metrics.

The remainder of this chapter will give more details about each of the steps in the assessment process outlined in Figure 9.1 and will provide some mathematical approaches for conducting initial trade studies.

9.2 Define the Baseline Aircraft

The first step is to define the baseline aircraft and its associated requirements set. A question of paramount importance is likely whether the aircraft is intended for military or civilian applications, as the requirements and FOMs for each will be drastically different. Military applications emphasize capability, stealth, thrust, mission viability, maneuverability, etc., whereas commercial applications value efficiency, reliability, safety, and affordability. Therefore, a clearly defined application and an understanding of its core requirements are essential prior to transitioning to the second phase of the assessment process. Without these, it will not be possible to accurately define an appropriate set of candidate EA architectures which might satisfy each of the requirements while achieving FOM improvement. Furthermore, it may be necessary – even in the conceptual design phase – to quantify certain key requirements (mass, size, etc.) so that those can be estimated at the EIS date and to verify that constraints are not violated during the sizing and synthesis step.

Table 9.1 Parallel hybrid aircraft power, specific power, and specific energy requirements for general aviation (GA) / commuter- and single-aisle-type passenger-class vehicles (adapted from the NAS [2]).

Parallel hybrid aircraft	Power [MW]	Specific power [kW/kg]	Specific energy [Wh/kg]
GA/commuter	<1	>3	>250
Single-aisle	1–6	>3	>800

9.2.1 Aircraft Requirements

For commercial applications, a key classifier for the type or class of airplane is the *passenger count* (PAX), the number of passengers that can be transported during standard operation. Unfortunately, PAX only identifies the number of passengers, not the cargo weight. Since the propulsion system and airframe cannot distinguish between passenger and cargo weight, for design purposes it is useful to generalize the weight parameter by specifying *payload weight*, which includes the weight of the passengers and the cargo. Generally, the vehicle size scales with payload weight, in part because of the structure and wing area required to carry the weight of additional passengers, but also because of the fuselage size and length extensions required to accommodate them comfortably. Due to physical issues of scale, it is generally advantageous for larger airplanes with high PAX to fly longer routes, since the fuel and additional weight required to fly is not much more, relatively speaking. For that reason, the maximum range and the size of large airliners tend to be correlated. Furthermore, there is a dividing line between twin-aisle (wide-body) and single-aisle (narrow-body) airliners, somewhere in the 180–250-passenger range (sometimes called the middle-of-market passenger class). Twin-aisle airplanes will typically fly long-haul flights while single-aisle airplanes support shorter domestic and somewhat longer transcontinental flights. Longer flights require further certification requirements like the US Federal Aviation Administration's (FAA's) Extended Operations (ETOPS) [1], which is required for flying twin-engine airplanes far away from a reachable airport. HE and AE propulsion architectures would have to meet such challenges for these longer-range airplanes.

Basic physics and aerodynamics dictate that the power required to lift an airplane and overcome the aerodynamic drag force during flight typically scales with its weight. As such, energy storage requirements will scale with weight and range. It is clear, then, that the first planes with a large percentage of their propulsion created electrically will be of the smaller variety; converting to EP on larger planes is a much bigger challenge. To illustrate this, the US National Academy of Sciences (NAS) reviewed the feasibility of HE airplanes and came to the conclusions in Table 9.1, specifically for the per-motor requirements of the parallel hybrid concept.

These results indicate that the passenger class, aircraft range, and other energy-related operational conditions should be clearly defined prior to defining the concept EP architecture. Other types of requirements then may extend from the concept, including several related to the propulsion system. Two of these – *climb rate capability* and *one-engine inoperative (OEI)* – are discussed in the following sections and are relevant to the system design created during the trade study and sizing and synthesis processes.

9.2.2 Climb Rate Capability

The *bypass ratio* (BPR) of a turbofan engine is the ratio of air flowing through the fan and bypassing the core (where combustion occurs) to the air flowing through the core. A high BPR is used to increase an engine's efficiency during cruise, but this comes at the expense of reduced *specific thrust*, the thrust produced per unit of mass flowing through the engine. This reduces the mean jet outlet velocity and increases the thrust *lapse rate*, the rate at which thrust decreases with increasing speed. This is particularly pronounced at higher altitudes with thinner air density. Thus, one issue with large turbofan engines with high BPRs is that they tend to lack sufficient top-of-climb thrust. The aircraft must maintain a certain *specific excess power* P_s, in W/N, which is usually specified as a *climb rate* dh/dt, in m/s. Climb rate (and, therefore, P_s) can be expressed in terms of *flight speed* V_∞, in m/s, and either *flight path angle* γ, in rad, or *thrust* T, *drag* D, and *weight* W, all in N, as follows:

$$P_s = \frac{dh}{dt} = V_\infty \sin(\gamma) = \left(\frac{T - D}{W}\right) V_\infty. \tag{9.1}$$

Different top-of-climb thrust requirements exist for each airplane, but the requirement is generally about 1.52 m/s (300 ft/min). From (9.1), it is clear that the thrust requirement is a function of the specific *weight-to-drag ratio* (in this case, close to L/D), *thrust-to-weight ratio* (which declines substantially at altitude), and V_∞. This requirement can become germane for certain HE or AE airplanes, which typically have heavier weights and therefore will require larger thrust-to-weight ratios to satisfy the necessary climb rate at altitude.

9.2.3 One-Engine Inoperative Operation

For reliability purposes, most large airliners must be able to take off in cases where a single engine is lost. This is because the probability of this occurrence is high enough that it would cause major safety concerns were it to routinely lead to catastrophic failures. The probability of two engines failing simultaneously is acceptably low, however, so flight safety regulations only require that the airplane must comply with the *one-engine inoperative* (OEI) condition. In the United States, airplanes must comply with the FAA regulations set forth in Part 25 [3] regarding OEI climb. These specify climb gradients that the aircraft must maintain while operating with the loss of an engine and are based on the total number of engines. While achieving the OEI regulatory gradient, the airplane must have enough residual power from the remaining engine(s) to overcome the additional drag produced by the inoperative engine and to power the vertical tail control surface deflections to maintain pitch and

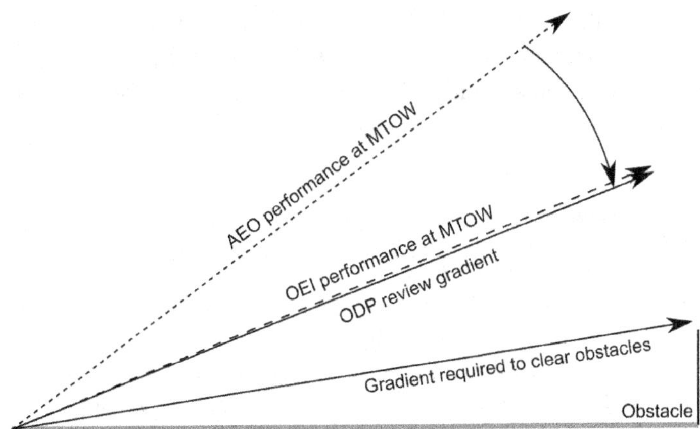

Figure 9.3 Notional flight path for AEO and OEI trajectories with required OEI gradient.

yaw control. This process is shown notionally in Figure 9.3, as the climb gradient is reduced from the nominal all-engines operating (AEO) trajectory to the planned OEI performance at a particular weight.

Many HE concepts have come to consider aircraft with a single gas turbine engine (so-called *single engine concepts*), while alternative sources of power are used for the OEI contingency. This selection contains various performance trades, but the OEI regulation is of primary concern for these types of architectures, and, of course, these novel aircraft concepts may require new regulations to safely certify them in the future. Part 25 technically only applies to airplanes with gas turbine engines of a certain number or count. For a HE airplane, the relative level of reliability for each of the electrical and gas/mechanical power paths must be determined, and specific gradient regulations for specific types of hybrid powertrains may have to be formulated. In the absence of these future regulatory updates, it is recommended that current regulations for the "two-engine" configuration are applied for assessments of large airliners.

9.3 Define the Entry-into-Service Date

The EIS date selected for an electrified propulsion concept aircraft greatly depends on the aircraft's intended application. For example, for single-aisle concept vehicles, it is impractical to consider EIS dates occurring within the next few years. One reason for this is that there have been recent revisions to the Boeing 737 and Airbus A320 single-aisle airplanes (which dominate the market) representing the current SOA. Another reason is the long lead times required to develop, design, and produce aircraft in the large airliner space. Therefore, single-aisle electrified propulsion concepts tend to be focused on the so-called single-aisle replacement that will likely occur in 15–20 years, such that an EIS date of 2035–2040 is more appropriate.

For smaller concepts, such as for commuter and regional aircraft that serve markets with "thin" passenger trip distributions – often called *thin-haul* – a significantly earlier EIS date may be chosen. This is because – relative to larger commercial airplanes – the aircraft development time is much shorter, the technology threshold is a bit lower, and the risk associated with airframe development is less. Also, smaller aircraft have not had as much investment in traditional engine technology as have the larger classes, and therefore, the relative benefit of electric technology may be greater. Often, the EIS date decision is considered an open question, which means that the enabling technologies on which the concept depends have yet to be determined. In these cases, the time required to fully mature the concept to the point of viability is determined as part of the assessment. The assessment process may then need to be conducted using various levels of technology assumptions at different EIS dates to attempt to determine where the crossover date would lie in the future. All of this should be done with thought given to the innate uncertainty involved in these determinations and should look at multiple scenarios and the technical risks involved in the reaching the technology targets assumed.

9.4 Define the Concept Electric Architecture

After the baseline and EIS date have been chosen, the EP concept is defined. The fundamental assessment questions are considered at the highest level, prior to performing preliminary performance calculations during the trade study phase. In comparison with traditional turbofan-based propulsion, HE and AE propulsion allow many more powertrain and airframe options. A sample EA combination space, or *morphological matrix*, is given in Table 9.2. There are, of course, many more design options than those shown available during the more detailed design

Table 9.2 Example EA morphological matrix.

		Propulsor			Airframe Design		Battery	
Architecture	Power plant location	Type	Location	No.	Wing	Tail	Location	Charging
Parallel HE	Wing tips	Turbofan	Rear fuselage	4	Swept	T-tail	In-floor	In-flight
TE	In-board wing	Ducted fan	Wing tips	3	Blended	Conventional	Wing	At airport/ replaceable
Series HE	Center line (top)	Combo	Front fuselage	6	Truss-braced	Pi-tail	Pods	At airport/ unreplaceable
Partially TE	Center line (tail)	Non-ducted fan	Wings	8+	Box	Canard	Pylon	In-flight
AE	Center line (aft)	Propeller	Over wing	2	Straight	V-tail	Fuselage	At airport/ replaceable

phase of an aircraft, but during concept architecture definition, only those options that specifically pertain to the configuration of the propulsion system are considered. Creating a morphological matrix (potentially with expanded categories) provides a systematic process through which particular options are eliminated based on the intent of the concept, basic engineering common sense, or known study results in literature. During this process, one key question must be considered when deciding on each option in the configuration: "What is the core benefit provided by this architecture?" If a particular option does not provide a substantial benefit for the application in mind, then it can easily be eliminated from the process.

9.4.1 Electrified Propulsion Concept Potential Benefits

General potential benefits provided by EP systems and a few example airplanes in which these have been demonstrated are summarized in the following sections. Efficiency improvements in thermal conversion and propulsion and reduced emissions and costs are discussed.

9.4.1.1 Improved Thermal Conversion Efficiency

The *thermal efficiency* of a gas generating core is the ratio of the amount of available useful energy produced to the chemical energy consumed by the device. An improved thermal efficiency can be achieved with a hybrid propulsion system through the inclusion of various forms of non-hydrocarbon energy storage, which generally have much higher thermal efficiencies. This must be traded against any weight penalty of the additional hybrid equipment and generally lower specific energy of batteries and fuel cells, etc. An EP concept can also increase thermal conversion efficiency by facilitating a more efficient overall aircraft design or by enabling the gas engine (in HE concepts) to be operated more efficiently. This can be done by offsetting gas engine power with alternative energy sources during high-demand portions of the mission to enable more efficient engine operation closer to its "on-design" peak [6]. For certain small-aircraft configurations, this can also be achieved by reducing the number of gas engines to a single unit, thereby increasing the power output and efficiency of the single engine. However, this configuration would require alternative energy storage to supply backup power in the event of an OEI condition.

9.4.1.2 Improved BPR and Propulsive Efficiency

Increases in propulsive efficiency are generally driven by higher BPRs, as discussed previously. For EA concepts, further increases can be achieved by enabling configurations that have a higher "effective" BPR than can be achieved by a conventional system. For some applications, turboelectric (TE) distributed propulsion configurations can achieve higher BPR while incurring lower penalties due to nacelle wetted area or the weight increases normally associated with very-high-BPR turbofan engines. Moreover, aero-propulsive interaction concepts, such as boundary layer ingestion (BLI) [7] or over-the-wing propulsors [8], can enable further increases in efficiency. Higher BPR can also be achieved by alleviating nacelle outer mold line constraints typically encountered on some airplanes.

9.4.1.3 Improved Emissions Characteristics

EA propulsion has the potential to improve emissions characteristics, especially with respect to carbon dioxide (CO_2) and nitrous oxide (NO_x). For HE propulsion, CO_2 reductions may be achieved by utilizing the potential for alternative forms of energy storage enabled by HE configurations. These can have lower net carbon intensities (grams of CO_2 per kWh of useful work produced) and therefore release overall less carbon into the atmosphere during their lifecycle of production, distribution, and use. For an electrical storage device, such as a battery, carbon intensity strongly depends on the source of the electrical charging energy. The carbon intensity of non-fossil-fuel-based forms of power production (wind, nuclear, solar, etc.) tends to be very low, while that of coal and natural gas tends to be much higher. The relative mix of energy sources varies widely by nation and region.

NO_x emissions may be reduced by using parallel HE propulsion to reduce the firing temperatures of the gas turbines. These temperatures strongly influence the concentrations of nitrous oxides produced. This can significantly improve the air quality in communities located near large airports.

9.4.1.4 Reduced Costs

Cost savings may be achieved by utilizing electric energy storage, which has the potential to be cheaper than jet fuel. The increasing cost effectiveness of renewable energy may make this a likely outcome, especially if fuel prices continue to increase. Other direct operating costs, such as maintenance, may see improvements with EA systems for some applications. For example, if the life of hot gas turbine components can be improved by significantly reducing their time at high temperature and stress, then large potential maintenance, repair, and overhaul (MRO) savings can be realized if the added costs for cycling and replacing the battery systems do not offset these reductions.

9.4.2 Electrified Propulsion Concept Architecture Considerations

Important features of several EP concept architectures that should be considered are given in the following sections. Considerations for parallel HE, TE, series HE, partially TE, and AE architectures are summarized.

9.4.2.1 Parallel Hybrid Electric

The parallel HE architecture can take on many forms; perhaps the most popular is the parallel hybrid turbofan. In fact, each major gas turbine engine manufacturer has its own research parallel hybrid concept engine for the single-aisle airplane class: GE has the hFan, as used on the Boeing Subsonic Ultra Green Aircraft Research (SUGAR)

Volt HE concept [9], the Rolls-Royce Electrically Variable Engine (EVE) [10], and UTRC, its parallel hybrid concept [6], which has also been considered for small unmanned aerial vehicles (UAVs) and helicopter applications. Generally, these concepts have a higher tank-to-wake thermal efficiency but have high electrical energy storage requirements, such that the evaluation of each concept is essentially an efficiency versus weight trade.

The breakeven analysis in Chapter 1 discusses the high sensitivity to battery specific energy of the parallel hybrid turbofan performance. Studies have shown that specific energy levels of at least 400–500 Wh/kg are required to break even at short range (less than 500 nmi), and even higher specific energies (greater than 800 Wh/kg) are required for a typical 900 nmi economic mission for the single-aisle airplane.

Single-aisle parallel HE concepts have generally included electric motors in the 1–3 MW range (per engine), which has yielded a degree of hybridization of power H_p in the range of 0–50 percent. This electric power can be used to offset power required from the gas turbine during high power points in order to optimize the engine's performance. It can also be used to enable substantial reliance on electrical energy throughout the mission, which may reduce emissions.

9.4.2.2 Turboelectric

In the turboelectric (TE) architecture, a gas turbine engine is used to produce torque and power to turn an electric generator that provides a large amount of electric power. The electric power generated by the turbine engine can be distributed throughout the vehicle and in many concepts is used to power a set of distributed propulsor fans for what is termed "distributed EP."

Distributed propulsion is a design concept in which power generated from a turbine core engine is distributed to propulsors located remotely from that engine; in other words, the mechanical propulsor is decoupled from its power source. While this can be accomplished via mechanical or electrical connections, the electrical option provides some benefit in that it comprises static hardware, is generally lighter weight per meter of distribution, and, thus, should trade well if the power is desired to be distributed along greater distances, as it is in the NASA N3-X concept [11]. When compared to an electric machine (EM), a simple mechanical linkage, such as a shaft and gearbox arrangement, is generally be more reliable and efficient due to the relative simplicity and lower losses. However, in the case of a highly distributed transmission, the distributed propulsion option is much more competitive due to the increasingly complex nature of the mechanical option. The downside to the distributed propulsion concept relative to traditional turbofan engine propulsion is that the distribution path creates additional losses and substantial weight in the form of the electrical or mechanical conversion and transmission components. The propulsion system overall becomes much heavier and less efficient if the losses are not negligible, as is shown by the breakeven analysis in Chapter 1.

Part of the benefit of distributed electric power comes from the flexibility afforded by the decoupling of the power generation and thrust systems. The gas turbines can operate at speeds that are independent from those of the propulsors, and the propulsors can be located in the airframe to produce aero-propulsive interactions. Aero-propulsive interaction for the TE airplane includes the potential for BLI, blown wings, or overwing nacelles that might increase propulsive efficiency. These propulsors may also have much lower fan pressure ratios, and, therefore, the system equivalent BPR is quite large in comparison to what a turbofan engine can reasonably achieve, such that the overall propulsive efficiency is increased due to both the aero-propulsive influence and the increased BPR. To achieve the benefits of distributed electric power, the challenge is to produce lightweight, high-power, and efficient electric distribution components.

The turboshaft engines used in these distributed configurations primarily generate electric power but also produce some residual nozzle thrust. However, the H_p parameter for this system is close to unity because the residual thrust is small. Since the energy source is the fuel inserted into the gas turbine, the H_e parameter is zero. The decoupled speeds of the turboshaft engines and propulsors enable some efficiency optimization for engine on- and off-design conditions.

There are several examples of TE distributed propulsion concepts. These include ES-Aero's ECO-150, the NASA N3X, and various other pure TE systems.

9.4.2.3 Series Hybrid Electric

An evolution of the TE system is the series HE. This hybrid powertrain contains a significant TE generation component but also has a battery system that can be used for various purposes. The battery can be used to supplement the energy system of the airplane in a way that is equivalent to that of the parallel hybrid system. Many of the trades involved in using the battery in this way are analogous to those discussed for the parallel hybrid system (i.e., essentially a weight vs. efficiency trade). In fact, from a mathematical perspective, the system can be viewed as identical to a parallel hybrid system with additional efficiency terms in the conventional power series chain due to the additional electrical conversion and transmission equipment. However, these systems are very different physically, as the electric motor is remotely located from the gas turbine engine and driven by electric power from both powertrain systems. The speeds of the gas generator and electric motors are not specifically coupled as they are in a parallel hybrid turbofan. Furthermore, this system enables swapping between various power sources to operate a distributed propulsor array.

There is added complexity for this type of architecture because the battery current must be tied into the electrical system via a bus to be distributed to the propulsors. This potentially adds weight and reliability concerns but enables detailed power management approaches for the optimization of the energy system. Furthermore, if the battery can be made large enough with high enough specific power, the turbo-shaft engine size can be reduced (in multiengine architectures, it may be possible to eliminate one or more engines). If the architecture can cut down to a single large gas turbine with a battery backup power, there is likely a large fuel burn reduction that can be achieved since gas turbine engines tend to be more efficient at larger scales. Additionally, reducing the number of gas generators leads to MRO cost reductions.

There are many hybrid concepts in the public literature that consider series HE concepts with a single gas generator. In such cases, the primary concern is the reliability of the system, OEI power requirements for the battery system (which drives weight), and the battery specific energy and power that are required for the system to buy its way onboard.

9.4.2.4 Partially Turboelectric

In the partially TE airplane, thrust is produced by both the main gas turbine engine and a remote (or multiple remote) propulsor(s) not tied to the main engine. The example of the single-aisle TE aircraft with aft boundary layer propulsor (STARC-ABL) aircraft is discussed in Chapter 1, where two underwing engine/generators produce a large amount of thrust and generate enough electricity to power an aft fan, which ingests the fuselage boundary layer [12]. This architecture enables the reduction of the underwing engine size by moving some of the captured propulsive air to the aft of the airplane. This enables a higher BPR for both the underwing engine itself and the system as a whole, while providing additional propulsive synergy through fuselage BLI. All of these have the potential to improve propulsive efficiency.

The negative aspects of this architecture are the weight (specific power) requirements of the components (as discussed in Chapter 1) and, in the case of the STARC-ABL arrangement, the challenges involved with integrating the aft fan with the tail cone. This represents a somewhat complicated structural layout, creating issues of interference with the auxiliary power unit (APU) placement, and creating concerns about tail-strike on the bottom of the fan nacelle. As with most BLI systems, the fan experiences flow that is not "clean" but heavily distorted, which imposes aero-mechanical and noise concerns for the aft fan.

9.4.2.5 All Electric

Like the parallel HE, the AE configuration is heavily dependent on the required range and available battery specific energy. Typically, AE airplanes are confined to short ranges and only practical in very small configurations (e.g., for general aviation). In the very small UAV regime, the AE architecture dominates because of the relative inefficiency of gas engines at this scale.

An AE airplane requires a high-reliability electric power system because this system is mission critical and the only power source on the aircraft. Substantial reserve power is required to safely fly the airplane, such that the system sensitivity to battery specific energy is exacerbated even further beyond that of the other hybrid configurations, since some percentage of the energy must be reserved for emergency contingencies. Finally, thermal constraints must be carefully considered when defining the flight trajectories that dictate the battery power demand. For example, the NASA X-57 demonstrator has shown the critical importance of this type of optimization for somewhat larger AE configurations [13].

9.4.3 Summary of Architectures and Selection Criteria

The benefits detailed in Section 9.4.1 are assessed for each of the architectures in Section 9.4.2 and summarized in Table 9.3. When choosing a concept, it is necessary to consider the main performance benefits that each architecture might provide

Table 9.3 Potential benefits of electrified propulsion architectures.

Architecture	Thermal efficiency	BPR increase	Aero-propulsive	Emissions	Cost**
Parallel HE	High	Medium	Low	High*	Medium
TE	Low	High	High	Medium	Low
Series HE	Medium	High	High	High*	Medium
Partially TE	Low	High	High	Medium	Low
AE	High	Low	Medium	High*	Medium

* For CO_2, this depends on assumptions of where the battery energy comes from and life-cycle costs of batteries

** AE aircraft have the potential for cost reductions via efficiency improvements but also come with additional components, which can be costly. The history of electric cars shows that purchase costs will likely be higher, but energy and maintenance costs may offset that over the lifespan.

Table 9.4 Powertrain architecture and vehicle-class EIS heuristics.

Architecture	Thin-haul	Regional jet	Single aisle (narrow body)	Wide body
Parallel HE	N + 1	N + 2	N + 3	N + 4
TE	N + 1	N + 1	N + 3	N + 4
Series HE	N + 2	N + 3	N + 3	>N + 4
Partially TE	N + 1	N + 1	N + 2	N + 3
AE	N + 2	>N + 4	≫N + 4	Unlikely

from a general perspective as well as other ancillary effects such as reduced emissions and cost. The matrix Table 9.3 provides some guidance when thinking about how the propulsive layout of an airplane might synergize with an EP concept.

The preceding considerations, along with the numerical breakeven analyses described in Chapter 1, provide some clear heuristics for selecting an appropriate powertrain. An example is provided in Table 9.4, wherein N represents the current generation of each vehicle class. The number added to N represents the predicted number of generations away from the EIS date of the EP architecture is. For example, the EIS date for a single-aisle TE aircraft is expected to be three generations away.

There is no guarantee that the architecture chosen in this initial down-selection will be globally optimal due to the difficulty in predicting how the concept powertrain will integrate with the airplane application chosen. Furthermore, there may exist many possible combinations of propulsor layouts that are compatible with the basic powertrain system, each having its own performance potential. Each of these feasible concepts must be scrutinized with a higher-fidelity analysis to determine its true benefit. Due to this, several potentially competitive architectures may remain after the initial concept definition exercise described in this section.

9.4.4 Concept of Operations

The discussion thus far has focused on the potential performance benefits and emission and cost reductions of each EP architecture. Going one level deeper, the unique features of the powertrain are considered vis-à-vis how they might be used to enable further performance benefits by augmenting the operation of the airplane and power systems.

Each airplane architecture has a concept of operations (CONOPS) associated with it. A CONOPS is defined by specifying how the EP system is used during each phase of a mission and how that might change with the operational characteristics of the airplane (payload and range). CONOPS for EP architectures attempt to utilize the hybrid systems in phases where the gas turbine is weakest, such as ground idle, high altitude climb, and descent. This eliminates some of the inefficiency of the gas turbine system at extreme off-design conditions, and with high enough specific energy in the battery system, this concept can buy its way onto the airplane. Additional operations that are potentially synergistic with hybrid configurations include AE taxi (eTaxi), core shutdown configurations [9], and top-of-climb assist (by enabling higher cruise altitude).

Going a little deeper on the CONOPS, specific hybrid powertrain uses can be defined. For example, within the parallel hybrid turbofan architecture, there are actually several ways to define the system CONOPS. The first can be called *power offset*, while the second is an *energy offset* CONOPS. The power offset CONOPS uses the hybrid system to reduce the gas turbine engine size, improve its thermal efficiency, and potentially perform e-Taxi with a large electric motor and high-power-density battery pack. The energy offset CONOPS focuses on inserting an optimal amount of battery energy at various ranges and especially at short range. This concept is shown in Figure 9.4. At short range, the lower fuel load creates a margin between the maximum weights of the airplane and where it is operating, which creates an opportunity to use larger battery weights at short range to offset further fuel quantities. At longer ranges, smaller batteries or no batteries would be used to reduce the impact of the battery weight on sizing the airplane for a max range capability. However, this means that different battery sizes would have to be loaded on the airplane for various airport pairings and payload sizes (PAX). This poses an infrastructure question for the airport and integration issues for the airplane, since the battery packs would have to be accessible and interchangeable in a modular way such that the required change could be performed in a reasonable amount of time during airport turnaround. It is very important to define these kinds of details in terms of how the energy system CONOPS would look (even during the conceptual design phases) so that the assumptions made about infrastructure requirements would be consistent with a far enough EIS date.

9.5 Baseline Technology Projection

The next step in the performance assessment process diagrammed in Figure 9.1 is the baseline technology projection (BTP), which follows concept definition and in parallel with electric technology selection. An example of BTP for gas turbine engines was shown with the S-curves in Figure 9.2. In this section, the BTP process is detailed, focusing on turbine engine technology progression in Section

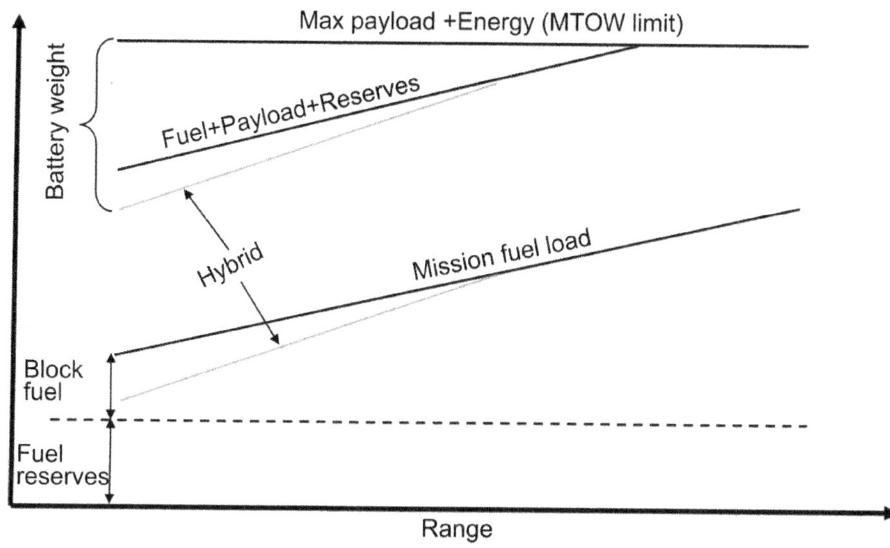

Figure 9.4 Example Energy-Offset CONOPS for the parallel hybrid configuration.

9.5.1 and predicting future technologies and technology readiness level (TRL) in Section 9.5.2.

9.5.1 Turbine Engine Technology Progression

The difficulty of predicting both propulsion and airplane technology levels increases significantly as the EIS date is moved farther into the future. The intent of turbine engine technology progression, then, is to provide a systematic method to extrapolate the baseline gas engine performance forward in time. Amongst many approaches to accomplish this, the most common is engine thermodynamic cycle analysis, whereby the actual engine thermodynamic performance is modeled and progressed in time.

The process is shown notionally in Figure 9.5 and is comprised of a few essential steps: (1) construction of an appropriate cycle model, (2) calibration to open-source SOA engine information, (3) inclusion of historical technology regressions into the cycle model, and (4) selection of the advanced conventional gas turbine cycle. The result of (4) is used for determining the application and selecting EIS date. The forecasted cycle (and its associated performance) that arises from this process can be tailored later to the meet the thrust requirements of the projected baseline aircraft as model data becomes available. The primary data source used to supplement this process is the International Civil Aviation Organization (ICAO) Aircraft Engine Emissions Databank [4], which serves as a collection of performance data for existing engines, including rated thrust, fuel flow, emissions indices, pressure ratios, and BPRs that can be used to tune the

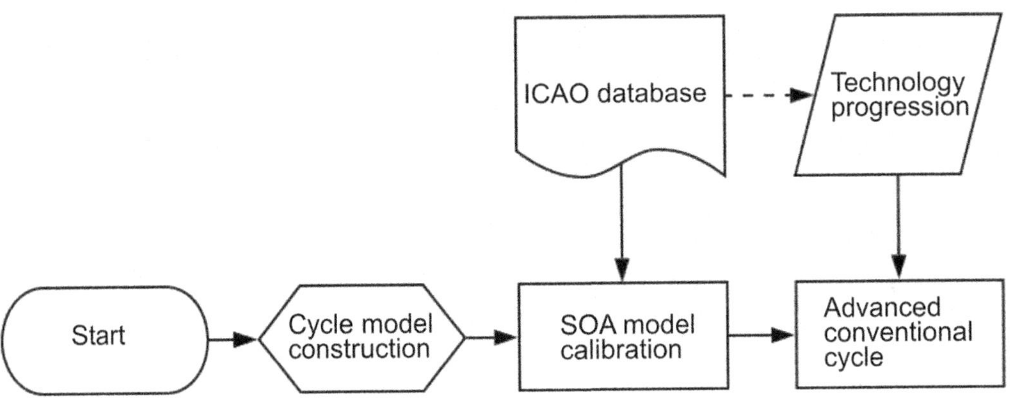

Figure 9.5 Notional turbine engine technology progression process.

model. With a calibrated model, the engine performance can be updated to that of a future engine through the technology progression exercise. The purpose of this is to forecast technology and cycle enhancements and their associated performance improvements out to the EIS date.

Conventional engine technology can be progressed in several key areas, and these are discussed in Sections 9.5.1.1–9.5.1.5.

9.5.1.1 Turbine/Rotor Inlet Temperature

The turbine/rotor inlet temperature (TIT/RIT) is the hottest temperature encountered in the engine thermodynamic cycle. Thermodynamics dictates that a higher TIT/RIT generally leads to a more efficient engine; thus, technologies are sought that enable this temperature to be as high as possible. As such, progress in cooling technologies and advanced materials lead to increases over time in the maximum temperature the cycle can achieve. Unfortunately, TIT/RIT data from engine manufacturers is largely unavailable to the public and is not typically published in the ICAO databank. To construct a cycle model in the absence of this data, the temperature parameter can typically be adjusted to match the bypass ratio, thrust, and fuel flow values. This is part of the cycle calibration exercise shown in Figure 9.5.

9.5.1.2 Bypass Ratio

BPR is a key driver of propulsive efficiency and values have steadily increased over time, as demonstrated in Figure 9.6. Current gas turbine engines have sea-level static (SLS) BPRs between 10 and 11 for larger engines and slightly lower for lighter aircraft applications.

9.5.1.3 Overall Pressure Ratio

The overall pressure ratio (OPR) is the ratio of the highest total pressure in the engine to the ambient pressure. This is typically on the order of 20–50 for most gas turbines and increases with the rated output (thrust) of the engine. OPR values have steadily

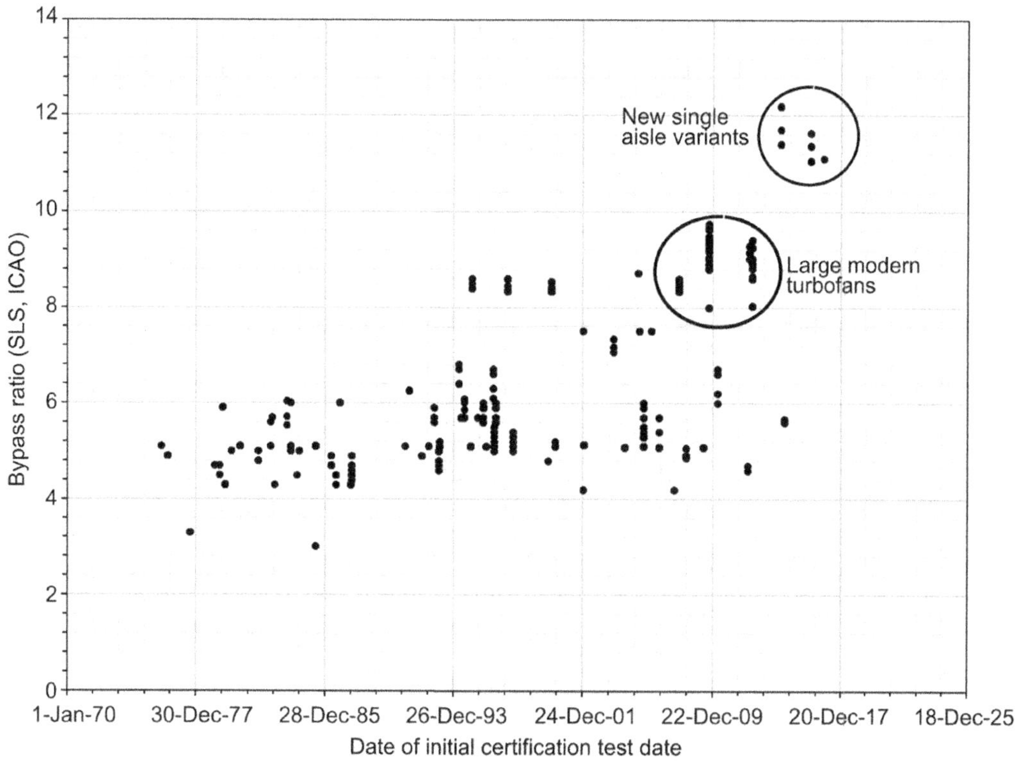

Figure 9.6 SLS BPR vs. certification test date for large turbofan engines. Data from ICAO databank, [4]

increased over time due to advances in compressor design techniques and materials. Figure 9.7 demonstrates this progression of OPR versus rated output and regression lines for different series and classes of engines.

9.5.1.4 Component Efficiencies

The efficiencies and pressure drops of its constituent components determine the actual gas turbine engine cycle efficiency in relation to the Brayton cycle ideal efficiency. Like TIT/RIT, these values are protected fiercely by the engine companies and are not available in the ICAO databank; however, they can be estimated during the calibration process. Many technologies target improvements in component efficiencies, especially for the turbomachinery in smaller engines where efficiencies are generally lower.

9.5.1.5 Engine Weight

The engine weight can be estimated based on historical data and is best modeled by regressing the thrust-to-weight ratio and extrapolating forward in time from that. The thrust-to-weight ratio of engines using gas turbines has progressed significantly over the course of gas turbine technology development, primarily due to advances in

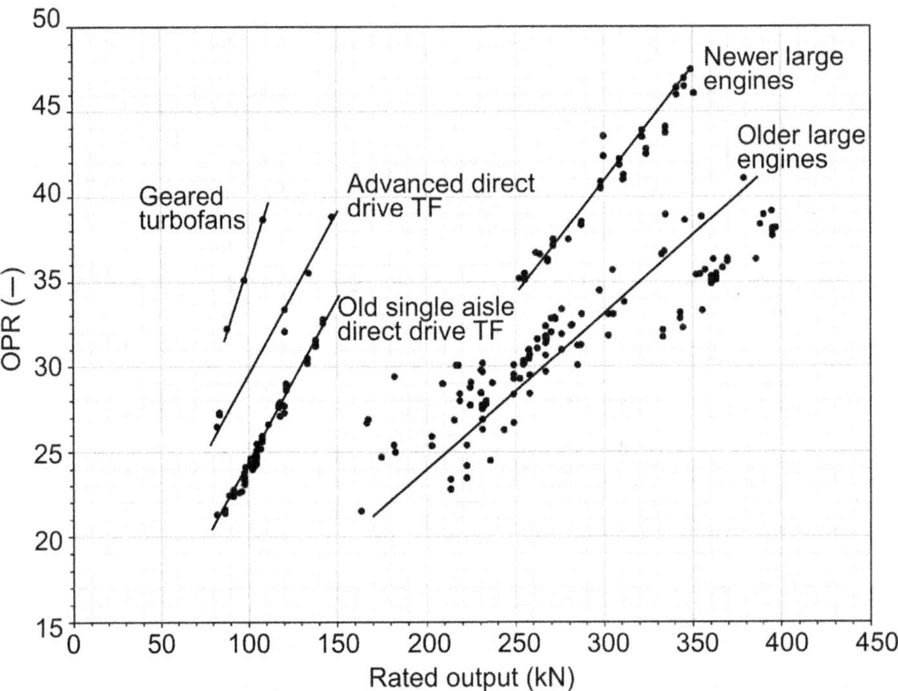

Figure 9.7 Rated output vs. OPR (SLS uninstalled test conditions) for large turbofan engines with greater than 60 kN of thrust. Data taken from ICAO databank, [4]

materials and through structural design that minimizes the weight required for the turbomachinery components and casings.

9.5.2 Future Technologies and Technology Readiness Level

One metric for defining the level of "readiness" of a technology for implementation/deployment in the market is the *technology readiness level* (TRL) [5]. The concept of TRL, originally developed in the 1970s with six or seven levels of readiness, is a means to compare different technologies with respect to their maturity levels. Since its inception, the TRL scale has gradually expanded to include two additional levels for a total of nine, and, generally speaking, a technology must achieve TRL 9 before it is considered flight ready.

Any new concept technology must be at or nearing at least TRL 6 at the beginning of the preliminary design of the airplane – which can generally take 5–10 years – in order to justify the risk of including it in early designs. This means that the EIS date will have a definite impact on which technologies are selected for the baseline engine and airplane concept, as selecting technologies that are too aggressive for a near-term EIS date will yield overly optimistic results for the baseline platform. This fact means that the EIS date choice can either amplify or suppress the impacts of future propulsion technologies, such as EP, if future combustion engine technology is overestimated or underestimated.

The same statements apply to the airframe technology, where future aerodynamic improvements – with the potential to increase the L/D of the base airplane – might offset the performance penalties due to weight paid by the typically heavier EP systems. Ancillary considerations like cost, reliability, and other concerns may also impact the decision of whether or not specific technologies are included for the baseline configuration for the chosen EIS date. It is clear that great care must be taken when choosing which technologies are feasible and can thus be included in the baseline application.

9.6 Electric Technology Selection

Once a powertrain architecture is chosen and a concept defined, the powertrain details must be defined in order to conduct detailed modelling. The essential component categories required to define the powertrain for each standard architecture are shown in Table 9.5. Certain categories – transmission, power electronics, and protection systems – are present in all architectures. Others are optional, such as the additional propulsive devices in the case of parallel HE and the different energy storage options, each of which may be present depending on the choice of secondary energy storage.

In the following sections, several of these component categories are discussed.[1] Prior to conducting a preliminary breakeven analysis, it is necessary to state assumptions for the standard parameters presented. If possible, the assumed parameter values should be based on technology progression trends (i.e., they should be reasonable and attainable) so that the estimated benefit at the concept EIS date is as accurate as possible.

Table 9.5 Powertrain component categories and their use in EP-based architectures.

	Component								
Architecture	**Engine**	**Battery**	**Capacitor**	**Fuel cell**	**Electric mach.**	**Prop. device**	**Trans-mission**	**Power electronics**	**Protection**
Parallel HE	Y	O	O	O	Y	O	Y	Y	Y
TE	Y	N	N	N	Y	Y	Y	Y	Y
Series HE	Y	O	O	O	Y	Y	Y	Y	Y
Partially TE	Y	N	N	N	Y	Y	Y	Y	Y
AE	N	O	O	O	Y	Y	Y	Y	Y

* O: Optional; Y: Yes; N: No

[1] Protection devices, power electronics, and fuel cells are covered in earlier chapters. A detailed analysis of the transmission – a mechanical component – is outside the scope of this text.

9.6.1 Engine

There are many types of gas turbine engines, the choice of which depends on the application and airplane size (i.e., passenger class). Regardless of the architecture, the gas turbine core will produce power that can either be used to produce high-velocity gas within a nozzle and/or mechanical shaft power that can be converted to electricity or used to drive a propulsor or rotor. The gas turbine engine will have a rated power and rated thrust, which depend on the rating structure of the engine and on the temperature and physical limits that define the ratings and at what conditions they occur. The SLS-rated thrust and power output are usually quoted in lbf or kN and hp or kW (MW), respectively.

It is also important to know the assumed fuel type and its specific energy, which is often considered to be the lower heating value of the fuel in BTU/lbm or kW/kg-h. It is then useful, at a minimum, to quote a performance value at a cruise Mach number (or speed) and altitude. Typical efficiency values are quoted in terms of the *brake-specific fuel consumption* (BSFC) or *thrust-specific fuel consumption* (TSFC). The former is used in the case where shaft power P_{sh} is being produced, while the latter is used if it is thrust F_{n}. The equations for BSFC and TFSC (for which the variable c is used) are as follows, where $\dot{m}_f$ is the *fuel mass flow rate.*

$$\mathrm{TSFC} \equiv c = \frac{\dot{m}_f}{F_{\mathrm{n}}} \tag{9.2}$$

$$\mathrm{BSFC} = \frac{\dot{m}_f}{P_{\mathrm{sh}}}. \tag{9.3}$$

Although SI units should be used in these equations, the TSFC is commonly reported in g/N-s or lbm/lbf-hr, and BSFC replaces N with W, and lbf with hp. Appropriate unit conversions are necessary. For the shaft-power-producing engines, the overall thermal efficiency can be computed directly from the TSFC.

Finally, the weight of the gas turbine needs to be included in the study. Often, this is quoted in terms of a *core-specific power*, in kW/kg or hp/lb, or in terms of the engine thrust-to-weight ratio for turbofan/turbojet engines.

9.6.2 Battery

The battery plays a key role in many HE or AE systems; thus, it is imperative that the targeted technology level for the battery is carefully selected based on the EIS date and that the parameters of the battery are consistent with its desired use on the system. The fundamental identifying characteristic for a battery system is its battery chemistry, which determines the specific energy/power and efficiency achievable at the selected EIS date. Chapter 7 outlines the current SOA and potential future directions for batteries, which can be used as a basis for determining the appropriate assumptions, based on EIS date, for the battery chemistry type, packaging factors, etc.

There are several things to consider when deciding whether to use batteries in an aircraft design; perhaps the three most important are the range, type, and size of the aircraft. Table 7.3 lists estimated battery pack–specific energy values required for conventional takeoff and landing (CTOL) and AE vertical takeoff and landing (eVTOL) applications derived from the literature. These are useful in a first-order assessment, and the first step in the process is to identify the concept vehicle mission requirements and cross-check these against Table 7.3 to determine the energy density threshold. The second step is to evaluate whether that threshold can be met by the intended EIS date for the airplane of interest. This can be done by simple extrapolation from the current SOA battery chemistry to the intended EIS date to determine the year-over-year improvement rate required to reach the threshold. If the improvement rate is much greater than both the historical average and what is deemed reasonable, then batteries should be excluded from the consideration, barring some other compelling reason to proceed with a hybrid battery configuration. If the mission or concept is not covered by Table 7.3, simple range equations – e.g., Equation (7.1) – can be used to determine the energy requirements, but more trade studies may also be needed to determine if the concept is amenable to battery integration. If batteries are deemed feasible at the chosen EIS date for the intended application, then the designer can proceed to formal trade studies to determine the degree to which batteries should be used on the concept.

For AE concepts, the primary concern is the need for high power and C-rate during takeoff and dynamic maneuvers and the associated high power density needed. These concerns are amplified by the need to keep considerable state-of-charge margin within the battery to maintain an ample reserve for contingency considerations. Generally, aircraft designers require a reserve energy portion of between 5 and 25 percent of total mission energy depending on the application. Additionally, the designer of the AE configuration should consider the number of required battery cycles per year and should consider limiting the state-of-charge range of each cycle to extend the life of the battery and prevent the need for early replacement. Additional concerns for AE configurations are thermal management and its impact on mission performance, which requires a multidisciplinary optimization of potential flight paths and power usage profiles to keep the battery within its safe operating limits and to reduce risk of thermal runaway and other heat-related challenges.

As opposed to AE concepts, in which all the energy must come from the batteries, HE concepts provide the designer with the added freedom to choose the degree of hybridization of energy. For example, the HE concept could use a battery system, (1) to a limited degree, to provide some targeted benefit, or (2) to a large degree, for renewable energy insertion during flight, such as with the parallel HE turbofan concept. Obviously, the energy density requirements are much higher in the latter. Additionally, due to the presence of the backup gas engine, there is freedom in choosing when to use the battery during the mission. Although this provides opportunities for energy-use optimization, it adds complexity to the design and analysis problem. A CONOPS must be defined to identify when the battery system is allowed to be discharged and recharged. Only then can meaningful trade studies or optimizations be performed to determine the optimal battery size, usage profile, and any associated C-rate and thermal limits that might impact the system.

9.6.3 Capacitor

Capacitors used in HE and AE applications are of the ultra/supercapacitor variety and are characterized by high power density and low energy density. Because of this, they can discharge at very high C-rates, making them ideal for high-power-burst applications. Supercapacitors can be used as temporary energy storage devices to provide transient energy during peak-power-demand situations. This is sometimes called "peak shaving" because the supercapacitor shaves the power peaks from the load seen by the engine. NASA researchers have investigated using a supercapacitor-driven device to power an operability augmentation system for a turbofan engine [14].

The critical point to consider when deciding whether a supercapacitor provides a net benefit to a vehicle powertrain is whether the need for high power density and high discharge rate justifies the (fairly low) weight penalty. Powertrains that require significant peak power spikes over short duration time periods can reasonably consider this option for energy storage or as a potential supplement to other forms of energy storage in a hybrid system. Other advantages of a supercapacitor system may include engine size reduction and cycle improvements due to peak power shaving, reduced maintenance – and the associated cost savings – due to decreased engine peak loads, and the enabling of additional missions requiring high-power ramp rates that exceed the capabilities of standard aircraft engines.

9.6.4 Electric Machine

The EM is the crucial piece of any EP configuration because it is the primary conversion device from electric to mechanical power (and, thus, thrust). EMs are used to power propellers, ducted fans, and possibly turbofan engine shafts. Therefore, the first task in the process of selecting an EM for an aircraft concept is to determine the power and speed (and, from these, torque) requirements for the EM. This will bracket the basic machine operational space. The second task is to determine the required gravimetric specific power and efficiency requirements for the concept. The equations from Chapter 1 for various powertrain types can be used to determine specific power and efficiency requirements. In general, systems with a high degree of electrification (such as fully TE systems) tend to have much higher breakeven requirements on the EMs, since more of the propulsive power is electric, and more of the powertrain weight falls upon the EMs and their associated components.

Chapters 3 and 4 discuss many of the details of EM design, topology selection, sizing considerations, and the comparison of superconducting to conventional machines for various applications. By following the first two steps, enough information should be available about the EM to utilize the information in Chapters 3 and 4 to perform a basic motor topology and cooling scheme selection and to determine an appropriate machine technology. From an aircraft modeling perspective, depending on the level of analysis the EM specific power, efficiency, cooling scheme, and maximum torque/speed limits should be included within the powertrain model. If more details about the relationship between torque, speed, and efficiency are known (such as an electric motor map), then this should also be included to supplement an aircraft mission analysis, which does have some degree of off-design variability in torque and speed over time.

Thus far, the EMs have been considered as propulsive motors. However, EMs also serve as generators to provide primary electric power. Thus, a similar initial assessment of the requirements for the generators should be conducted. For example, the number of generators required for redundancy should be considered, especially for high-power and EP applications. Multiple generators may be required for safety (i.e., redundancy) and to lower the design power of the machines. Generally speaking, the generators will be driven by some other power-producing device, such as an internal combustion or gas turbine engine. The rotational speed of an internal combustion engine is typically one or two orders of magnitude slower than that of a power turbine on a turboshaft engine, so gearboxes may also need to be included in the analysis. To perform an aircraft concept assessment, it is necessary to determine many of the same characteristics for the generator as in the case of the electric motor, and therefore, the guidelines described previously should suffice for choosing a generator technology level.

9.6.5 Power Electronics

AEA concepts contain power electronics as part of the powertrain analysis. In Chapter 5, various power electronics components, technologies, topologies, and basic modeling approaches relevant to EA propulsion were discussed, and in Chapter 6, superconducting technology was discussed. For the aircraft designer performing a concept assessment, the primary need is to determine the nature of the transmission system (i.e., ac direct drive or dc distribution). These are choices that should be determined prior to performing the analysis, as they are necessary to define the powertrain components and have implications for the overall system specific power.

At the design stage, the key power system variable is the distribution voltage, for which there are various existing options, such as the latest +/–270 Vdc system on the Boeing 787 Dreamliner. Systems with high degrees of electrification, such as the fully TE, are more sensitive to the distribution voltage because there is a strong sensitivity of power electronics and cable distribution weight to voltage. The aircraft designer should decide on the appropriate voltage level based on the concept's EIS date and should perform sensitivity studies using the equations from Chapter 1 to determine whether or not superconducting power electronics are required to achieve the necessary performance.

9.6.6 Propulsive Device

A propulsive device is any machine that can convert mechanical shaft power, such as that provided by an EM, into thrust. While turbofan or turboshaft engines are certainly included in this category, this section is meant to cover electrically driven propulsive devices.

There are typically two types of propulsors: propellers and ducted fans. Propellers have been a staple of low-speed flight since the dawn of aviation, and they maintain a high efficiency at vehicle speeds up to very high Mach numbers. Ducted fans are more appropriate for high-speed applications, where ducting is required ahead of the propulsor to control the velocity of the air entering the blades. This duct-containing shroud also creates some noise shielding relative to an unducted configuration.

A typical propulsion performance metric is the *propulsor efficiency* η, defined in (9.4) as the ratio of thrust power (thrust T times velocity V_∞) to shaft power input to the propulsor P_{sh}:

$$\eta = \frac{TV_\infty}{P_{\text{sh}}}. \tag{9.4}$$

For a propeller, this efficiency includes losses in the shaft and propeller blading, as well as the basic propulsive efficiency associated with the remaining waste heat in the jet stream, which scales inversely with the jet velocity. For a ducted application, this efficiency also includes friction losses in the ducting and associated nozzle.

Another useful metric is the *thrust-to-power ratio* – which, as its name implies, is simply the ratio of thrust to power, typically expressed in lbf/hp. This parameter has the useful property of maintaining a meaningful definition at the static condition. The thrust-to-power ratio trends with the propulsive efficiency, but it can be difficult to work with because its range of possible values changes at different flight speeds, as shown in Figure 9.8. This parameter is useful in that it can provide a rough estimate of the ducted fan power requirement as a function of thrust by using a simple proportionality relationship.

Another important propulsive device-related variable is the *thruster specific mass*, typically expressed in lbm/lbf. As the units imply, this is the ratio of the propulsor's mass to its thrust. Thruster specific mass generally decreases with fan pressure ratio, implying a trade-off between efficiency and weight for thrust devices (devices with lower-pressure ratios are more efficient but heavier due to their larger disk diameters, which are required to accommodate the requisite mass flow). Larger disk areas require larger rotor blades, and the additional weight that comes with them. In the case of ducted fans, it also means that there are larger external drag effects and increased weight due to a larger nacelle. As pressure ratio drops, the blade tip diameter grows, and the rotational speed for optimal propulsor efficiency decreases in order to reduce the relative tip speed. This necessitates slower EMs or a geared configuration at very low pressure ratio values.

The choices of fan pressure ratio, propulsor type, and cruise speed are extremely important in the early stages of conceptual design because these propulsor requirements propagate through the rest of the powertrain and impact every component discussed thus far. The vehicle and propulsor thrust demands set the EP architecture

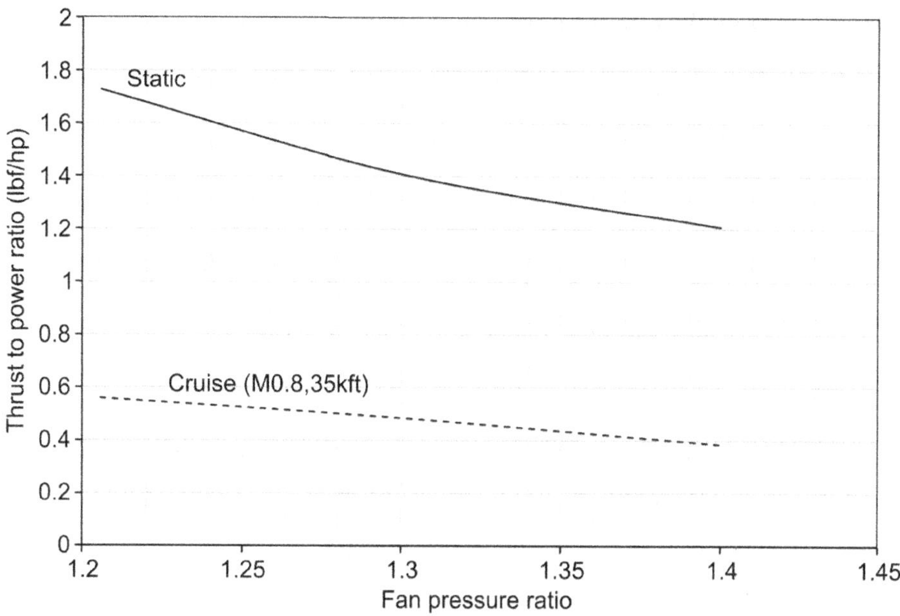

Figure 9.8 Uninstalled ducted fan thrust-to-power ratio vs. rotor fan pressure ratio for static and cruise conditions.

power rating and are the first quantities that must be determined in an initial trade study to set limits and requirements on the electric components.

Although it will not be detailed here, it is worth noting that there are several ways to improve the overall vehicle thrust-to-power ratio by using EP. One is by using BLI, which has been applied in marine propulsion engineering for many decades. The idea is that the vehicle propulsive efficiency can be improved by ingesting the trailing wake of a portion of its aerodynamic body such as the tail, wing, or fuselage, thereby improving the equivalent thrust. For example, the partially TE STARC-ABL configuration ingests nearly the entire fuselage boundary layer and has an electric propulsor thrust-to-power ratio of about 0.7 at the same cruise flight condition shown in Figure 9.8 and at a pressure ratio of 1.25. This is significantly higher than the base curve and implies a range of improvement for the thruster itself of about 25 percent in thrust-to-power ratio.

9.7 Electrified Propulsion-Based Aircraft Design Process

After the baseline technology has been chosen and progressed to the EIS date and the concept electric technology has been selected, the EP-based aircraft design can begin. The process consists of preliminary trade studies followed by detailed vehicle sizing and synthesis, which are both shown in the flowchart in Figure 9.1. Each of these is described in the following sections.

9.7.1 Preliminary Trade Studies

An essential part of the EP assessment process is a preliminary trade study to determine general feasibility of the concept prior to conducting higher-fidelity analyses on various parts of the system. The goals of such trade studies are as follows:

- To determine the relative potential performance benefits of various concepts; in particular, vehicle range impacts are key.
- To identify potential viable ranges of the key variables that dictate the size and design of the system – e.g., propulsor fan pressure ratio, degree of hybridization, and power-to-weight ratio.
- To compute the key technology breakeven points and regions of achievable benefit by the target EIS date.

For some EP-based architectures, the KPP equations from Chapter 1 can be used in these trade studies. A base technology level should be selected for consideration at the EIS date chosen, and sensitivities using these equations should be performed for each architecture considered. For many of the equation inputs, more detailed models may be required to improve the assessment during the vehicle sizing and synthesis step, which will be discussed in the next section.

9.7.2 Vehicle Sizing and Synthesis

After preliminary trade studies, it is necessary to conduct a sizing and synthesis exercise for both the baseline and the EP architectures, a generic process for which is shown in Figure 9.9. The process consists of three main *disciplinary analysis* steps: (1) assembling a propulsion system powertrain model, (2) computing empty weights, and (3) constructing aerodynamic buildups. Each of these consists of additional substeps, as indicated in Figure 9.9, and their results feed into *constraint analysis* and *generic mission analysis*. In principle, aspects of the EP system that also impact the aerodynamics should find their way into each of the three disciplinary analyses, though it is not shown explicitly in the diagram. For instance, any potential synergistic airframe–propulsion interaction effects should be considered during the generation of the aerodynamic data or otherwise bookkept within the propulsion system model. Each architecture will affect each of the three main analyses in a unique way.

Hybridization parameters are those that are part of the EA design space, such as degree of hybridization, level of TE distribution, or other indicators of the way power is divided amongst the various parts of the system. These must be defined in the powertrain sizing routine, as component sizing depends on how much electrical and mechanical power are generated. They can then be varied or optimized during the mission analysis to optimize the fuel burn and energy use of the airplane during flight.

It is now possible to conduct detailed constraint and mission analyses for the airplane. Constraint analysis is necessary to ensure that mission constraints are satisfied, and the thrust-to-weight ratio and wing loading are the most commonly used variables for this. Some applicable constraints that impact the wing and propulsion

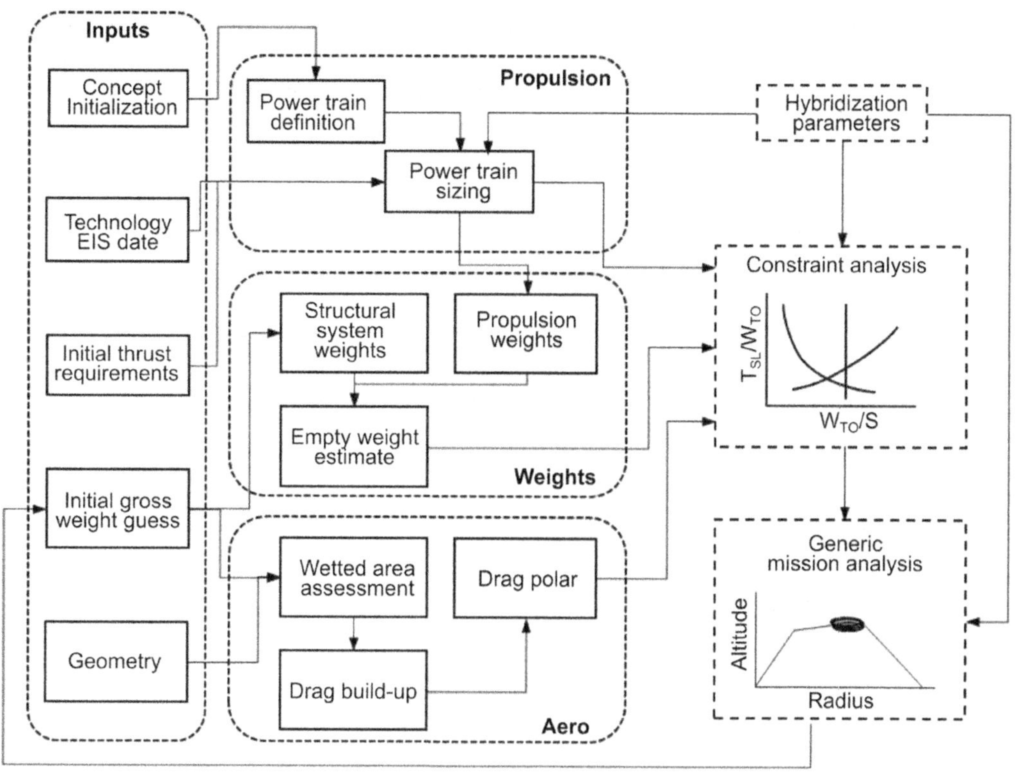

Figure 9.9 Vehicle sizing and synthesis data process.

system sizing are the takeoff and landing field length, minimum climb rate at altitude, OEI climb rate, fuel volume, and maximum velocity or ceiling. The degree of hybridization and the propulsion system performance can significantly impact these constraints. For example, hybrid systems that improve the maximum achievable lift coefficient will ease the takeoff and landing requirements on the gas engine but may also add additional weight to the airplane, thus counteracting some of their benefit and shrinking the feasible design space.

Using the weight, aerodynamic variables, and propulsion data in the constraint analysis yields the information necessary for EP mission analysis. At this point, the vehicle is most easily/commonly modeled as a point mass. Of course, more detailed and higher-fidelity approaches are possible including three- or six-DOF models, but a point mass analysis is sufficient to determine the overall vehicle energy use. In such an analysis, the mission of interest is discretized, and the point performance and fuel burn rate of the vehicle are integrated over each segment. At the end of each segment, the weight is updated along with the lift, drag, and required thrust; this propagates to the fuel flow model for the next segment and so on.

If electrical energy sources such as batteries or capacitors are considered, they must be sized appropriately. Unlike fuel, these do not lose weight as their energy is consumed, and they must have their state-of-charge and other related operational limits taken into consideration. For HE systems, detailed optimization must be conducted during the mission analysis to find the optimal way to utilize available electrical power during the mission to minimize the fuel burn or total energy consumed. Although this is a complex problem, simple approaches can suffice as a first step, leaving more detailed optimization algorithms to the later, more detailed phases of design [15].

9.8 Metrics for Electrified Propulsion Architecture Evaluation

In this section, metrics to evaluate performance as well as environmental and economic impact (cost) for an EP concept are presented. These represent a small sampling of what is possible; it is ultimately up to the designer to decide which metrics are most effective for evaluating a concept for its intended application.

9.8.1 Performance Metrics

There are several high-level metrics that provide insight into the overall airplane performance. These are somewhat standard performance metrics well known to aircraft designers and engineers that broadly measure/quantify a design's quality with respect to efficiency and overall capability. The *fuel specific air range* is one of these, as defined as follows, where R is the *range*, in m, and V is the *instantaneous aircraft speed*[2], in m/s:

$$-\frac{dR}{dW} = \frac{V}{cT} = \frac{1}{W}\frac{V}{c}\frac{L}{D}. \tag{9.5}$$

Equation (9.5) only applies to the case where a single fuel source is used to power the airplane. The fuel specific air range is analogous to the "mile-per-gallon" metric commonly reported for automobiles. This can be extended to a more generic definition for the hybridization case using the *energy specific air range* (ESAR):

$$\text{ESAR} = \frac{dR}{dE} = \frac{V\frac{L}{D}}{cW} = \frac{\eta}{W}\frac{L}{D}, \tag{9.6}$$

where dE is the differential electrical energy storage, in J, and η is the so-called *blockchain efficiency*, which is the ratio of effective thrust supplied to the time rate of change in energy storage that is consumed in the process. Note that this equation

[2] In practice, there is very little difference between the instantaneous aircraft speed, V, used here, and the flight speed. V_∞, used earlier. The difference is that V is the forward flight speed, whereas V_∞ is the relative free-stream air speed.

can be computed for both conventional systems and hybrid systems, taking into account the specific degrees of hybridization for the system. It takes into account weight and aerodynamic effects through the W and L/D terms. To correctly compute ESAR requires an understanding of the EP system's impact on the weight of the propulsion system, the blockchain efficiency, and the aerodynamic performance. Note that this parameter is calculated at a specific flight altitude and does not include climb, taxi, and other flight segments that contribute to the final fuel burn of the airplane itself.

To compute η, it is necessary to consider both parallel and series component arrangements with n components connected in series. These components can be either mechanical, with efficiencies related to mechanical loss sources (e.g., friction), or electrical, with efficiencies related to resistive and other related losses. The total efficiency of this "chain" is

$$\eta = \prod_{i=1}^{n} n_i. \tag{9.7}$$

Systems connected in parallel combine two series chains, each of which has its own value of η. In this case, the total blockchain efficiency is

$$\eta = \frac{P_1\eta_1 + P_2\eta_2}{P_1 + P_2} = \frac{P_1\prod_{i=1}^{n_1}\eta_i + P_2\prod_{j=1}^{n_2}\eta_j}{P_1 + P_2}, \tag{9.8}$$

where P_n is the power supplied by chain n and η_n is its corresponding series blockchain efficiency. Define H_p, the *hybridization of power* as

$$H_p \equiv \frac{P_2}{P_1 + P_2}, \tag{9.9}$$

where it is assumed that chain 2 is powered by electrical energy. Rewriting (9.8) in terms of H_p reveals that η is the average blockchain efficiency weighted by the proportion of power from the two power paths and their associated efficiencies:

$$\eta = \frac{P_1\eta_1 + P_2\eta_2}{P_1 + P_2} = (1 - H_p)\prod_{i=1}^{n_1}\eta_i + H_p\prod_{j=1}^{n_2}\eta_j. \tag{9.10}$$

If both power paths contain electrical components, then (9.9) no longer applies and (9.8) must be used.

9.8.1.1 Block Fuel Burn and Block Energy

The *block fuel burn* and *block energy* are the most commonly cited benefits (or detriments) of an EP configuration. The block fuel burn is the fuel required for the mission, ignoring reserves and taxiing, and is sometimes called *trip fuel*. The *block energy*, E_{block}, is the total mission energy required, fuel-based and otherwise. This is generally a weighted sum of energy from liquid hydrocarbon fuels and any other storage such as batteries, capacitors, or hydrogen for fuel cells.

For example, the block energy for a battery-based HE mission can be calculated as follows:

$$E_{\text{block}} = h_{lhv} m_{\text{fuel}} + E_{\text{batt}}. \tag{9.11}$$

Here, E_{batt}, in J, is the battery energy used; h_{lhv}, in J/kg, is the lower heating value of the fuel; and m_{fuel}, in kg, is the total mass of the liquid hydrocarbon fuel burned during the mission. Fuel burn and energy terms are sometimes given as "deltas" (percent difference) relative to the baseline configuration because most decision makers do not know raw fuel burn or energy consumption values by heart for a given airplane.

9.8.1.2 Off-Design Performance

It is also important for the concept assessment to consider many different missions that the aircraft might fly, apart from its "Max PAX" design mission. Every aircraft has a maximum range that is a function of payload, but most fly a variety of missions of differing ranges during their lifespan. One approach to quantifying this variance for a commercial airplane is to consider the performance at all of the possible mission ranges and then to perform a weighted average based on the actual distribution of flights that this class of airplane tends to fly. For example, the flight distribution of single-aisle class (150 PAX) airplanes is shown in Figure 9.10, which is based on data from the US Bureau of Transportation Statistics (BTS). Note that 50 percent of flights are less than 1,000 nmi, and 90 percent of flights are less than 1,900 nmi.

The distribution in Figure 9.10 can be used as a weighting function in performance assessments comparing the concept to a baseline configuration. These assessments yield a metric that is intended to represent a "fleet-wide" average for the family of missions considered. One way to compute such a metric is to divide the range variation region into discrete intervals (e.g., 100 nmi increments) between R_{low} and R_{high}, the lowest and highest expected mission ranges, respectively. At each discrete range, the block energy is multiplied by the number of flights, n_{flight}, at that range point and accumulated:

$$E_{\text{cum}} = \sum_{i=R_{\text{low}}}^{R_{\text{high}}} (n_{\text{flight}})_i (E_{\text{block}})_i. \tag{9.12}$$

9.8.2 Environmental Metrics and Indices

A commonly cited benefit of air vehicle electrification is its potential to use renewable or decarbonized energy. A key related metric is the aircraft *total life cycle carbon emissions* (LC-CO_2), which is like a carbon footprint, but for airplanes. The difficulty in computing LC-CO_2 lies in the fact that there are many sources of carbon emissions, even for electric vehicles, stemming from various sources including battery manufacturing and recycling, electricity generation, grid losses, etc.

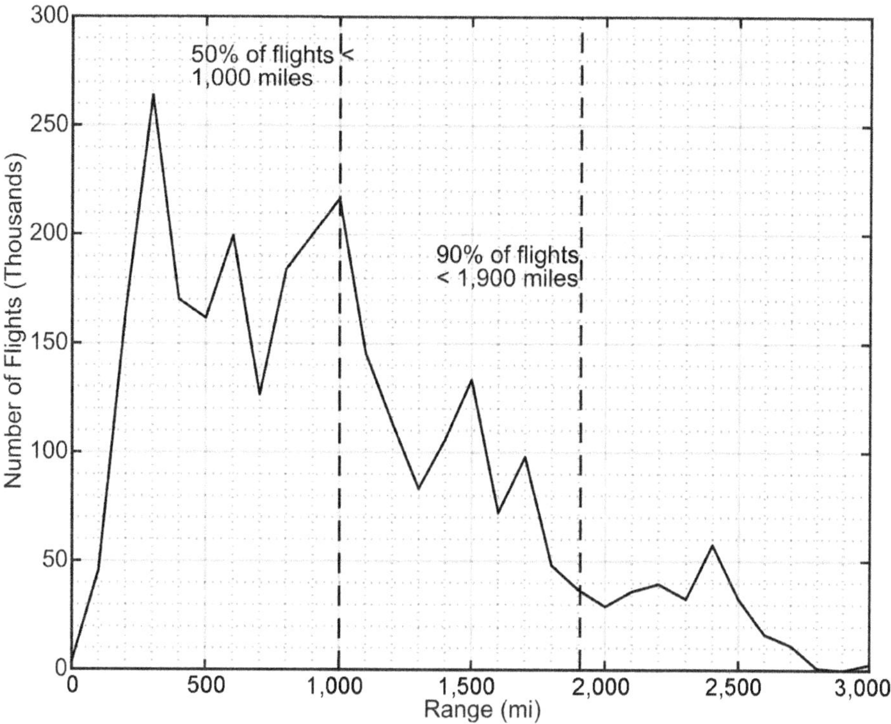

Figure 9.10 Range variability for single-aisle airplanes [16]. Chart is based on data from the US BTS [17].

There is a myth that circulates in popular culture (due to various advertising efforts) that all "zero-emission" vehicles are, by default, environmentally friendly. The truth is that zero-emission vehicles are only low carbon if their energy is derived from a non-hydrocarbon power source such as nuclear, solar, wind, or hydroelectric. Even then, these so-called renewables have some residual carbon footprint, albeit much less than that of coal or natural gas. It is therefore difficult to make a carbon reduction assessment, since the result ultimately depends on where the grid power is assumed to come from. Another approach is to simply consider the grid average across the world or across various regions.

Figure 9.11 compares the grams of CO_2 emitted per kWh of shaft power provided to a propulsor for a gas engine using Jet-A fuel and an equivalent electric system that is driven by grid power. On the horizontal axis is the thermal efficiency – i.e., how efficient the machine is at turning stored energy into useful shaft work.

Currently, the best large single-aisle engine is approximately 55 percent thermally efficient, but electrical systems can take on a range of thermal efficiencies that are much higher (80–100 percent). These numbers account for average grid losses and battery charging efficiencies. Despite this apparent disadvantage, using current grid numbers (even in the United States), the best gas turbine engine is actually better than an

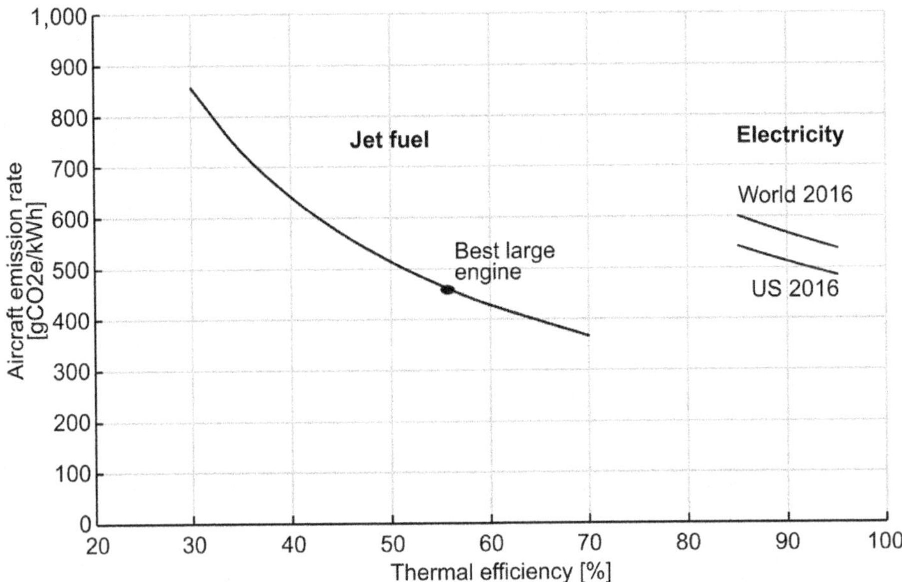

Figure 9.11 Average LC-CO_2 emissions for gas engine and electric power at varying thermal efficiencies.

equivalent electrical propulsor supplied with energy from the grid – by a significant margin. However, a closer look at the numbers yields some interesting trends. First, the primary driver of the grid gCO_2/kWh metric is the relative use of coal within a state or region. In the United States, states like Kentucky and West Virginia rely heavily on coal power and therefore would have very high rates of emissions for EVs and HEVs that utilize large amounts of grid power. However, coal use has been waning due to regulation and competitive pricing trends from renewables and natural gas, and therefore the overall grid emissions index is dropping by about 9.9 gCO_2/kWh per year. This implies that grid-based electrical energy storage on an airplane can or will be able to reduce CO_2 emissions if basic performance levels for energy storage devices are improved to the point of being competitive with gas turbines (see Chapter 7 for details).

To perform a concept environmental impact assessment, any CO_2 metric computation should consider the source of the grid electricity and use reliable data sources such as from the US Environmental Protection Agency (EPA) or other government agencies. It should also be noted that the prior analysis has not considered the effects of battery manufacturing and replacement over the life span of the airplane, which may make a substantial difference. Many studies neglect this fact.

Another environmental metric that could be impacted by electrification is nitrous oxide (NO_x) emissions. These particulates pose a significant hazard to public safety, contribute to urban air pollution, and are regulated by government agencies via internationally recognized metrics such as those provided by ICAO. Aircraft engines typically are certified via testing of their NO_x emissions by use of the dp/Foo metric

[4] (ICAO Databank). Each engine type and class has a limit of dp/Foo over which it cannot exceed that is a function of the engine overall pressure ratio (OPR). The NO_x emissions of an engine are a function of the pressure and temperature entering the burner and scale with temperature and combustor design. Obviously for AE aircraft, the NO_x production is significantly reduced, though one should consider whether those emissions are being shifted from the aircraft exhaust to the grid power plant smokestack. Even so, removing emissions from population centers can have a large impact on reducing aviation-related pollution. Certain parallel hybrid configurations have been looked at to reduce engine firing temperatures during takeoff, climb out, and landing in order to significantly reduce landing and takeoff NO_x. If these benefits are to be considered during a concept assessment, some degree of combustor and engine knowledge must be known, and they must be weighed against any performance or cost impacts.

9.8.3 Economic Metrics

Although computing the true economic impacts of aircraft EP applications is a heavily researched topic, the impacts are still not fully understood. The difficulty arises from the fact that EP propulsion systems typically utilize novel technologies that add weight and complexity to the powertrain while improving efficiency and fuel burn. Therefore, as in the HEV world, there is a trade-off between the achievable operating cost reductions versus the initial price of designing and purchasing the airplane. Also, engine maintenance tends to consume a large portion of an airline's direct operating costs, but the impact of EP architectures on maintenance costs is not yet defined in the public literature. Nor have the effects of EP architectures on the net present value of the various classes of vehicle been fully investigated within the public domain. As such, most studies tend to define the benefits in terms of fuel burn per passenger as a surrogate for the direct airline operating costs while assuming that other factors remain neutral in the absence of better information.

9.9 Summary

This chapter has outlined a methodology for conducting performance assessments of HE and AE aircraft. The information provided serves as a guide to aircraft designers – new and old – who are just delving into this exciting field and starting to explore the large design space enabled by EP configurations. Combining information from the preceding chapters rounds out the performance assessment process described herein, addressing component-, physics-, and technology-related considerations to help in conducting the aircraft system-level assessment.

The material in this chapter should well prepare the reader to begin concept generation, providing numerous important factors to consider when choosing an EP-based architecture. Methods are also detailed for aircraft sizing and computation of system metrics for the chosen concept.

As a final takeaway, it is imperative that significant consideration and careful thought be given to the aircraft assessment process. The methodologies and assumptions used must be reasonable, and an "apples-to-apples" comparison to a baseline that itself is given a fair chance at technology progression must be included. Without this, it is impossible to gauge the true benefit (or cost) of replacing a traditional architecture with an EP-based one. It is the hope of the authors that the information in this chapter, together with the previous chapters, will lead to more structured, careful, and reasonable processes for the generation and evaluation of concept designs and that this will accelerate the safe electrification of the aircraft industry.

Abbreviations

AE, AEA, AEV	all-electric, AE aircraft, AE vehicle
AEO	all engines operating
APU	auxiliary power unit
BPR	bypass ratio
BSFC	brake-specific fuel consumption
BTP	baseline technology projection
BTS	Bureau of Transportation Statistics
CO_2	carbon dioxide
CONOPS	concept of operations
CTOL	conventional takeoff and landing
DOFs	degrees of freedom
EA	electrified/electric aircraft
EIS	entry into service
EM	electric machine
EP	electrified/electric propulsion
EPA	Environmental Protection Agency
ETOPS	extended twin operations
eVTOL	AE vertical takeoff and landing
EVE	electrically variable engine
FAA	Federal Aviation Administration
FOM	figure of merit
GA	General Aviation
GE	General Electric
HE, HEV	hybrid-electric, hybrid-electric vehicle
ICAO	International Civil Aviation Organization
ICE	internal combustion engine
LC-CO_2	life-cycle carbon dioxide
LTO	landing and takeoff
MRO	maintenance, repair, and overhaul
NAS	National Academy of Sciences

NASA	National Aeronautics and Space Administration
NOx	nitrous oxide
OEI	one-engine inoperative
OPR	overall pressure ratio
PAX	number of passengers, or passenger class
RIT	rotor inlet temperature
RPM	revolutions per minute
SLS	sea-level static
SOA	state of the art
STARC-ABL	single-aisle turboelectric aircraft with aft boundary layer propulsor
SUGAR	Subsonic Ultra Green Aircraft Research
TE	turboelectric
TIT	turbine inlet temperature
TRL	technology readiness level
TSFC	thrust-specific fuel consumption
UAV	unmanned aerial vehicle
VTOL	vertical takeoff and landing

Variables

c	thrust-specific fuel consumption (TSFC), [kg/N-s]
D	drag, [N]
dh/dt	aircraft climb rate, [m/s]
dR/dW	fuel-specific air range, [m/kg]
E_{batt}	battery energy, [J]
E_{block}	block energy, [J]
E_{cum}	cumulative fleet block energy, [J]
dR/dE	energy-specific air range (ESAR), [m/J]
h_{lhv}	lower heating value of fuel, [J/kg]
H_p	degree of hybridization of power
L/D	lift-to-drag ratio
m_{fuel}	total fuel mass, [kg]
n_{flight}	number of flights
P_n	power required for path n, [W]
P_s	specific excess power, [W/N]
R	aircraft range, [m]
R_{high}	high-end of aircraft range, [m]
R_{low}	low-end of aircraft range, [m]
T	aircraft thrust, [N]
V_∞	flight speed, [m/s]
W	aircraft instantaneous weight, [N]
η	blockchain efficiency
γ	flight path angle, [rad]

References

[1] Federal Aviation Administration, "Extended Operations (ETOPS and Polar Operations)," Washington DC, 2008, FAA AC 120-42B.

[2] National Academies of Sciences, Engineering, and Medicine, *Commercial Aircraft Propulsion and Energy Systems Research: Reducing Global Carbon Emissions*, Washington, DC: National Academies Press, 2016.

[3] Code of Federal Regulations, "Title 14 – Aeronautics and Space, Chapter 1 – Federal Aviation Administration Department of Transportation, Subchapter C – Aircraft, Part 25 – Airworthiness Standards: Transport Category Airplanes."

[4] International Civil Aviation Organization, "ICAO Aircraft Engine Emissions Databank," 2019. [Online]. Available: www.easa.europa.eu/easa-and-you/environment/icao-aircraft-engine-emissions-databank

[5] J. C. Mankins, "Technology readiness and risk assessments: A new approach," *Acta Astronaut.*, vol. 65, pp. 1208–1215, November 2009.

[6] C. E. Lents, L. W. Hardin, J. Rheaume, and L. Kohlman, "Parallel hybrid gas-electric geared turbofan engine conceptual design and benefits analysis," presented at the 52nd AIAA/SAE/ASEE Joint Propulsion Conference, Salt Lake City, UT, July 2016, Paper AIAA 2016-4610.

[7] D. Hall et al., "Boundary layer ingestion propulsion benefit for transport aircraft," *J. Prop. Power*, vol. 33, No. 5, pp. 1118–1129, September 2017.

[8] R. de Vries, M. Hoogreef, and R. Vos., "Preliminary sizing of a hybrid-electric passenger aircraft featuring over-the-wing distributed-propulsion," presented at the 57th AIAA Aerospace Sciences Meeting, San Diego, CA, January 2019, Paper AIAA 2019-1811.

[9] M. Bradley et al., "Subsonic ultra green aircraft research: Phase II – Volume II – Hybrid electric design exploration," NASA, Cleveland, OH, Tech. Rep. NASA/CR–2015-218704/Volume II, 2015.

[10] C. Perullo et al., "Cycle selection and sizing of a single-aisle transport with the electrically variable engine (EVE) for fleet level fuel optimization," presented at the 55th AIAA Aerospace Sciences Meeting, Grapevine, TX, January 2017, Paper AIAA 2017-1923.

[11] J. L. Felder, G. V. Brown, D. Hyun, and J. Chu, "Turboelectric distributed propulsion in a hybrid wing body aircraft," presented at the 20th International Society for Airbreathing Engines Meeting, Gothenburg, Sweden, September 2011, Paper ISABE-2011-1340.

[12] J. R. Welstead and J. L. Felder, "Conceptual design of a single-aisle turboelectric commercial transport with fuselage boundary layer ingestion," presented at the 54th AIAA Aerospace Sciences Meeting, San Diego, CA, January 2016, Paper AIAA 2016-1027.

[13] N. K. Borer et al., "Design and performance of the NASA SCEPTOR distributed electric propulsion flight demonstrator," presented at the 16th AIAA Aviation Technology, Integration, and Operations Conference, Washington, DC, June 2016, Paper AIAA 2016-3920.

[14] J. L. Kratz, D. E. Culley, and G. L. Thomas, "A control strategy for turbine electrified energy management," presented at the 2nd AIAA/IEEE Electric Aircraft Technologies Symposium, Indianapolis, IN, August 2019, Paper AIAA 2019-4499.

[15] D. Trawick, K. Milios, J. C. Gladin, and D. N. Mavris, "A method for determining optimal power management schedules for hybrid electric airplanes," presented at the 2nd AIAA/IEEE Electric Aircraft Technologies Symposium, Indianapolis, IN, August 2019, Paper AIAA 2019-4500.

[16] J. C. Gladin, C. Perullo, J. C. Tai, and D. N. Mavris, "A parametric study of hybrid electric gas turbine propulsion as a function of aircraft size class and technology level," presented at the 55th AIAA Aerospace Sciences Meeting, Grapevine, TX, January 2017, Paper AIAA 2017-0338.
[17] RITA/BTS, Office of Airline Information, "Air Carrier Statistics Database (T-100)," [Online]. Available: www.transtats.bts.gov.

Index

Printed in the United States
by Baker & Taylor Publisher Services